Materials Challenges in Alternative and Renewable Energy

Materials Challenges in Alternative and Renewable Energy

Ceramic Transactions, Volume 224

A Collection of Papers Presented at the Materials Challenges in Alternative and Renewable Energy Conference February 21–24, 2010, Cocoa Beach, Florida

Edited by
George Wicks
Jack Simon
Ragaiy Zidan
Edgar Lara-Curzio
Thad Adams
Jose Zayas
Abhi Karkamkar
Robert Sindelar
Brenda Garcia-Diaz

A John Wiley & Sons, Inc., Publication

Published by John Wiley & Sons, Inc., Hoboken, New Jersey.
Published simultaneously in Canada.

For general information on our other products and services or for technical support, please contact our Customer Care Department within the United States at (800) 762-2974, outside the United States at (317) 572-3993 or fax (317) 572-4002.

Wiley also publishes its books in a variety of electronic formats. Some content that appears in print may not be available in electronic format. For information about Wiley products, visit our web site at www.wiley.com.

Library of Congress Cataloging-in-Publication Data is available.

ISBN 978-1-1180-1605-3

Printed in the United States of America.

10 9 8 7 6 5 4 3 2 1

Contents

BATTERIES AND ENERGY STORAGE MATERIALS

Preface

Materials Challenges in Alternative & Renewable Energy (Energy 2010) was an important meeting and technical forum held in Cocoa Beach, Florida, on February 21–24, 2010. This represented the second conference in a new series of inter-society meetings and exchanges, with the first of these meetings held in 2008, on "Materials Innovations in an Emerging Hydrogen Economy." The current Energy Conference- 2010 was larger in scope and content, and included 223 participants from more than 25 countries and included more than 160 presentations, tutorials and posters. The purpose of this meeting was to bring together leaders in materials science and energy, to facilitate information sharing on the latest developments and challenges involving materials for alternative and renewable energy sources and systems.

Energy 2010 marks the first time that three of the premier materials organizations in the US have combined forces, to co-sponsor a conference of global importance. These organizations included The American Ceramic Society (ACerS), ASM International, and the Society of Plastics Engineers (SPE), representing each of the materials disciplines of ceramics, metals and polymers, respectively. In addition, we were also very pleased to have the support and endorsement of important organizations such as the Materials Research Society (MRS) and the Society for the Advancement of Material and Process Engineering (SAMPE), in this endeavor.

Energy 2010 was highlighted by nine "tutorial" presentations on leading energy alternatives provided by national and international leaders in the field. In addition, the conference included technical sessions addressing state-of-the art materials challenges involved with Solar, Wind, Hydropower, Geothermal, Biomass, Nuclear, Hydrogen, and Batteries and Energy Storage. This meeting was designed for both scientists and engineers active in energy and materials science as well as those who were new to the field.

We are very pleased that ACerS is committed to running this materials-oriented conference in energy, every two years with other materials organizations. We believe the conference will continue to grow in importance, size, and effectiveness and provide a significant resource for the entire materials community and energy sector.

GEORGE WICKS
Savannah River National Laboratory
Energy Conference-2010 Co-Organizer/President-Elect of ACerS

JACK SIMON
Technology Access
Energy Conference-2010 Co-Organizer/Past President ASM International

Acknowledgments

Conference Co-Chairs

Dr. George Wicks, Savannah River National Laboratory, Aiken, SC
Dr. Jack Simon, Technology Access, Aiken, SC

Advisory & Technical Planning Committee:

Dr. Jack Simon, Technology Access and Alpha Sigma Mu Honorary Society
Dr. George Wicks, Savannah River National Lab
Dr. Thad Adams, Savannah River National Lab
Dr. Joel Ager, Lawrence Berkeley National Lab
Dr. Ming Au, Savannah River National Lab
Dr. Amir Farajian, Wright State Univ.
Ms. Rita Forman-House, ASM International
Dr. Brenda Garcia-Diaz, Savannah River National Lab
Dr. Frank Goldner, U.S. Dept. of Energy
Prof. Hong Huang, Wright State Univ.
Dr. M. Ashraf Imam, Naval Research Lab
Dr. Natraj Iyer, Savannah River National Lab
Prof. Puru Jena, Virginia Commonwealth Univ.
Dr. Enamul Haque, Bostik, Inc.
Dr. Abhi Karkamkar, Pacific Northwest National Lab
Dr. Gene Kim, Cookson Electronics
Ms. Lesley Kyle, Society for Plastics Engineers (SPE)
Dr. Edgar Lara-Curzio, Oak Ridge National Lab
Mr. Richard Marczewski, Savannah River National Lab
Dr. Rana Mohtadi, Toyota Technical Center NA
Dr. Ali Raissi, Florida Solar Energy Center Univ. of Central Florida
Dr. Bhakta Rath, Naval Research Lab
Dr. Robert Sindelar, Savannah River National Lab

Prof. Rick Sisson, Worcester Polytechnic Institute
Ms. Hidda Thorsteinsson, U.S. Dept. of Energy
Ms. Agatha Wein, U.S. Dept. of Energy
Mr. Jose Zayas, Sandia National Labs
Dr. Kristine Zeigler, Savannah River National Lab
Dr. Ragaiy Zidan, Savannah River National Lab
Mr. Mark Mecklenborg, The American Ceramic Society
Mr. Greg Geiger, The American Ceramic Society

Conference Sponsors

Institute of Metal Research, Chinese Academy of Sciences
Oak Ridge National Laboratory
Pacific Northwest National Laboratory
National Energy Technology Laboratory
National Renewable Energy Laboratory
Sandia National Laboratories
Savannah River National Laboratory
Solar Solutions

Hydrogen

HYDROGEN STORAGE TECHNOLOGIES – A TUTORIAL WITH PERSPECTIVES FROM THE US NATIONAL PROGRAM

Ned T. Stetson
U. S. Department of Energy
Washington, DC, US

Larry S. Blair, Consultant
1550 Bridger Road
Rio Rancho, NM 87144

ABSTRACT

While the demand for electrical power generated by clean, efficient hydrogen fuel cells is rapidly growing, one of the key technical issues that remains to be resolved is the storage of hydrogen, or hydrogen-bearing fuels, to be available to the fuel cell within the design and performance constraints of the total power system. Criteria such as hydrogen storage capacity, weight, volume, lifetime and cycle-life, and certainly cost, become important factors in determining the best storage system for a particular application. In this paper we review the various storage approaches that are currently under investigation and provide a brief materials science tutorial on the storage mechanism for each approach.

Physical storage approaches store hydrogen as a compressed gas, a cryogenic liquid or as a cryo-compressed gas. Materials-based storage systems are based on storing hydrogen by adsorption, absorption or chemical bonding to various materials such as reversible or regenerable hydrides. Each of these storage systems will be discussed and the particular materials science challenges involved will be noted. At the present time no hydrogen storage approach meets all volume, weight and cost requirements for automotive fuel cell power systems across the full range of vehicle platforms. It is clear that materials science will play a key role in the ultimate solution of the hydrogen storage challenge.

INTRODUCTION

Hydrogen fuel cells are emerging as a leading candidate in the search for a clean, efficient alternate energy source. Fuel cells fueled with hydrogen are coming out of the Laboratory and moving toward commercialization in a variety of important applications. Initially fuel cells provided high-value power for both manned and unmanned spacecraft, but more recently they are being developed for "down to earth" applications such as back-up power for telecommunications and uninterrupted power systems (UPS), stationary power for residential, commercial and industrial uses, and portable power for hand-held instrumentation and military applications. Longer term transportation deployments are targeted toward the personal automobile market with specialty vehicles (e.g., forklifts), transit buses, and fleet vehicles leading with early market entry. In 2008 world-wide cumulative shipments of fuel cells exceeded 50,000 units (see Figure 1).

As hydrogen fuel cells become a viable contender in the alternative energy arena, attention is being focused on overcoming the major technical challenges that may ultimately impact introduction in potential early markets. For example, fuel cell cost is a significant factor that must be addressed for this technology to be competitive with conventional, petroleum-based power systems. Likewise the availability of hydrogen to fuel the system is a technical challenge. For the ultimate transportation application – the consumer automobile – a sufficient amount of hydrogen must be stored on-board the vehicle to allow a 300-mile driving range.

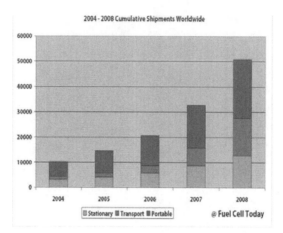

Figure 1. Worldwide Cumulative Fuel Cell Shipments. (Source Fuel Cells Today)

Hydrogen continues to receive intense study and support as a leading candidate to provide clean, safe and efficient power as an alternative to petroleum/hydrocarbon sources. Like all potential fuels hydrogen has both advantages and disadvantages. It is the lightest of all the elements. Based on its lower heating value (LHV) hydrogen has a very attractive specific energy of 120 kJ/g or 33.3 kWh/kg – approximately three times that of gasoline. Of course, with a normal boiling point of 20 K, hydrogen is a gas in its normal state with a density of ~0.09 g/L or 11 L/g. So while hydrogen has a high specific energy, due to its low density it has a normal energy density of only 10 kJ/L compared to gasoline at ~32,000 kJ/L. Therefore the challenge for hydrogen storage is to increase its normal energy density, thus it is normally stored either at high pressure or as a cryogenic liquid. Its storage is further problematic due to its ability to diffuse through many containment materials and can cause embrittlement, resulting in diminished material strength and lifetime challenges. On the other hand hydrogen combustion products from fuel cells are only water and heat making it a non-polluting energy carrier. Additionally hydrogen can be derived from various liquid fuels that can be reformed either internally within or externally to the fuel cell, and it can also be produced using alternative energy sources (such as solar, wind, nuclear, etc.).

The U. S. Department of Energy has identified several key characteristics for viable hydrogen storage systems. Key storage system characteristics include: gravimetric and volumetric capacities (i.e. system weight and volume per unit H_2); operating temperature and pressure; transient response (start-up and shut-down times and load following); refill time; dormancy (i.e. length of idle time before H_2 loss occurs); cycle life and costs (capital, maintenance and refueling). The actual values of these characteristics will vary depending on the specific needs of the particular application. However the most challenging requirements, by far, are those for the ≥300-mile range, on-board hydrogen storage system for automobiles. A complete list of the DOE system performance targets for vehicular, on-board hydrogen storage can be found on the DOE website.[1] The performance targets are system targets and must be achieved simultaneously. Presently extensive research, development and testing are underway to address the challenge of hydrogen storage for fuel cell power systems. Material science is the key to the long-term development of practical hydrogen storage systems that meet the established performance and cost targets. Hydrogen storage concepts are based on physical storage systems and materials-based approaches; each is summarized in the following sections of this paper.

PHYSICAL STORAGE

Physical storage techniques generally involve storing hydrogen as a compressed gas or as a cryogenic liquid in a qualified container. High pressure storage vessels are the present state-of-the-art in hydrogen storage. Most commercially available fuel cell power systems operate on high pressure compressed hydrogen stored in certified tanks. Storage at cryogenic temperatures allows hydrogen to be stored at liquid densities and cryo-compressed storage concepts attempts to take advantage of both high pressure and cryogenic temperatures.

Compressed Storage

For compressed gas storage the higher the pressure the higher the density of stored gas. While merchant hydrogen is typically delivered for industrial uses in the pressure range from 150 to 250 bar, automotive storage systems commonly operate at 350 bar with the goal of increasing the operating pressure to 700 bar. Clearly for on-board storage the higher the pressure the greater quantity of hydrogen that can be contained in a fixed volume. However as the pressure increases the cost and weight of the storage tank increases and ultimately a point of diminishing returns is reached. The walls of all-metal storage tanks (Type I) must contain all of the stress from the high pressure, thus the wall thickness of the containment vessel increases rapidly with pressure. Since the wall thickness relates to the operating pressure and ultimate tensile and yield strength of the metal, higher strength metals could lead to lighter cylinders, however current standards and regulations limit the ultimate tensile strength of steels used in hydrogen service to 950 MPa due to hydrogen embrittlement issues.[2] The materials R&D challenges for all-metal hydrogen storage cylinders therefore include the development of high strength metals that are not susceptible to hydrogen embrittlement. In addition there is a need to more fully understand cycle fatigue failure under hydrogen storage operating conditions.

Fiber reinforced composite cylinders are also being developed for hydrogen storage. These include hoop-wrapped (Type II) and fully wrapped with either metal liners (Type III) or non-metal liners (Type IV). These composite tanks can either share the strain load between the liner and fiber layers (Type II and III) or have the fiber layer fully bear the strain load (Type IV). Composite cylinders generally allow higher pressure operation resulting in higher gravimetric capacities (>5 wt.%) compared to more conventional Type I metal vessels (typically <2 wt.%). However cost is an important issue with composite tanks and current analyses indicate approximately 75% of the cost is due to the carbon fiber layer.[3] The key material R&D challenge for composite storage vessels is the development of low-cost, high-strength carbon fiber suitable for reinforcing these vessels.

Liquid Hydrogen Storage

Hydrogen for industrial applications is often transported and stored as a cryogenic liquid. Several automotive manufacturers have incorporated liquid hydrogen storage into fuel cell concept vehicles. The cryogenic temperatures (33 K hydrogen critical temperature, 20 K normal boiling point) required for liquefying hydrogen necessitates double-walled containment with multi-layer vacuum super insulation (MLVSI). These vessels are designed to minimize conductive, convective and radiative heat transfer between the inner and outer vessel walls to maximize the dormancy before pressure buildup due to boil-off causes venting and loss of hydrogen. Storage system capacities in the range of 5-6 wt.% have been projected for liquid hydrogen storage. In addition to a problem with dormancy, the energy required for hydrogen liquefaction results in an efficiency penalty that must be addressed; the total liquefaction energy is approximately 30% of the stored hydrogen energy. The development of low-cost materials of construction including super insulation is a material R&D challenge.[3]

Cryo-Compressed Storage

Cryo-compressed hydrogen gas storage uses temperature along with pressure to increase the density of stored hydrogen. At temperatures slightly above the critical temperature, hydrogen density increases rapidly with pressure. Densities greater than the liquid density (71 g/L) are possible with sufficiently low temperature and high pressure. Storage capacities of >6 wt.% are projected to be achievable with a doubled-walled tank with a high-pressure Type III inner vessel and MLVSI.[4] Cryo-compressed hydrogen storage systems are similar in design to liquid storage tanks with the inclusion of a high-pressure capable inner vessel. If filled using liquid hydrogen, this storage concept still has the liquefaction energy penalty and system costs are still a significant issue. Cryogenic compatible materials of construction, including high-pressure seals, are material R&D challenges for this storage concept.

MATERIALS-BASED STORAGE

Hydrogen can be stored on the surfaces of solids (adsorption) or within solids (absorption). In adsorption, hydrogen attaches to the surface of a material either as hydrogen molecules (H_2) or hydrogen atoms (H). In absorption, hydrogen molecules dissociate into hydrogen atoms that are incorporated into the solid lattice framework. Finally, hydrogen can be strongly bound within molecular structures, as chemical compounds containing hydrogen atoms. These materials-based approaches may make it possible to store larger quantities of hydrogen in smaller volumes at low pressure and at temperatures closer to room temperature than is possible through physical storage methods. Materials for hydrogen storage being investigated include high surface area adsorbents, intermetallic hydrides, complex hydrides and chemical hydrogen storage materials, each of which will be discussed in the following sections of this paper.

Figure 2 shows the density of hydrogen at several temperature and pressure configurations and in various materials. For the materials, the densities are at near-room temperature and relatively low-pressures (0 to 1 MPa). The high hydrogen densities at low to moderate conditions of the materials shown in the middle portion of the figure clearly indicate why they are very attractive candidates for hydrogen storage.

The nature of how hydrogen is bound within materials is an important property for potential hydrogen storage materials. Storage characteristics such as hydrogen uptake and release temperatures and pressures are strongly dependent on the binding energies. For sorbents, the primary binding type is weak van der Waals attractions between the sorbent substrate and the diatomic hydrogen molecule. For carbon-based substrates the binding energy is typically 4-6 kJ per mole hydrogen, meaning there is only significant adsorption at cryogenic temperatures. The goal is to increase the binding energy to around 15-20 kJ per mole hydrogen for ambient temperature adsorption. For interstitial or intermetallic hydrides, the binding energy type is primarily metallic and a range of binding energies is possible; materials developed for most practical applications have a range of about 20 to 40 kJ per mole hydrogen. Complex hydrides generally have hydrogen covalently bonded to a metal forming a complex anion with the charged balanced by a metal cation. Some complex hydrides have additional ionic-bound hydrogen as well. The strong covalent bonds are often greater than 40 kJ per mole hydrogen, which will typically lead to high hydrogen release temperatures. Like complex hydrides, chemical hydrogen storage materials have primarily strongly covalent bound hydrogen. Since with chemical hydrogen storage materials, hydrogen is not released through equilibrium processes, the strong binding is not necessarily a disadvantage. For materials-based hydrogen storage, the key

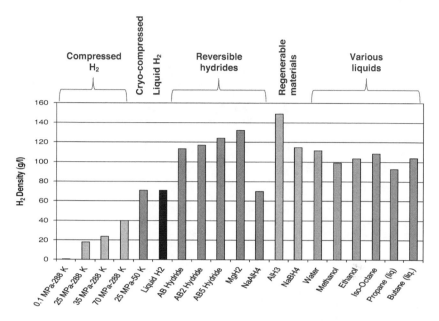

Figure 2. Hydrogen density at various temperatures and pressures or in various materials.

research areas include increasing the hydrogen capacity (both by weight and volume) and optimizing both the energetics and rates of hydrogen uptake and release.

Hydrogen Sorbents

Hydrogen adsorption is typically characterized by the formation of a dense physisorbed layer of diatomic hydrogen molecules on the surface of the sorbent material; thus the adsorption capacity is usually a function of the material's surface area. The increased hydrogen content of the dense surface layer over and above hydrogen gas at the corresponding pressure is known as the excess surface capacity. The total capacity is the sum of surface excess capacity plus the gas stored in the void space of a sorbent material. Research to discover effective hydrogen sorbents focuses on high surface area materials, such as various graphite nanostructures and metal-organic frameworks (MOF), as good potential candidate sorbents. Graphite nanostructures (nanotubes, nanohorns, fullerenes and aerogels), due to their lightweight and open structure (high void volume), have been studied extensively. MOFs are synthetic, highly porous, crystalline materials composed of metal clusters linked by organic molecules to form highly ordered three dimensional structures. The resulting material can have very large surface areas. Figure 3 shows a schematic of hydrogen adsorbed within MOF-74, as determined by neutron diffraction experiments, with an exceptionally high hydrogen density.[5] In addition to traditional MOFs, there are a number of related highly ordered synthetic materials (e.g. covalent organic frameworks (COFs), porous coordination networks (PCNs), zeolitic imidazolate frameworks (ZIFs)) that are being investigated as well.

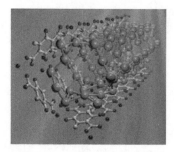

Figure 3.   Hydrogen Filling in MOF-74 (only part of the MOF-74 backbone shown, hydrogen atoms represented by the green balls).[6]

Figure 4 shows H_2 uptake as a function of pressure for a typical MOF (MOF-177). Note that the adsorption at 77 K is shown; low temperatures are required to optimize adsorption due to the low binding energy between hydrogen and the sorbent material.

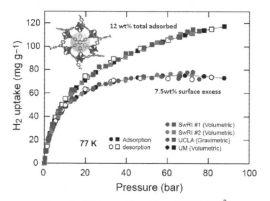

Figure 4. Hydrogen Capacity of MOF-177.[7]

One area of active research with hydrogen sorbents is to increase the binding energy of hydrogen to the substrate from the typical binding energy of 4-6 kJ/mol of hydrogen on carbon materials to 15 kJ/mol or higher. Several approaches that are being investigated include doping of carbon structures with metals or elements such as boron; developing materials that contain unsaturated metal centers and developing materials with void spaces with narrow size distribution of specific size.[8] While the first two approaches have been found to increase the binding energy of the initial hydrogen adsorbed, the binding typically is found to become lower as more hydrogen is adsorbed. More constant binding energy for a greater portion of the absorbed hydrogen has been found with materials with narrow pore size distribution of appropriate size.[9] The materials R&D challenge is the development of new sorbent materials tailored to have high surface area and high pore volume with a narrow size distribution.

Reversible Metal Hydride Storage

Reversible interstitial or intermetallic metal hydrides are a class of metal alloys that absorb hydrogen under pressure and then release the absorbed hydrogen when the pressure is released. As shown in Figure 5, the hydriding process involves several steps: hydrogen adsorption, disassociation of the hydrogen molecule and dissolution of the hydrogen atoms into the metal lattice. Initial dissolution of hydrogen into the metal forms what is known as the -phase. As further hydrogen dissolves into the metal, a -phase nucleates and grows. The co-existence of the two phases results in a plateau in plots of isothermal measurements of the equilibrium hydrogen concentration as a function of pressure (known as PCT or PCI plots). Since an equilibrium condition exists between the free hydrogen gas and hydrogen within the solid, the hydrogen is released from the solid when the pressure is lowered or heat is applied. Due to the relatively heavy weight of the metal atoms compared to hydrogen, the stored hydrogen capacity is fairly low, approximately 2 wt.% for many interstitial hydrides. The - phase is typically characterized by an increase in the distance between metal atoms, and thus results in a swelling of the solid. Many interstitial hydrides will swell by 20-25%, which often leads to the

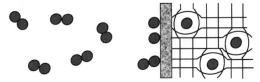

Figure 5. Schematic of Hydrogen Diffusion into the Lattice of a Reversible Metal Alloy.

breakdown of the solid into fine powders, a phenomenon called decrepitation. Along with system engineering challenges, material R&D challenges include the discovery of new lightweight, low-cost hydride formers that are tolerant to air exposure.

Complex Hydride Storage

Complex hydrides are materials with hydrogen covalently bonded to a metal to form a multi-atom (complex) anion with an ionically-bound cation to balance the charge. For example sodium alanate (Na AlH$_4$) can be written $Na^+(AlH_4)^-$. The strong covalent bonds between hydrogen and the

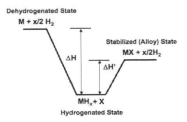

Figure 6. Generalized enthalpy diagram showing transition between hydrogenated and dehydrogenated states and how enthalpy change is reduced through formation a binary alloy upon dehydrogenation.

metal result in high hydrogen release temperatures. To achieve reversible hydrogen release at low to moderate temperatures requires other strategies to facilitate release from the complex hydrides. An approach that has shown promise is to alter the reaction pathway by adding additional phases so that products closer in energy to the complex hydride phase are formed on release of hydrogen. This approach is shown schematically in Figure 6. In addition to high release temperatures, complex hydrides tend to have slow sorption kinetics. The use of additives to catalyze the hydrogen uptake or release reactions has been successful in a number of complex hydrides. One of the best examples demonstrating the positive effect that an additive can have on hydrogen release from complex hydrides is the addition of titanium to sodium alanate.[10] The enhanced release of hydrogen from titanium-doped sodium alanate at relatively moderate conditions has opened up new interest and activity in the discovery of new complex hydride materials along with development of appropriate additives and/or catalysis. Figure 7 shows an example of the additive effect on a metal hydride resulting in improved hydrogen release

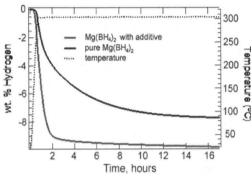

Figure 7. Example of the Use of Additives to Destabilize Complex Hydrides[11]

The following table compares important characteristics of representative interstitial and complex hydrides. As illustrated by the table complex hydrides demonstrate substantially higher stored hydrogen capacity and lighter weight which are attractive attributes for hydrogen storage systems. Unfortunately the complex hydrides suffer from slow hydrogen sorption rates and require

	Interstitial Hydrides	Complex Hydrides
Gravimetric capacity	1-4 wt.%	5-18 wt.%
Material bulk density	4-9 g/cc	<5 g/cc
Enthalpy (ΔH)	<20 - >40 kJ/mol	>35 kJ/mol
Sorption kinetics	fast	Slow
Cycle-life	moderate-good	poor-moderate
Other issues	decrepitation	volatile species (e.g. B_2H_6, NH_3)

Table I. Comparison of Interstitial Metal and Complex Hydrides for Hydrogen Storage.

higher temperature for hydrogen release. On the other hand the interstitial metal hydrides display faster sorption kinetics and release stored hydrogen at lower temperature but suffer from low storage capacity and higher weight. Advances in materials science may improve the feasibility of both of these types of hydride materials for hydrogen storage in the future. The material R&D challenges for complex hydrides are the development light-weight, high capacity hydride materials with low sorption enthalpies and fast kinetics. Additives/dopants, and catalysis are needed that will complement the hydrogen release and sorption kinetics of the complex hydride storage materials are also needed.

Chemical Hydrogen Storage

In this concept hydrogen is released irreversibly from a hydrogen-rich storage material. Hydrogen is stored in these materials via strong chemical bonds and is released by thermal decomposition or other chemical reaction such as the catalyzed reaction of sodium borohydride with water to produce sodium borate and gaseous hydrogen. The reaction by-products must then be regenerated back to the original hydrogenated material requiring a separate process. The hydrogen release mechanisms and thermodynamics of candidate chemical hydrogen storage materials vary widely and can range from strongly exothermic (e.g., hydrolysis reactions of LiH or $NaBH_4$) to endothermic decomposition of liquid or solid compounds assisted by catalysis and heating. Due to their high gravimetric storage capacities (>10 wt.%) and fast desorption kinetics at near-ambient temperatures, these systems are ideal for single-use applications where disposal of the spent fuel is acceptable or in situations where efficient fuel regeneration schemes exist. The configurations of chemical hydrogen storage systems are thus highly dependent on the nature of these reactions as well as the characteristics of the materials.[12]

Research and development of these materials has focused on catalysts to control hydrogen release rate and to increase the usable hydrogen yield from the chemical reaction. Figure 8 shows catalyst development progress on a typical chemical hydrogen storage material – ammonia borane (NH_3BH_3) (AB).

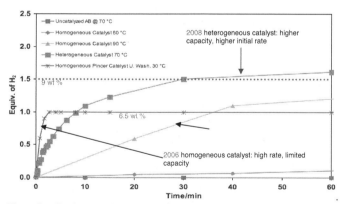

Figure 8. Catalyst development has increased the kinetics for hydrogen release and enhanced the usable hydrogen obtained for AB.[13]

Likewise solvents are being studied as a means to achieve full thermal decomposition at reduced temperatures (See Figure 9). In this example the hydrogen evolution rate of the chemical hydrogen storage material, alane – AlH_3, is shown first as a dry powder. Next the AlH_3 is dissolved in the hydrocarbon solvent, $C_{10}H_{22}O_2$, and an increase in hydrogen release is observed, particularly at lower temperature. Finally titanium is added to the mixture and a significant increase in the rate of hydrogen release is obtained. This example showing both the effect of solvents and additives on the performance of hydrogen storage materials moves toward the goal for chemical hydrogen storage materials of full decomposition of the material at reasonable, acceptable temperatures.

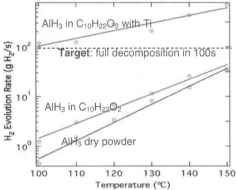

Figure 9.   The Effect of Added Solvent on Hydrogen Release in Chemical Hydrogen Storage Materials.[14]

The technical issues associated with chemical hydrogen storage materials include the total cycle energy efficiency of the entire reaction process including the regeneration of the hydrogenated starting materials. Likewise the overall cost of hydrogen production/storage, again including regeneration, must be reduced so that the costs are competitive with conventional fuels. The many material R&D challenges include new high-capacity hydrogen materials, new release catalysts, and regeneration methods that are energy efficient and cost effective.

CONCLUSIONS

Figures 10 and 11 summarize the status of the state-of-the-art in hydrogen storage systems. The capacities shown are for complete systems based on experimental capacities for the basic materials. Thus engineering design and analysis has been applied to project system characteristics and properties based on the basic material measurements. The system analyses have been based on a total system capacity of 5.6 kg of hydrogen – the estimated nominal amount of hydrogen required for a 300-mile range for a conventional-type passenger automobile. The performance targets for system capacities depend strongly on the particular fuel cell system and the specific application (portable, stationary CHP, back-up, auxiliary, etc.).

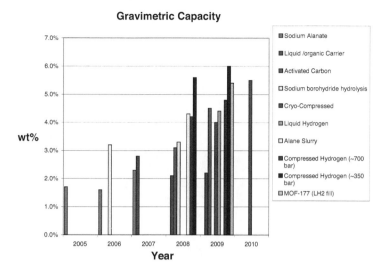

Figure 10. Estimates of gravimetric capacities projected for on-board storage systems that can supply 5.6 kg of usable hydrogen (based on engineering analyses at Argonne National Laboratory).

The system capacities shown in Figures 10 and 11 include the hydrogen, storage media (i.e., metal hydrides, chemical hydrides, or sorbents), the containment vessel, as well as associated plumbing, valves, and auxiliary components required to supply hydrogen to the inlet of the fuel cell power system. The engineering assessments of these hydrogen storage systems were based on information from prototypes (when publicly available), publicly released reports and documents on candidate storage materials and schematic designs of proposed system configuration.[15]

The figures show a steady trend of increasing capacities resulting from the dedicated efforts of many researchers and organizations, however it is also clear that there is still considerable work to be done. While significant progress has been made in the development of hydrogen storage systems, none currently meet all of the stringent performance requirements for the most demanding applications, such as on-board vehicle storage. The materials-based approaches require further development to improve sorption kinetics, less extreme operating temperatures and pressures, and higher volumetric and gravimetric capacities. Additionally there are significant engineering challenges for the materials-based and physical storage methods, not the least of which is to reduce the costs. Material science is the key to answering the challenge of hydrogen storage and enabling the commercialization and economic and societal benefits of fuel cells as a credible alternative energy source.

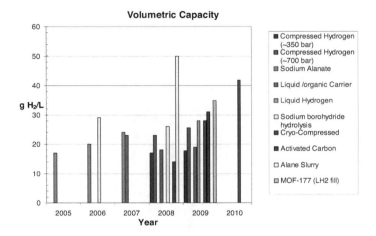

Figure 11. Estimates of volumetric capacities projected for on-board storage systems that can supply 5.6 kg of usable hydrogen (based on engineering analysis at Argonne National Laboratory).

ACKNOWLEDGEMENTS

The authors wish to thank and acknowledge the contributions of the DOE Hydrogen Storage Team – Sunita Satyapal, Carole Read, Grace Ordaz and Monterey Gardiner. Also thanks are in order to the Storage Team technical support network – Bob Bowman, George Thomas, Gary Sandrock, John Petrovic and Anita Vanek. Tremendous thanks and acknowledgements are also due to the many researchers whose work through the US DOE National Hydrogen Storage Project was drawn upon in preparing this paper.

REFERENCES

1. http://www1.eere.energy.gov/hydrogenandfuelcells/storage/pdfs/targets_onboard_hydro_storage.pdf

2. R. S. Irani, Hydrogen Storage: High-Pressure Gas Containment. *MRS Bulletin* 27(9):680-682 (2002).

3. S. Lasher, K. McKenney, J. Sinha, and P. Chin, Analyses of Hydrogen Storage Materials and On-Board Systems, *2009 DOE Hydrogen Program Review*, Arlington, VA. (2009). http://www.hydrogen.energy.gov/pdfs/review09/st_12_lasher.pdf .

4. R. K. Ahluwalia, T. Q. Hua, J-K. Peng, and R. Kumar. System Level Analysis of Hydrogen Storage Options, *2009 DOE Hydrogen Program Review*, Arlington, VA. (2009). http://www.hydrogen.energy.gov/pdfs/review09/st_13_ahluwalia.pdf .

5. Y. Liu, H. Kabbour, C. M. Brown, D. A. Neumann, and C. C. Ahn, Increasing the Density of Adsorbed Hydrogen with Coordinatively Unsaturated Metal Centers in Metal-organic Frameworks, *Langmuir* 24, 4772 (2008).

6. Image credit: NIST-NCNR, http://www.nist.gov/ncnr/hydrogen_040108.cfm
7. Christian Doonan and Omar Yaghi, Hydrogen Storage in Metal-Organic Frameworks, *2009 DOE Hydrogen Program Review, Arlington*, VA. (2009). http://www.hydrogen.energy.gov/pdfs/review09/st_33_doonan.pdf
8. See for instance reports from the Hydrogen Sorption Center of Excellence in the DOE 2009 Annual Progress Report, http://www.hydrogen.energy.gov/annual_progress09_storage.html#c
9. Channing Ahn, Enhanced Hydrogen Dipole Physisorption. *DOE Hydrogen Program Annual Progress Report* (2009). http://www.hydrogen.energy.gov/pdfs/progress09/iv_c_11_ahn.pdf
10. B. Bogdanovic, and M. Schwickardi. 1997. Ti-doped Alkali Metal Aluminum Hydrides as Potential Novel Reversible Hydrogen Storage Materials. *J Alloys Compds*. 253/254:1-9 (1997).
11. V. Stavila, Sandia National Laboratories, publication in preparation.
12. C. L. Aardahl and S. D. Rassat, Overview of Systems Configurations for On-Board Chemical Hydrogen Storage, *Int. J. Hydrogen Energy* 34, 6676-6683 (2009).
13. Anthony Burrell, Chemical Hydrogen Storage R&D at Los Alamos National Laboratory, *2009 DOE Hydrogen Program Review* (2009). http://www.hydrogen.energy.gov/pdfs/review09/st_17_burrell.pdf]
14. Jason Graetz, and et. al., Aluminum Hydride Regeneration, *2009 DOE Hydrogen Program Review* (2009). http://www.hydrogen.energy.gov/pdfs/review09/st_05_graetz.pdf
15. S. Satyapal, J. Petrovic, C. Read, G. Thomas and G. Ordaz, The U.S. Department of Energy's National Hydrogen Storage Project: Progress towards meeting hydrogen-powered vehicle requirments, *Catalysis Today*, 246-256 (2007).

STRUCTURAL STUDY AND HYDROGEN SORPTION KINETICS OF BALL-MILLED Mg-10 wt%Ni ALLOY CATALYSED BY Nb

Sima Aminorroaya[1,3], Abbas Ranjbar[1,3], Younghee Cho[2,3], Hua Liu[1,2] , Arne Dahle[2,3]
1. The Institute for Superconducting and Electronic Materials, University of Wollongong, NSW 2522 Australia
2. Materials Engineering, the University of Queensland, Brisbane, QLD 4072, Australia
3. CSIRO National Hydrogen Materials Alliance, CSIRO Energy Centre, 10 Murray Dwyer Circuit, Steel River Estate, Mayfield West, NSW 2304, Australia.

ABSTRACT
    It is well known that Mg_2Ni facilitates hydrogen molecule dissociation in the composite Mg-Mg_2Ni and enhances hydrogen diffusion at phase boundaries. To further improve the hydrogen sorption kinetics, Nb has been introduced into Mg-Mg_2Ni composite. In the present study, the eutectic structure of cast Mg-10wt%Ni was refined by addition of 1wt% Nb into the melt during casting. Chips of cast Mg-10%Ni and Mg-10%Ni-1%Nb were ball milled separately and then with 5 wt% multi-walled carbon nanotubes (MW-CNTs), and 0, 1.5, 3, and 5 mol% Nb. Scanning electron microscope (SEM) analysis was carried out on cast and ball-milled samples to study the microstructure and distribution of Nb and Mg_2Ni. The absorption and desorption kinetics of samples were measured at 350, 250, and 200°C by Sievert's method apparatus. The results showed that 1%Nb addition during casting accelerates the hydrogen diffusion compared to cast Mg-10 wt%Ni without Nb. Moreover, addition of 5% MWCNTs to ball-milled powders improved the activation characteristics of the samples significantly. The absorption/desorption kinetics, as well as the hydrogen capacity of the sample containing 1.5% Nb were improved considerably even at temperatures as low as 100°C. Thermogravimetric analysis/ differential scanning calorimetry (TGA/DSC) results showed that the releasing temperature decreased by approximately 100°C compared to the ball-milled Mg-10 wt%Ni sample with no Nb.

INTRODUCTION
    Magnesium has been one of the most promising metals for hydrogen storage. Magnesium has low cost, high hydrogen capacity, excellent reversibility, and is relatively safe to handle. However, the very low hydriding/dehydriding kinetics of magnesium has been the main drawback for practical application [1]. Several attempts have been made to enhance the hydrogenation kinetics of magnesium by mixing and/or mechanical alloying with certain metals. Nickel has been the main candidate for this purpose. Nickel in magnesium is able to form Mg_2Ni compound, which has high hydrogen capacity and facilitates hydrogen molecule dissociation in the composite Mg-Mg_2Ni [2-5]. Mg_2Ni in magnesium is prepared by ball-milling of Ni with magnesium [2, 3], by casting [1, 6], or through the milling of Mg_2Ni and Mg [5]. Yim et al. [6] showed that eutectic Mg-23.5 wt% Ni alloy has the highest hydrogenation kinetics and hydrogen capacity among Mg-x wt% Ni (x = 13.5, 23.5 and 33.5) alloys. However, Mg-10wt% Ni has shown the highest hydriding rate and a relatively high dehydriding rate among Mg-Ni alloys prepared by mechanical alloying [2, 3]. Moreover, the hydrogen capacity of eutectic Mg-Ni alloy is less than 6 wt%, and therefore, cast Mg-10 wt% Ni alloy (with hydrogen capacity of 6.7 wt%) has been employed in the current study to explore the catalytic effect of Nb on its hydrogenation properties.
    Transition metals such as V, Ti, and Nb have been ball milled with magnesium as a catalyst to improve the hydrogenation kinetics [7]. Niobium and magnesium are immiscible, and therefore, ball milling provides heterogeneously catalysed materials. Various attempts have been made in the past to enhance the hydrogenation properties of magnesium hydrides with metallic Nb [7, 8], niobium oxide [9-11], niobium fluoride [12, 13], or the hydrogenation that occurs during milling of magnesium and niobium [14]. In the current study, less than 1 wt% Nb was added to Mg-10 wt% Ni alloy during casting to investigate the effects of intermetallic Nb-Ni phases on hydrogenation/dehydrogenation properties of the alloy, followed by milling of various amount of

metallic niobium with the alloy. We [15] recently found that the addition of 5 wt% multi-walled carbon nanotubes (MWCNTs) to Mg-Ni alloy offers excellent activation characteristics and leads to a pronounced improvement in the hydrogen storage properties. Therefore, 5 wt% MWCNTs was added to all ball-milled samples in the present study.

EXPERIMENTAL

Mg-10wt% Ni alloy with an average weight of 1 kg was melted in a steel crucible in an electric resistance furnace at 760°C under a cover gas (CO + SF$_6$) atmosphere. Commercial purity magnesium (99.8%) and nickel (99.8%) were used for alloying. Additions of Nb lower than ~1 wt% were made using Ni-60wt% Nb and NbF$_5$. The chemical composition of the cast alloys is summarized in Table I.

Chips of cast alloys were prepared by drilling the cast ingot. Then, the chips that were created were ball milled in a laboratory high energy mill, QM-3SP2 (from Nanjing University Instrument Plant) in a stainless steel vial and a ball-to-powder ratio of 20:1 for 60 hours at 400 rpm, which was followed by further milling with 5 wt% multi-walled carbon nanotubes (MWCNTs) for 2 hours. The handling of samples during milling was carried out under purified argon atmosphere in a glove box. Ball-milled powder of Mg10NiNb2 (see Table I) was ball milled with 1.5, 3, and 5 mol% Nb, respectively, for 20 hours prior to addition of 5 wt% MWCNTs.

The microstructure of cast samples, and the morphology and distribution of ball-milled powders were analysed using a JEOL 7500 high resolution scanning electron microscope (HR-SEM) equipped with an energy dispersive spectrometer (EDS). X-ray diffraction (XRD) patterns were obtained from ball-milled powders using a X-ray diffractometer with Cu K$_\alpha$ radiation (λ = 1.544 Å, 40 kV, 30 mA) from GBC Scientific Equipment Ltd. The hydrogen storage properties of the powders were measured by a volumetric method using Sievert's apparatus designed by Advanced Materials Corporation, PA, USA at 370°C and 20 atm. Calorimetric measurements were performed with a differential scanning calorimeter (DSC, Mettler Toledo) under a high purity Ar flow of 20 ml/min. The samples were heated from 25 to 400 °C with a 5 °C/min heating rate.

Table I: Chemical composition of cast Mg-Ni-Nb alloys.

Alloy name	Alloying description	Ni (wt%)	Nb (ppm)
Mg10Ni	Pure Mg and Ni powders	9.41	-
Mg10NiNb1	Nb addition using mechanically milled Ni60Nb master alloy	9.38	240
Mg10NiNb2	NbF$_5$ was added to Mg10NiNb1 for a consecutive increase in Nb amount	9.45	370

RESULTS AND DISCUSSION

Figure 1 shows the typical microstructure of Mg-10 wt% Ni cast alloy, consisting of a primary Mg dendrite and eutectic Mg-Mg$_2$Ni mixture. The cooling rate increment of a few tens of degree per second provides a high growth rate of the eutectic and refines its structure dramatically, with inter-lamellar spacing of a few hundred nanometres. Furthermore, refined eutectic Mg-Mg$_2$Ni structure is associated with a significant increase in interphase interfaces along which hydrogen migration occurs with a high diffusivity [16].

Transition metals, such as Ti, V, and Nb, are normally immiscible in magnesium, while they are able to form intermetallic phases with Ni during solidification. Figure 2 shows an Nb-Ni intermetallic phase as an example, which is frequently observed in the inter-dendritic region in grains approximately 10 μm in size.

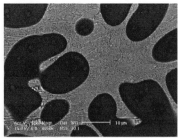

Figure 1: Backscattered electron micrograph showing α-Mg matrix and well refined eutectic structure of non-faceted Mg phase (in dark gray) and faceted Mg$_2$Ni (in white).

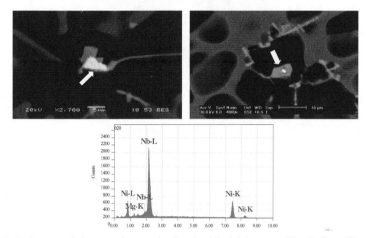

Figure 2: Backscattered electron micrographs showing Nb-rich intermetallics (indicated by arrows) formed in (a) Mg10NiNb1 alloy, and (b) Mg10NiNb2 alloy. (c) EDS analysis corresponding to Nb-rich intermetallic.

Figure 3 shows the hydrogen absorption kinetic curves measured for the Mg10Ni, Mg10NiNb1, and Mg10NiNb2 chips at an absorption temperature of 250°C. All samples were activated at 350°C and 2 MPa hydrogen pressure prior to the hydrogenation experiments. The addition of Nb in the range of 240–370 ppm to Mg-10wt% Ni alloys does not show any significant influence on hydrogen absorption kinetics.

The hydrogen storage capacity for the chips and ball-milled powders of Mg10NiNb1 and Mg10NiNb2 samples are shown in Figure 4. Chips absorb 2.9 wt% hydrogen within 30 min of exposure to hydrogen at 250°C and 2 MPa pressure, while the ball-milled samples absorb 5.8 wt% hydrogen within 20 minutes exposure to hydrogen. Ball milling improves the hydrogenation kinetics as a result of a particle/crystallite size reduction [11]. The higher surface area and high concentration of crystal defects, such as dislocations, stacking faults, vacancies and the number of grain boundaries, provide more nucleation sites that are available for hydride [11]. However, Schimmel et al. [7] have proven that the defects in the magnesium hydride structure play no major role in the hydrogenation/dehydrogenation kinetics, as they are annealed out during cycling. The significant improvement that is shown in the hydrogenation kinetics of ball-milled samples in comparison to as-cast chips in Figure 4 is caused by the considerable increment in surface area available for hydride formation as a result of ball milling.

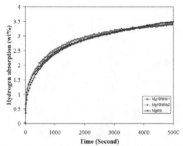

Figure 3: Hydrogen absorption curves of chips of Mg10Ni, Mg10NiNb1, and Mg10NiNb2 at 250°C under 2 MPa hydrogen pressure.

The ball-milled sample of Mg10NiNb2, which contains 370 ppm niobium, achieved 85% of the maximum hydrogen capacity within 2 minutes exposure to hydrogen, whereas ball-milled Mg10NiNb1 requires 10 minutes to absorb the same amount of hydrogen. Figure 4 demonstrates that an increment in niobium content from 240 to 370 ppm in the Mg-10wt% Ni significantly enhances the hydrogen absorption kinetics in ball-milled samples. However, its effect is negligible in the as-cast alloys (Figure 3). It is contended that the higher absorption kinetics in Mg10NiNb2 is due to the presence of intermetallic Ni-Nb phases, which were precipitated during solidification.

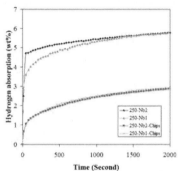

Figure 4: Hydrogen absorption curves of chips and ball-milled samples of Mg10NiNb1 and Mg10NiNb2, with the ball-milled samples containing 5 wt% MWCNTs, at 250°C under 2 MPa hydrogen pressure.

Figure 5 presents the hydrogen absorption kinetics curves measured for ball-milled powder of Mg10Ni, Mg10NiNb1, and Mg10NiNb2 samples, which were given an additional complementary ball milling with 5 wt% MWCNTs for 2 hours. The absorption measurements were performed at 200°C and 2 MPa hydrogen pressure. Figure 5 demonstrates that hydrogen capacity is increased by alloying of Mg10Ni cast samples, even by addition of 240 ppm niobium to the cast sample. It is proved that niobium addition to the cast alloy is able to effectively improve the hydrogenation kinetics as well as the maximum capacity.

The ball-milled Mg10NiNb2 sample revealed the best hydrogenation kinetics and capacity between three available alloys. Therefore, ball-milled powder of Mg10NiNb2 alloy was chosen to be alloyed by 1.5, 3, and 5 mol% niobium by further ball-milling processing. Niobium addition to Mg-10 wt% Ni alloy during the casting process forms Ni-Nb intermetallics. However, niobium and magnesium are immiscible, and therefore, niobium which is ball-milled with Mg10NiNb2 powder does not form any intermetallic phases.

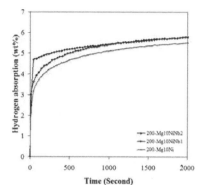

Figure 5: Hydrogen absorption curves of ball-milled samples of Mg10Ni, Mg10NiNb1, and Mg10NiNb2 at 200 and 250°C under 2 MPa hydrogen pressure.

The influence of milling with Nb on the structural characteristics of ball-milled Mg10NiNb2 alloy was further explored by X-ray diffraction. The X-ray diffraction patterns of the ball-milled powder and that of the powder ball-milled with 1.5, 3, and 5 mol% Nb for 40 hours, with all samples given a complementary milling with 5 wt% MWCNTs, are shown in Figure 6(a). Comparison of the XRD patterns reveals that the lattice parameters of magnesium and Mg_2Ni are identical to those of the starting materials, which indicates that no solid solubility occurs between the niobium, Mg_2Ni, and magnesium. The characteristic diffraction peaks of niobium appear on addition of 1.5 mol% niobium to the ball-milled Mg10NiNb2 sample, and their intensities are increased by increasing the amount of niobium added to the powder. The characteristic diffraction peak of MWCNTs does not appear in the XRD patterns. It can be concluded that the milling of MWCNTs with magnesium alloy, even for a short time (2 hours) is sufficient to disrupt the regular structure of the carbon nanotube layer and/or the amount of MWCNTs is lower than the detection limit of the X-ray diffractometer. Figure 6(b) presents a scanning electron micrograph of the surface of a particle after 2 hours of ball milling with MWCNTs, which clearly shows that the structure of the carbon nanotubes is not disrupted by ball milling. X-ray analysis of the samples was performed in air. and therefore, diffraction peaks of MgO are detected in all the histograms due to the thin layer of magnesium oxide which forms through exposure to the air.

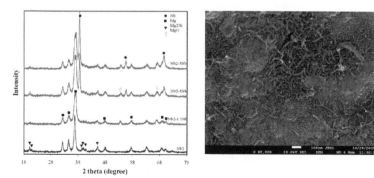

Figure 6: (a) X-ray diffraction patterns of Mg10NiNb2 powder, ball milled with 1.5, 3, and 5 mol% Nb for 40 hours, followed by milling with 5 wt% MWCNTs for 2 hours. (b) Scanning electron micrograph of the powder, illustrating the distribution of MWCNTs on the particle surface.

Figure 7 compares the hydrogenation kinetics of ball-milled Mg10NiNb2 with the same material ball-milled with 1.5, 3 and 5 mol% niobium for 40 hours, with all samples then ball milled for 2 hours with 5 wt% MWCNTs. Ball-milled powders of Mg10NiNb2 with 0, 1.5, 3, and 5 mol% Nb absorb 75%, 82%, 90% and 90%, respectively, of total hydrogen capacity of the sample within less than 2 minutes of exposure to hydrogen at 200°C. However, addition of Nb to the ball-milled sample reduces the total capacity of this sample. This might be due to the fact that magnesium is replaced by niobium, which is almost 4 times denser.

The desorption behaviour of the three alloys in the form of chips and ball-milled powder are compared in Figures 8(a) and 8(b), respectively, by constant heating rate differential scanning calorimetry (DSC). DSC samples were hydrogenated in the Sievert's apparatus. The endothermic peaks correspond to the decomposition of hydride phases. The DSC curves in Figure 8(a) show an endothermic peak associated with hydride decomposition, with onset at approximately 375°C for chips of all alloys, which is reduced to 345°C for ball-milled Mg10Ni alloy and approximately 325°C for the ball-milled Mg10NiNb1 and Mg10NiNb2 samples in Figure 8(b).

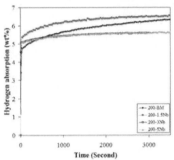

Figure 7: Hydrogen absorption curves of ball-milled Mg10NiNb2 alloy compared to those of the same material after additional ball milling with 1.5, 3 ,and 5 at% Nb for 40 hours. All samples were given a further 2 hours milling with 5 wt% MWCNTs at 200°C under 2 MPa hydrogen pressure.

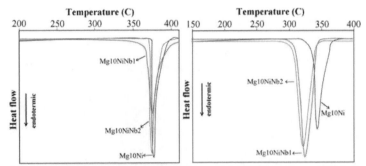

Figure 8: DSC curves of (a) Mg10Ni, Mg10NiNb1, and Mg10NiNb2 chips; and (b) ball-milled Mg10Ni, Mg10NiNb1, and Mg10NiNb2 alloys, which were additionally ball-milled with 5 wt% MWCNTs.

The desorption behaviour of Mg10NiNb2 alloy which is ball-milled with 1.5, 3, and 5 mol% Nb is compared with that of ball-milled Mg10NiNb2 powder with no additional Nb in Figure 9. The DSC curves in Figure 9 show an endothermic peak associated with hydride decomposition, with onset at approximately 325°C for ball-milled Mg10NiNb2, which is reduced to 263°C for Mg10NiNb2 that is ball-milled with 5 mol% Nb. In order to explain this phenomenon, the X-ray diffraction patterns of hydrogenated ball-milled Mg10NiNb2 alloy with 1.5, 3, and 5 mol% Nb powder are shown in Figure 13. All samples contain β-magnesium hydride, Mg_2NiH_4, and niobium hydride. Obviously, the intensity of the characteristic diffraction peaks of niobium hydride are increased by increasing the amount of niobium added to the original powder, which is consistent with the calorimetric measurements in Figure 9. It is contended that niobium hydride releases hydrogen at lower temperatures and acts as a gateway for dehydrogenation of magnesium nickel and magnesium hydrides.

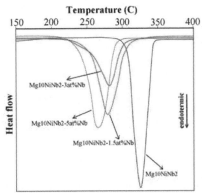

Figure 9: DSC curves of ball-milled Mg10NiNb2 alloy compared to that of the same powder ball-milled with 1.5, 3 and 5 atomic percent Nb for 40 hours. All samples were additionally ball-milled with 5 wt% MWCNTs.

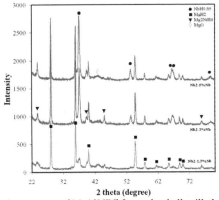

Figure 10: X-ray diffraction patterns of Mg10NiNb2 powder, ball-milled with 1.5, 3, and 5 mol% Nb for 40 hours, followed by milling with 5 wt% MWCNTs for 2 hours and hydrogenation in a Sievert's apparatus.

CONCLUSIONS

It is shown that Mg-10wt% Ni alloy contains a primary phase of magnesium, and a eutectic structure of $Mg-Mg_2Ni$. Nb-Ni intermetallic phase is also frequently observed in the inter-dendritic regions in approximately 10 μm in size.

Addition of Nb (less than 1 wt%) to Mg-10 wt% Ni alloy during casting improves the hydrogen absorption kinetics of ball-milled samples significantly. We contend that homogeneous distributions of intermetallic phases of Ni-Nb are responsible for the improvement. However, Nb addition in the cast samples has no significant influence on hydrogenation or dehydrogenation properties of the chips.

Ball milling of 1.5 mol% Nb with the ball-milled Mg10NiNb2 alloy improves the hydrogen absorption kinetics dramatically, but no further improvement is shown by addition of 3 and 5 mol% Nb. However, ball milling of various amounts of niobium with the ball-milled Mg10NiNb2 alloy does reduce the hydrogen desorption temperature..

It is proposed that niobium acts as a gateway for dehydrogenation of magnesium hydrides.

REFERENCES

[1]M.Y. Song, C.D. Yim, J.S. Bae, D.R. Mumm, and S.H. Hong: J. Alloys Compd., Vol. 463 (2008), p. 143-147.

[2]G. Liang, S. Boily, J. Huot, A. Van Neste, and R. Schulz: J. Alloys Compd., Vol. 267 (1998), p. 302-307.

[3]M.Y. Song: Int. J. Hydrogen Energy, Vol. 28 (2003), p. 403-408.

[4]L. Zaluski, A. Zaluska, and J.O. Strom-Olsen: J. Alloys Compd., Vol. 217 (1995), p. 245-249.

[5]A. Zaluska, L. Zaluski, and J.O. Strom-Olsen: J. Alloys Compd., Vol. 289 (1999), p. 197-206.

[6]C.D. Yim, B.S. You, Y.S. Na, and J.S. Bae: Catalysis Today, Vol. 120 (2007), p. 276-280.

[7]H.G. Schimmel, J. Huot, L. Chapon, F. Tichelaar, and F. Mulder: Journal of the American Chemical Society, Vol. 127 (2005), p. 14348-14354.

[8]J. Huot, J.F. Pelletier, L.B. Lurio, M. Sutton, and R. Schulz: J. Alloys Compd., Vol. 348 (2003), p. 319-324.

[9]G. Barkhordarian, T. Thomas Klassen, and R. Bormann: J. Alloys Compd., Vol. 407 (2006), p. 249-255.

[10]O. Friedrichs, T. Klassen, J.C. Sánchez-López, R. Bormann, and A. Fernández: Scr. Mater., Vol. 54 (2006), p. 1293-1297.

[11]M. Porcu, A.K. Petford-Long, and J.M. Sykes: J. Alloys Compd., Vol. 453 (2008), p. 341-346.

[12]Y. Luo, P. Wang, L.P. Ma, and H.M. Cheng: Scr. Mater., Vol. 56 (2007), p. 765-768.

[13]S.A. Jin, J.H. Shim, J.P. Ahn, Y.W. Cho, and K.W. Yi: Acta Mater., Vol. 55 (2007), p. 5073-5079.

[14]J.F.R. De Castro, S.F. Santos, A.L.M. Costa, A.R. Yavari, W.J.F. Botta, and T.T. Ishikawa: J. Alloys Compd., Vol. 376 (2004), p. 251-256.

[15]S. Aminorroaya, H.K. Liu, Y. Cho, A. Dahle: Int. J. of Hydrogen Enegry, Vol. 1 (2010), p. 4144-4153.

[16]J. Cermak and L. Kral: Acta Materialia, Vol. 56 (2008), p. 2677-2686.

MECHANICAL PROCESSING – EXPERIMENTAL TOOL OR NEW CHEMISTRY?

Viktor P. Balema
Aldrich Materials Science, Sigma-Aldrich Corp.
6000 N. Teutonia Av., Milwaukee, WI 53209 USA

ABSTRACT

The article highlights the preparation and modification of hydrogen-rich materials by mechanical processing of solids at ambient conditions in the absence of solvents. Possible mechanisms of mechanically induced chemical transformations in solids ranging from metal alloys to organic materials are discussed. Although exact mechanisms of particular processes have to be handled on a case-by-case basis, mechanically facilitated reactions of different solids share common features, which define mechanical processing as an enabling technique for a new type of chemistry - the Mechanochemistry.

INTRODUCTION

For more than a decade, hydrogen as an alternative to traditional energy sources such as oil, coal and natural gas has been the focus of research and development efforts in all technologically advanced countries of the world. It is strongly believed that hydrogen-based economy can resolve energy-related problems and slow down global climate change. Hydrogen can be produced from a variety of sources including renewables and photolytic water splitting.[1,2] It is non-toxic and, as an energy carrier, extremely environmentally benign. Water is the only product formed during the conversion of hydrogen into energy by oxidation.

Despite apparent benefits, an immediate incorporation of hydrogen into the world economy faces serious challenges. Unlike oil and natural gas, hydrogen has no large-scale infrastructure supporting its transportation. Although it is routinely used by chemical and refining industries, the cost of hydrogen storage and delivery is still too high for energy applications.

Storing hydrogen in solids - metal hydrides, composites or porous materials - offers a unique opportunity for its convenient and safe use in a variety of automotive, portable and stationary applications. Unfortunately, none of the materials currently on the market fully satisfies the needs of end users.[3] Therefore, search for new hydrogen storage media remains the focus of hydrogen-related research.

The article below provides a brief overview of an experimental approach, the mechanical processing, that has been used by practically every experimentalist involved in hydrogen storage research. The approach uses mechanical force in the form of milling or grinding for nano-scale design and modification of solid hydrogen storage media. The paper focuses on chemical events facilitated by mechanical processing and shows that mechanical milling and grinding have distinct chemical effect enabling a new type of chemistry – the Mechanochemistry.

MECHANICAL PROCESSING OF SOLIDS.

Although mechanical milling and grinding have been routinely used for the processing of solids for hundreds of years, their chemical effect was realized only in the second half of the last century.[4] By the beginning of the 21st century, mechanically induced chemical conversion of metals, also known as *mechanical alloying*, evolved into an experimental technique routinely used for the preparation of metal alloys, intermetallics, composites[5] as well as hydrogen energy storage media.[6-8]

Figure 1. A planetary mill (Pulverisette 5/4, Fritsch and milling containers,left and center), and a shaker mill (Spex 8000M, right). (Images provided by manufacturers).

Usually, mechanical alloying is performed by ball-milling of solids in tightly sealed containers (Fig.1), which can be loaded and unloaded under inert gas in a glove box. In a typical experiment, solids are charged into the milling container and ball-milled for a defined period of time. Once milling is complete, the material is recovered from the container and studied using solid-state analytical techniques such as X-ray or neutron powder diffraction, the magic angle spinning solid-state nuclear magnetic resonance (MAS NMR), IR/UV-spectroscopy, thermal analysis and microscopy.[4-6,9] It is worth noting that MAS NMR has recently proven particularly useful for monitoring chemical transformations in metal hydrides and other hydrogen-rich solids.[6,8-10]

An external cooling is often used to prevent the material from overheating during prolonged milling.[4-6,9] Unfortunately, the temperature inside the processed material is very difficult to manage and a significant effort has been invested into the better understanding of the conditions existing in solids during their milling.[4,5,11-13] Since it is usually carried out in nontransparent thick-walled vials with fast moving metal or ceramic balls (Fig.1), the direct temperature measurements inside the vial are an extremely challenging task. Thus, the temperature in the material during its processing has been estimated either by theoretical modeling or by milling substances that chemically transform or decompose at well-defined temperatures.

The temperature rise in metals, alloys and intermetallic compounds during mechanical alloying was estimated using a number of different theoretical models, which produced an extremely broad distribution of temperature values ranging from 50 °C to 1000°C.[4,5] At the same time, the calculations performed for the specific case of milling in a Spex 8000 mill revealed that the temperature increase inside the material is probably at the lower end on this range, i.e. it does not exceed 120°C.[11-13] According to the same studies, the pressure generated in the material trapped between colliding balls or balls and container walls can reach several GPa,[11-14] which is quite sufficient for triggering chemical reactions in solids.[15]

The temperature rise in molecular and ionic materials during their processing was also estimated experimentally by ball-milling thermolabile 7-triphenylmethylcycloheptatriene[16] and ammonium carbonate .[6] 7-Tiphenylmethylcycloheptatriene undergoes a quick radical dissociation and recombination above 60°C, while ammonium carbonate decomposes irreversibly above 60 °C (Fig. 2).

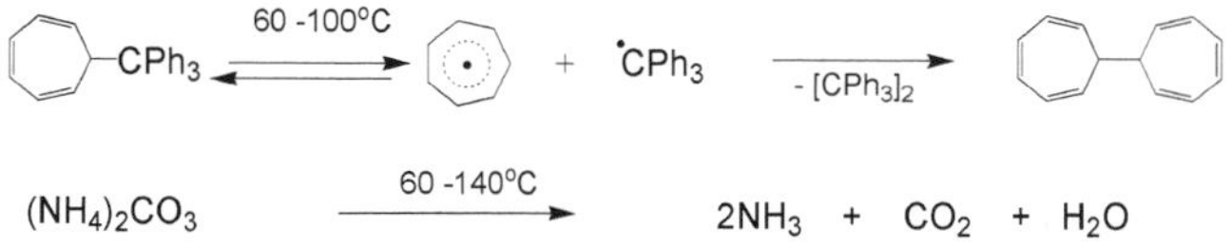

Figure 2. Thermally driven transformations of 7-triphenylmethylcycloheptatriene and ammonium carbonate

The ball-milling in a shaker mill for time periods ranging from between 30 min to 20 hours did not cause any noticeable changes in the tested materials indicating that the temperature in their bulk remained below 60 °C throughout the entire process.

PREPARATION AND MODIFICATION OF HYDROGEN STORAGE MATERIALS

Initially, mechanical processing was applied to metallic hydrogen absorbers and materials for nickel-metal hydride batteries, whereas it produced materials with hydrogen storage properties superior to those of the materials prepared using conventional metallurgical techniques.[17,18] In some cases, ball-milling of metallic hydrogen absorbers generated meta-stable phases, which ability to absorb and release hydrogen exceeded that of the parent materials.[19] It should be noted, however, that prolonged ball-milling of metallic phases often leads to their partial or complete amorphization[4,5], which may reduces the material's ability to retain hydrogen in the lattice.

While applied to stoichiometric metal hydrides, such as alkali metal aluminum hydrides (also known as alanates)[6,20], ball-milling does not alter their hydrogen content but, instead, reduces hydrogen release temperature by 20-50 °C. Also, mechanical processing proved capable of facilitating solvent-free chemical reactions between different metal hydrides and inorganic salts, and provided a unique insight into the solid-state chemistry of hydrogen storage materials (Eqs.1-11 and Fig.3).

Thus, ball-milling of alkali metal aluminohydrides with binary metal hydrides enabled the preparation of alkali metal hexahydroaluminates, which are hardly accessible through the conventional solution-based chemistry (Eq. 1). It became also evident that numerous chemical transformations in alkali metal – aluminum/boron – hydrogen systems do not, in fact, require a solvent and can be successfully run under solvent-free conditions (Eqs. 1-6).[20-27]

$$2MH \ + \ MAlH_4 \ \longrightarrow \ M_3AlH_6 \qquad (1)$$

$$MH \ + \ AlH_3 \ \longrightarrow \ MAlH_4 \qquad (2)$$

$$2MH \ + \ 2Al \ + \ 3H_2 \ \longrightarrow \ 2MAlH_4 \qquad (3)$$

$$M_3AlH_6 \ + \ 2AlH_3 \ \longrightarrow \ 3MAlH_4 \qquad (4)$$

M = Li, Na

$$TiCl_3 \ + \ 3LiBH_4 \ \xrightarrow{-3LiCl} \ Ti(BH_4)_3 \qquad (5)$$

$$LiH \ + \ H_3NBH_3 \ \xrightarrow{- H_2} \ LiH_2NBH_3 \qquad (6)$$

Solvent-free mechanical processing was also successfully applied to the preparation of a broad variety of binary and complex metal hydrides including the generation of alane and diborane. A few examples of such reactions are shown in Equations 7-11.[26,28-32]

$$2M \quad + \quad xH_2 \quad \longrightarrow \quad 2M(H)x \quad (7)$$

$$MgCl_2 \quad + \quad 3LiAlH_4 \quad \xrightarrow{-2LiCl} \quad LiMg(AlH_4)_3 \quad (8)$$

$$MgH_2 \quad + \quad 2B_2H_6 \quad \longrightarrow \quad Mg(BH_4)_2 \quad (9)$$

$$3LiAlD_4 \quad + \quad AlCl_3 \quad \xrightarrow{-3LiCl} \quad 4AlD_3 \quad (10)$$

$$SnCl_2 \quad + \quad 2NaBH_4 \quad \xrightarrow{-2NaCl} \quad B_2H_6 + H_2 + Sn \quad (11)$$

Furthermore, ball-milling experiments helped to identify possible intermediates, aluminum hydride species, which may be responsible for the complex set of transformations occurring in alkali metal alanates under solvent-free conditions (Fig. 3).[20-25]

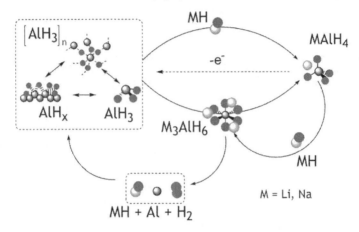

Figure 3. Mechanochemical transformations in the alkali metal – aluminum – hydrogen system.

Finally, mechanochemical approach helped to uncover the species likely responsible for the ability of Ti-salts to catalyze the decomposition of aluminum-based hydrides,[6-9, 33] including the conversion of LiAlH$_4$ into Li$_3$AlH$_6$, Al and H$_2$ upon ball-milling with catalytic amounts of TiCl$_4$ (Eq. 12 and Fig. 4) – the reaction that does not occur in the absence of mechanical processing.[34]

$$LiAlH_4 \quad \xrightarrow[- LiCl]{TiCl_4; \, 3 \, wt.\%} \quad Li_3AlH_6 + Al + H_2 \quad (12)$$

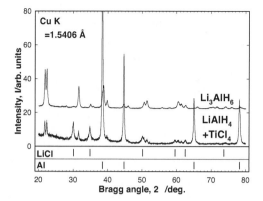

Figure 4. The transformation of $LiAlH_4$ in the presence of catalytic amount of $TiCl_4$ upon the ball-milling for 5 min.

It turned out that once added to alkali metal aluminum hydrides and ball-milled, titanium salts transform into a Al_3Ti alloy (Eq. 13)[6,9] and a minor amount of an instable TiH_x phase[35], which can act as a catalyst.[36-38] It is worth noting that titanium trichloride reacts with magnesium hydride under similar conditions giving rise to titanium hydride and magnesium dichloride (Eq. 14).[39]

$$n\,MAlH_4 \;+\; TiCl_n \xrightarrow[\substack{-\,nMCl \\ -\;Al}]{} TiAl_3/TiH_x \;+\; H_2 \quad (13)$$
$$M = Li,\,Na;\; n = 3,4$$

$$MgH_2 \;+\; TiCl_3 \xrightarrow[MgCl_2]{} TiH_2 \qquad (14)$$

The number of reports on the mechanochemical preparation of various materials for hydrogen storage applications is constantly growing. Further examples, including the preparation of metal hydrides, alloys, amides, nitrides and boron-based materials, can be found in research papers and reviews published elsewhere.[6-8, 40-44]

ABOUT THE MECHANISM OF MECHANICALLY INDUCED TRANSFORMATIONS IN SOLIDS

Even a quick look at the previous chapter is sufficient to realize the diversity of the processes that take place in solids trapped between steel or ceramic balls colliding in a tightly-closed milling container. Indeed, mechanical processing brings about a broad variety of defects such as cracks, pores, dislocations, vacancies and constantly creates new surface in the processed material. Prolonged milling or even grinding with mortar and pestle can destroy crystallinity of the material and lead to its partial or even complete amorphization.[4-6]

Changes in solids under mechanical stress (Fig. 5) start with elastic deformations, which can disappear once the load is lifted. If, however, the load increases, the elastic deformations transform into irreversible plastic deformations, including shear deformations, which are followed by the fracture and/or amorphization of the material.

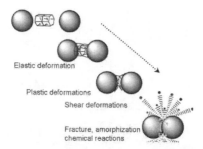

Elastic deformation

Plastic deformations

Shear deformations

Fracture, amorphization
chemical reactions

Figure 5. Changes in the material trapped between colliding balls during ball-milling

The constant formation of a fresh surface and the mass transfer facilitated by milling appear to be the major factors responsible for enhanced reactivity of solid metals and metal hydrides towards gaseous hydrogen or diborane (Eqs. 7 and 9).[6] However, these effects cannot be solely responsible for the solid to solid transformations mentioned in the previous section.

Prolonged milling of metals often produces amorphous phases, which crystallize upon subsequent heat treatment into metal alloys or intermetallic compounds.[4,5] In the case of non-metallic substances, metal salts or complex metal hydrides, mechanical processing can facilitate the formation of disordered solid solutions,[45,46] which may serve as media for chemical processes. This idea originates in mechanochemical experiments involving organic and metalorganic materials (Fig. 7), which account for a substantial fraction of known mechanically facilitated chemical transformations.

Similar to the transformations of metal hydrides, mechanochemical reactions of organic compounds do not require solvents and are as efficient as similar transformations in solution.[47-52] Often, they can be carried out in the same equipment that is used for mechanical processing of metal hydrides or alloys. For example, the transformations shown in the equations 4, 6, 12, 13 and in the figures 2 and 7 have been run using the same milling equipment under identical experimental conditions (Spex 8000 mill; helium atmosphere; forced air cooling at ambient temperature).[6,9,19,34,47-49]

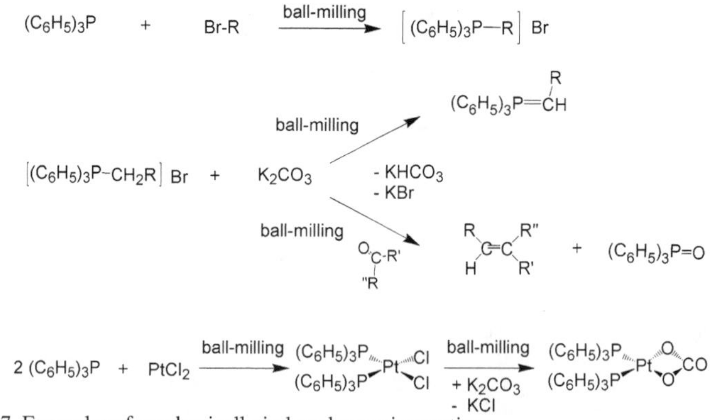

$$(C_6H_5)_3P \quad + \quad Br\text{-}R \quad \xrightarrow{\text{ball-milling}} \quad \left[(C_6H_5)_3P\text{—}R\right] Br$$

$$\left[(C_6H_5)_3P\text{-}CH_2R\right] Br \quad + \quad K_2CO_3 \quad \xrightarrow[\text{- KBr}]{\text{ball-milling}}^{\text{- KHCO}_3}$$

$(C_6H_5)_3P\!\!=\!\!\overset{R}{\underset{}{C}}H$

$$\xrightarrow{\text{ball-milling}} \quad \underset{"R}{\overset{O}{\underset{}{}}}C\text{-}R' \qquad \underset{H}{\overset{R}{}}C\!\!=\!\!C\overset{R''}{\underset{R'}{}} \quad + \quad (C_6H_5)_3P\!\!=\!\!O$$

$$2\,(C_6H_5)_3P \quad + \quad PtCl_2 \quad \xrightarrow{\text{ball-milling}} \quad \underset{(C_6H_5)_3P}{\overset{(C_6H_5)_3P}{}}Pt\overset{Cl}{\underset{Cl}{}} \quad \xrightarrow[\text{- KCl}]{\text{ball-milling}}^{\text{+ K}_2\text{CO}_3} \quad \underset{(C_6H_5)_3P}{\overset{(C_6H_5)_3P}{}}Pt\overset{O}{\underset{O}{}}CO$$

Figure 7. Examples of mechanically induced organic reactions.

Until recently, mechanically induced chemical reactions of organic compounds were regarded as true solid-state processes.[51,52] However, the latest studies have revealed a distinct possibility that they really occur in a liquid phase.[49] For example, the reaction between o-vanillin (oV) and p-toluidine (pT), when they are ground together at about 0 °C and slowly warmed up to room temperature, appears to be a solid-state process (Fig.8). In reality, it takes place in a liquid eutectic, which forms upon mechanical processing and remains hidden behind solid reactants and the reaction product.[49]

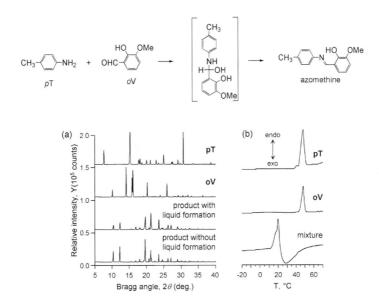

Figure 8. The solvent-free reaction of o-vanillin (oV) with p-toluidine (pT): (a) XRD patterns of pT, oV, and the reaction product, azomethine (b) DSC trace of pT, oV, and their mixture grinded at 0°C; the eutectic melts at ~10°C.

Ball-milling or grinding of organic solids, which are incapable of reacting with each other, also generates mixed amorphous phases that can further crystallize as eutectics[48,50] or mixed crystals, co-crystals.[53,54] Mixed phases can also serve as reaction media for those organic materials and metal hydrides, which do not form low-melting eutectics or are instable in the melt. An insight into the forces driving mechanochemical transformations in the solid state is provided by the photochemical dimerization of anthracene (Fig. 9).[55,56] It readily occurs in solution but does not take place in the solid state because of an unfavorable orientation of anthracene molecules in the crystal. No photochemical reaction is also observed if crystalline anthracene is subjected to the hydrostatic pressure up to 10 GPa.[55] However, once an external pressure is combined with a shear deformation, the photochemical dimerization reaction becomes possible.[56] Apparently, high pressure and shear deformation not only reduce the distance between molecules in the solid but also change their orientation thus enabling this and other similar solid-state transformations.[57,58]

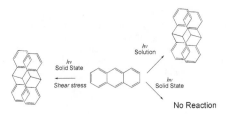

Figure 9. Photochemical dimerization of anthracene.

As mentioned above, both high pressure and shear deformations are generated in the solid particles trapped between colliding balls during ball-milling. If these chemically active events are combined with the amorphization and intimate mixing of reacting materials, they would create conditions sufficient for triggering mechanochemical transformations in solids without an intermediate formation of liquid phases.

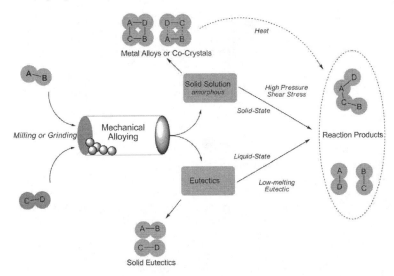

Figure 10. Mechanically induced transformations of solids - likely scenarios.

Summarizing the facts available to date (Fig.10), it is reasonable to assume that the formation of mixed amorphous and/or solid solution phases-like phases, which represents the mechanical alloying step of mechanical processing, is a common feature for all mechanically driven chemical transformations regardless of the materials involved. In the case of metals, the mechanical processing concludes at the alloying stage, while milling or grinding of ionic or molecular (organic) materials may produce low melting eutectics and facilitate chemical reactions in the melt. High pressure and shear stress generated in solids disordered by milling are probably the factors responsible for the chemical conversion of those materials that do not form liquid eutectics but, nevertheless, react with each other upon mechanical processing.

CONCLUSIONS

Mechanical processing has proven to be an extremely useful tool for the preparation and modification of hydrogen storage materials. Although the exact mechanism of each particular mechanochemical transformation has to be determined on a case-by-case basis, it seems quite possible that mechanochemical transformations of different solids share common features and the knowledge acquired in one area of mechanochemistry can help to better understand chemical events in other areas of chemistry and materials science. The mechanical processing is a crucial element of mechanically driven chemical reactions because it enhances interactions between reacting materials, enables their intimate mixing, provides mass transfer and generates high pressure and shear deformations facilitating such transformations.

ACKNOWLEDGEMENTS

The author would like to express his gratitude to his current colleague Dr. Shashi Jasty (Aldrich Materials Science) and the former co-workers and current collaborators Dr. Vitalij Pecharsky, Dr. Karl Gschneidner, Jr., Dr. Jerzy Wiench, Dr. Marek Pruski and Dr. Alex Dolotko (all the Ames Laboratory of the US DOE) for valuable contributions and support that enabled this publication.

REFERENCES

[1] D. Nocera, *Chem. Sus. Chem.*, **2**, 387 (2009).

[2] http://www.hydrogen.energy.gov/annual_progress08.html (The author does not assume liability for the content or availability of this URL.)

[3] C. Read, G. Thomas, G. Ordaz, and S. Satypal, *Material Matters*, **2.2**, 3 (2007).

[4] G. Heinicke, *Tribochemistry*; Academie Verlag: Berlin, 1984.

[5] C. Suryanarayana, *Progr. Materials Sci.*, **46**, 1 (2001).

[6] V.P. Balema, L. Balema, *Phys. Chem. Chem. Phys.* **7**, 1310 (2005).

[7] S. Orimo, Y. Nakamori, J.R. Eliseo, A. Züttel, and C.M. Jensen, *Chem. Rev.* **107**, 4111(2007).

[8] W. Grochala, P. Edwards, *Chem. Rev.* **104**, 1283 (2004).

[9] V.P. Balema, J.W. Wiench, K.W. Dennis, M. Pruski and V.K. Pecharsky, *J. Alloys Compd.* **329**, 108 (2001).

[10] R.C. Bowman, S.J. Hwang, *Material Matters, 2.2*, 29 (2007).

[11] D.R. Maurice, T. H. Courtney, *Metall. Trans. A* **21A**, 289 (1990).

[12] D.R. Maurice, T. H. Courtney, *Metall. Trans. A* **26A**, 2432 (1995).

[13] D.R. Maurice, T. H. Courtney. *Metall. Trans. A,* **27A**, 1981 (1996).

[14] C. C. Koch, *Int. J. Mechanochem. Mech. Alloying*, **1**, 56 (1994).

[15] *High Pressure Molecular Science, NATO Science Series, E358*; Eds. Winter R.; Jonas J.; Kluwer Academic Publishers: Dordrecht, Boston, London, 1999.

[16] K. Komatsu, *Top. Curr. Chem.* **254**, 185 (2005).

[17] S.Orimo, H.Fujii, *Appl.Phys. A.* **72**, 157 (2001).

[18] P.G. McCormick in Handbook *on the Physics and Chemistry of Rare Earth*, K.A. Gschneidner, Jr., and L. Eyring (Eds.), Vol. 24, Elsevier Science B.V. 1997, p. 47.

[19] V.P. Balema, A.O. Pecharsky, and V.K. Pecharsky, *J. Alloys Compds* **307**, 184 (2000)

[20] V.P. Balema, V.K. Pecharsky, and K.W. Dennis, *J. Alloys Compds*. **313**, 69 (2000).

[21] T.N. Dymova, D.P. Aleksandrov, V.N. Konoplev, T. Silina, and N.T. Kuznetsov, *Russian J. Coord. Chem.* **19**, 491 (1993).

[22] S. Chaudhuri, J. Graetz, A. Ignatov, J.J. Reilly, and J.T. Muckerman, *J. Am. Chem. Soc.*, **128**, 11404 (2006).

[23] J.M. Bellosta von Colbe, M. Felderhoff, B.Bogdanovic, F. Schüth, and C. Weidenthaler, *Chem. Commun.* 4732 (2005).

[24] Y. Kojima, T. Kawai, T. Haga, M. Matsumoto, and A. Koiwai, *J. Alloys Compds.* 441, 189 (2007).

[25] J. Hout, S. Boily, V. Gunther, and R. Schulz, *J. Alloys Compds.* **283**, 304 (1999).

[26] V.V. Volkov, K.G. Myakishev, *Inorg. Chim. Acta*, **289**, 51 (1999).

[27] Zh. Xiong, Ch. K. Yong, G. Wu, P. Chen, W. Shaw, A. Karkamkar, T. Autrey, M.O. Jones, S.R. Johnson, P.P. Edwards, and W. I. F. David, *Nature Materials*, **7**, 138 (2008).

[28] S. Chen, J. Williams, *Mater. Sci. For 225/ 227*; TransTech Publications Inc.: Stafa-Zurich, 1996; p 881.

[29] M. Mamathab, B. Bogdanović, M. Felderhoff, A. Pommerin, W. Schmidt, F. Schüth, and F. Weidenthaler, *J. Alloys Compds*, **407**, 78 (2006).

[30] K. Chlopek, C. Frommen, A. Leon, O. Zabara, and M. Fichtner, *J. Mart.Chem.* **17**, 3496 (2007).

[31] H.W. Brinks, A. Istad-Lem, and B.C. Hauback, *J. Phys. Chem.* **110**, 25833 (2006).

[32] J. Chen, H.T. Takeshita, D. Chartouni, N. Kuriyama, and T. Sakai, *J. Mater. Sci.* **36**, 5829 (2001).

[33] J.L. Herberg, R.S. Maxwell, and E.H. Majzoub, *J. Alloys Compds.* **417**, 39 (2005).

[34] V.P. Balema, K.W. Dennis, and V.K. Pecharsky, *Chem. Commun.* 1665 (2000).

[35] P.E.Vullum, M.Pitt, J.Walmsley, B.Hauback, and R. Holmestad, *Appl.Phys. A: Mat. Sci. Proc.* **94**, 787 (2009).

[36] P. Wang, X.-D. Kang, and H.-M. Cheng, *J. Phys. Chem. B*, **109**, 20131 (2005).

[37] A. Marashdeh, R.A. Olsen, O. M. Løvvik and G.-J. Kroes, *J. Phys. Chem. C*, **112**, 15759 (2008).

[38] S. Li, R. Ahuja, C. M. Araújo, B. Johansson, and P. Jena, *J. Phys. Chem. Solids* doi:10.1016 / j.physletb.2003.10.071 (2010) .

[39] J. Charbonnier, P. de Rango, D. Fruchart, S. Miraglia, L. Pontonnier, S. Rivoirard, N. Skryabina, and P. Vulliet, *J. Alloys Compds.* **383**, 205 (2004).

[40] V. Iosub, T. Matsunaga, K. Tange, M. Ishikiriyama, and K. Miwa, *J. Alloys Compds.* **484**, 426 (2009).

[41] J. Zhang, W. Yan, C. Bai, and F. Pan, *J. Mat. Res.* **24**, 2880 (2009).

[42] C. Kim, S.J.Hwang, R.C. Bowman, J. W. Reiter, J.A.Zan, J. G. Kulleck, H. Kabbour, E.H. Majzoub, and V. Ozolins, *J. Phys. Chem. C* **113**, 9956 (2009).

[43] T. Ichikawa, H.Y. Leng, S. Isobe, N. Hanada, and H. Fujii, *J. Power Sources* **159**, 126 (2006).

[44] R.J. Newhouse, V. Stavila, S.-J.Hwang, L.E. Klebanoff, and J.Z.Zhang, *J. Phys. Chem. C*, **114**, 5224 (2010).

[45] V. Iosuba, T. Matsunagaa, K. Tangea, M. Ishikiriyamaa, and K. Miwab, *J. Alloys and Compds* **484**, 426 (2009).

[46] L. M. Arnbjerg, D.B. Ravnsbæk, Y. Filinchuk, R.T. Vang, Y. Cerenius, F. Besenbacher, J.- E. Jørgensen, H. J. Jakobsen⊥ and T. R. Jensen, *Chem. Mater.* **21**, 5772 (2009).

[47] V.P. Balema, J.W. Wiench, M. Pruski, and V.K. Pecharsky. *J. Am. Chem. Soc.* **124**, 6244 (2002)

[48] V.P. Balema, J.W. Wiench, M. Pruski, and V.K. Pecharsky, *Chem. Commun.* 724 (2002).

[49] A. Dolotko, J.W. Wiench, K. Dennis, V.K.Pecharsky, and V.P. Balema, *New J. Chem.* **34**, 25 (2010)

[50] G. Rothenberg, A.P. Downie, C.L. Raston, and J.L. Scott, *J. Am. Chem. Soc.* **123**, 8701 (2001).

[51] K. Tanaka, F. Toda, *Chem. Rev.* **100**, 1025 (2000).

[52] G. Kaupp, *Top. Curr. Chem.* **254**, 95 (2005).

[53] R. Kuroda, J. Yoshida, A. Nakamura, and Sh.-i. Nishikiori, *CrystEngComm* **11**, 427 (2009).

[54] D. Braga, F. Grepioni, *Chem. Commun.* 3635 (2005).

[55] M. Oehzelt, R. Resel, *Phys. Rev. B.* **66**, 174104 (2002).

[56] A.A. Politov, B.A. Fursenko, V.V. Boldyrev, *Doklady Phys. Chem.* **371**, 28 (2000).

[57] Zharov in *High Pressure Chemistry and Physics of Polymers*, Ed.: A.L. Kovarskii, CRC Press, Boca Raton, 1994, 267 p.

[58] J.J. Gilman in *High-pressure shock compression of solids VI: old paradigms and new challengesy* Eds.: Y. Horie, L. Davison, N.Thadani,, Springer, Heidelberg, New York, 2003, p. 111.
[59] M. Matsuoka, K. Danzuka, *J. Chem. Eng. Japan* **42,** 393 (2009).

PRODUCTION OF HYDROGEN AND CARBON MONOXIDE FROM WATER AND CARBON
DIOXIDE THROUGH METAL OXIDE THERMOCHEMICAL CYCLES

Eric N. Coker, Andrea Ambrosini, Mark A. Rodriguez, Terry J. Garino and James E. Miller
Sandia National Laboratories, PO Box 5800, Albuquerque, NM 87185-1349, USA.

ABSTRACT

Two-step thermochemical cycles using ferrite-based materials to split water and carbon dioxide are promising routes for the production of H_2 and CO (syngas). To aid in the design of highly efficient materials for H_2 and CO production, this work aims to identify the metal oxide phases present during thermochemical cycling and how they change as a function of temperature and gas composition. High-temperature X-ray diffraction (HT-XRD) was used to monitor the structure of iron oxides supported on YSZ (10 wt.-% Fe_2O_3 basis) and cobalt-substituted ferrites during thermochemical cycling. HT-XRD showed dynamic behavior as iron migrated into and out of YSZ at elevated temperatures, monitored by the lattice parameter of the YSZ. Iron oxides were seen to thermally reduce stepwise from Fe_2O_3 to Fe_3O_4 and finally FeO as the temperature increased from ambient to 1400 °C under He with a low background of O_2. Between 800 and 1100 °C no iron species were detected, indicating that all iron was in solid solution with YSZ. Similar cycles were performed with a cobalt-substituted ferrite which exhibited similar phase evolution. Exposure of $Fe_xCo_{1-x}O$ to CO_2 or air resulted in re-oxidation to $Fe_{3x}Co_{3-3x}O_4$. Thermogravimetric analysis corroborated the reduction/oxidation behavior of the materials during thermal reduction and subsequent re-oxidation by H_2O or CO_2. A complimentary study on diffusion of iron oxide into YSZ revealed a steep increase in diffusion rate once temperatures exceeded 1475 °C. Fusion and vaporization of iron species at these high temperatures occurs.

INTRODUCTION

The primary goal of this work is to lay the foundation to enable the synthesis of hydrocarbon fuels from carbon dioxide and water using concentrated solar power as a heat source to drive a two-step solar-thermochemical cycle. This process can be described as a way to "re-energize" CO_2 and H_2O, which are the thermodynamically stable products of hydrocarbon combustion (Figure 1A). Once CO_2 and H_2O have been re-energized (reduced) to CO and H_2, traditional syngas chemistry can be applied to convert these products into hydrocarbon fuels.

Solar-driven two-step ferrite (e.g., Fe_3O_4) thermochemical cycles are promising as a method for producing H_2 and CO via H_2O- and CO_2-splitting,[1,2,3,4,5] as illustrated in simplified form in Figure 1B. The basic cycles consist of a thermal reduction step (TR; reaction (1)) in which solar thermal energy reduces Fe^{III} to Fe^{II}, i.e., spinel transforms to wüstite, followed by a water-splitting step (WS; reaction (2)), or carbon dioxide-splitting step (CDS; reaction 3) wherein the ferrite spinel is regenerated:

$$Fe_3O_4 \quad \rightarrow \quad 3FeO + 0.5\,O_2 \tag{1}$$
$$3FeO + H_2O \quad \rightarrow \quad Fe_3O_4 + H_2 \tag{2}$$
$$3FeO + CO_2 \quad \rightarrow \quad Fe_3O_4 + CO \tag{3}$$

However, hydrogen production using Fe_3O_4, originally proposed by Nakamura,[6] is not practical since the TR requires 1800 °C resulting in sintering or fusion that must be undone by, e.g., mechanical crushing or milling in order to activate the material for successive cycles; supporting Fe_3O_4 on zirconia, or yttria-stabilized zirconia (YSZ) reduces this problem.[7,8] Alternative redox systems using $A_xFe_{3-x}O_4$ where $A \neq Fe$ enable reduced temperature TR, e.g., A = Mn, Co, Ni, Zn, and are now receiving considerable attention.[9] The TR can be driven as low as 1100 °C, although kinetics usually dictate that temperatures above 1300 °C are used, which are readily achievable using concentrated

37

solar-thermal energy. Yields of H_2 and CO are maximized in the range 1080 – 1230 °C. The requirements of large temperature swings as well as spatial/temporal isolation of the TR and WS/CDS reactions to avoid energetic re-combination of O_2 and H_2/CO, were addressed in the design of Sandia's Counter-Rotating-Ring Receiver Reactor Recuperator (CR5), described in detail elsewhere[10] and shown in concept in Figure 2. The work described here focuses on some of the materials fundamentals in an effort to understand reaction pathways and enable efficient H_2 and CO production using the CR5.

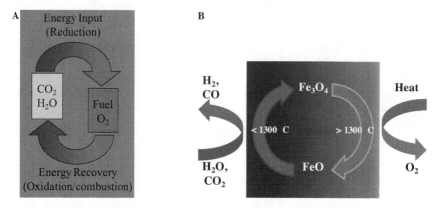

Figure 1. Process fundamentals. By re-energizing CO_2 and H_2O, we can effectively close the carbon cycle and recycle combustion products into fuel (1A); a simplified representation of a thermochemical cycle to convert CO_2 or H_2O to CO or H_2 using a ferrite material (1B).

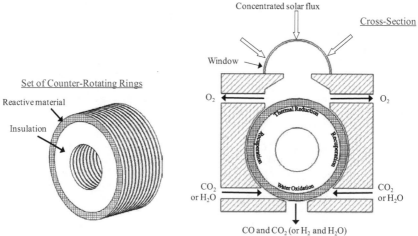

Figure 2. Design of Sandia's prototype CR5 reactor for conversion of H_2O to H_2, and CO_2 to CO using concentrated solar power.

Significant advances have been made in the field of solar thermochemical H_2O- and CO_2- splitting technologies using metal oxides; however a lack of fundamental research into the behavior of the metal oxides under the high temperature conditions present in these cycles has hampered materials development. Basic questions relating to oxygen transport, surface chemistry, structural changes vs. redox reactions, materials synthesis methods, effects of thermochemical cycling on the material, and the role of supports have still to be addressed.

While it is known that the ad-mixing of a high temperature-stable support, such as ZrO_2 or YSZ to the ferrite is necessary in order for the process to be cyclable,[4] the interaction between the reactive material (e.g., ferrite) and the support is largely un-explored in the high temperature environment relevant to thermochemical processing.

The current work has probed the complex interaction between the ferrite and the support, with particular emphasis on investigating solid solubility and phase evolution during thermal processing. The iron oxide/YSZ system was chosen for study based on a combination of its effectiveness in CDS and WS, and because an understanding of the basic Fe_3O_4 system can be used as a basis to understand the more complex systems such as $Co_xFe_{3-x}O_4$ and other substituted ferrites which are currently under investigation. A considerable body of published work exists describing the physico-chemical properties of iron oxide-zirconia and iron oxide-YSZ, including their in-situ characterization at temperatures up to ~ 1200 °C [11,12,13,14,15,16,17,18,19,20,21,22,23,24]; however, the behavior of these materials at temperatures up to ~1500 °C has not been evaluated in-situ, and is of importance for the development of solar thermochemical cycles. While many studies have used post mortem analysis of high temperature-treated materials, the data obtained may not accurately reflect the properties of a material under high temperature conditions. Thus, our work has attempted to monitor the materials during simulated thermochemical cycle operation up to 1500 °C. In addition, some preliminary observations of the behavior of un-supported cobalt-substituted ferrites under CO_2-splitting conditions are reported here. Finally, we present some data exploring the fusion and volatility of iron species in Fe_2O_3/8YSZ samples at various temperatures and under various dynamic atmospheres. The principal techniques employed in the current work include high-temperature X-ray diffraction (HT-XRD) and thermogravimetric analysis (TGA) under conditions simulating TR/WS and TR/CDS reactions.

EXPERIMENTAL DETAILS
Materials synthesis
The yttria (Y_2O_3) content of YSZ is defined as mole-% Y_2O_3 in ZrO_2; thus 8YSZ has the composition $(ZrO_2)_{0.92}(Y_2O_3)_{0.08}$. Unless otherwise described in the text, Fe_2O_3/8YSZ samples were prepared by physically mixing Fe_2O_3 (Fisher, 99.5%) with 8YSZ (Tosoh Corporation, 99.9%) and calcining the mixture in air to 1500 °C. The calcination protocol involved ramping the temperature from ambient to 1375 °C at 5 °C min^{-1}, holding there for between 2 and 48 hours, then ramping to 1500 °C at 5 °C min^{-1}, and holding for 2 hours. The temperature was then ramped back to ambient at 5 °C min^{-1} (nominal). The soak at 1375 °C was to allow iron species to diffuse into the YSZ and form a solid solution prior to the final sinter at 1500 °C.

Cobalt-substituted ferrites were prepared by co-precipitation of cobalt and iron nitrate salts using excess ammonia. The washed and dried precipitate was then calcined to 1400 °C to yield a product with nominal composition $Co_{0.95}Fe_{2.05}O_4$. This composition was chosen based on results of thermodynamic modeling of phase formation and stability as a function of spinel composition, temperature, and gas composition.[25] The modeling results suggested that Co : Fe ratios between 0.95 : 2.05 and 1.0 : 2.0 offer the optimal spinel fraction in the material, while minimizing the formation of additional iron or cobalt phases.

Single-crystalline 9YSZ platelets, 10mm x 10mm x 0.5mm, with <100> orientation (MTI Corporation) were used as substrates during the in-situ XRD experiments.

Characterization

Figure 3 shows a simplified diagram of the HT-XRD experimental setup. The experiments were performed using a Scintag PAD X diffractometer (Thermo Electron Inc.; Waltham, MA). This diffractometer is equipped with a sealed-tube source (Cu Kα, $\lambda = 0.15406$ nm), an incident-beam mirror optic, a peltier-cooled Ge solid-state detector, and a Buehler hot-stage with Pt/Rh heating strip and surround heater. The hot stage lies within a sealed cell with X-ray-transparent beryllium windows, and is operable from ambient temperature to 1600 °C, and at gas pressures from 10^{-9} to 10^3 Torr. An all-metal gas manifold was attached to the inlet of the reaction cell allowing the controlled flow of helium, air, or carbon dioxide through the cell. An oxygen getter furnace (Centorr TM 1B) was installed in the helium inlet line to remove trace levels of oxygen, and an oxygen- and moisture-specific adsorbent purifier bed was used in the CO_2 line. An oxygen analyzer (Ametek CG1000) attached to the exit line from the cell verified that the oxygen background, when running He or CO_2, was below 0.1 ppm (limit of detection). Samples of typically 20 – 30 mg material were analyzed as thin films (ca. 50 – 100 μm) of powder on top of single-crystal <100> 9YSZ platelets. In-situ HT-XRD experiments were conducted at atmospheric pressure, under gas flow rates of 150 ml min^{-1} (corrected to STP). Experiments investigating the solubility and phase evolution of iron oxides in YSZ typically involved purging the reaction cell with He, then ramping the temperature stepwise to 1400 °C and back down to ambient. The temperature intervals were typically 200 °C (ambient to 600 °C), 100 °C (600 to 1000 °C) and 50 °C (1000 to 1400 °C) for both the up- and down-ramps. Diffraction patterns were recorded at each step during a 30 minute isothermal hold. Heating and cooling ramp rates were set to 20 °C min^{-1}. For the study of phase transformations during TR/CDS cycling, the sample was typically heated stepwise to 1400 °C under He, held at 1400 °C under He for 2 hours, stepped down to 1100 °C under He, then exposed to CO_2 at 1100 °C for 5 hours, or until no further change in XRD pattern was observed. Using this experimental set-up, phase fractions as low as ~ 1 wt.-% could be reliably detected. The temperature calibration was performed using the thermal expansion behavior of known materials (e.g., alumina or Pt) to an accuracy of ±5 °C. Diffraction patterns were collected at 40 kV and 30 mA using fixed slits over a scan range of 20 – 80 °2θ at a step-size of 0.04 °2θ and a count time of 1 s. Subtle displacement of the stage was observed in the process of instrument calibration. The displacement effects (stage moved up upon heating) were repeatable and therefore mapped out through structure refinement of various materials under differing atmosphere conditions. An established sample displacement table as a function of temperature enabled the removal of this systematic instrumentation error from the structural results obtained on each sample.

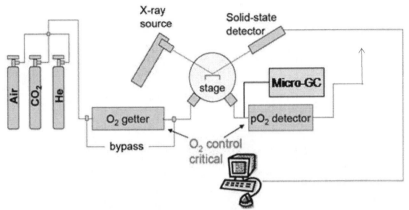

Figure 3. Schematic layout of high-temperature XRD apparatus.

Room temperature XRD patterns were recorded for finely-ground powders on either a PANalytical X-Pert Pro diffractometer, or a Siemens D500 diffractometer using Cu Kα radiation. Thermogravimetric analysis (TGA) was carried out using a TA Instruments SDT Q600. Argon and CO_2 gas streams both passed through purification beds to remove traces of oxygen prior to entering the TGA. Samples for TGA analysis were sintered bars of reactive material (approx. 1 x 1.7 x 6 mm^3) placed across the top of a Pt crucible. Thermochemical cycling experiments in the TGA were conducted under heating and cooling rates of 20 °C min^{-1}, and gas flow rates of 100 ml min^{-1}. The sample was heated under Ar to 1400 °C, held isothermally for 2 hours (thermal reduction), cooled to 1100 °C, exposed to CO_2 and held isothermally for 2 hours (oxidation).

RESULTS AND DISCUSSION
Iron oxide / Yttria-stabilized Zirconia system
A detailed investigation into the 10 wt.-% Fe_2O_3 / 8YSZ system was carried out using in situ HT-XRD. Figure 4 shows a contour plot summarizing the results of a HT-XRD study of the TR process under helium carried out on a material previously calcined to 1500 °C in air. At room temperature, the diffraction pattern consisted of a strong YSZ signature and a weak peak for Fe_2O_3 only; the remainder of the iron species being in solid solution with the YSZ. Once the sample temperature reached 800 °C, the Fe_2O_3 peak began to disappear, and was no longer detected by 900 °C. Between 900 and 1100 °C, no diffraction peaks were detected for iron compounds, and at 1100 °C the first observation of Fe_3O_4 was made. The solubility of Fe in 8YSZ in the temperature range 900 – 1100 °C is therefore assumed to be at least 14 mol-% (equivalent to 9 wt.-% Fe_2O_3, allowing for a phase detection limit of 1 wt.-%), which is significantly higher than the values reported from post-mortem measurements.[26,27,28] Upon further heating, FeO appeared at about 1250 °C and coexisted with Fe_3O_4 briefly. By 1300 °C the Fe_3O_4 was no longer detected, and FeO was the only observable iron species. The FeO peaks remained at approximately constant intensity throughout the remainder of the experiment (i.e., heating up to 1400 °C, then cooling stepwise to ambient temperature, still under an inert environment). The YSZ remained cubic throughout the temperature cycling. The fact that iron species were again detected at temperatures above 1200 °C indicates that iron solubility is maximal between 900 and 1100 °C. Similar phenomena have been observed for the solubility of iron in ZrO_2 during crystallization of iron-zirconia co-gels at various temperatures.[16,20]

The (111) reflection of the 8YSZ exhibited significant asymmetry during the 400 – 900 °C temperature range during sample heating. This temperature range corresponds to the migration of Fe into the YSZ as Fe_2O_3 disappeared, and the asymmetric peak shape may be due to asymmetry of the structure caused by local ordering, or to the coexistence of two YSZ lattices with differing Fe contents. The observed peak asymmetry is not consistent with a distortion of the YSZ from cubic to tetragonal symmetry, since the (111) is not expected to split in the tetragonal phase. There is a chance that a minor component of monoclinic lattice forms, leading to the observed asymmetric (111) peak, but the evidence for this is weak. At temperatures below 400 °C and above 900 °C, and also at all temperatures during cooling, the peak shape appeared symmetrical.

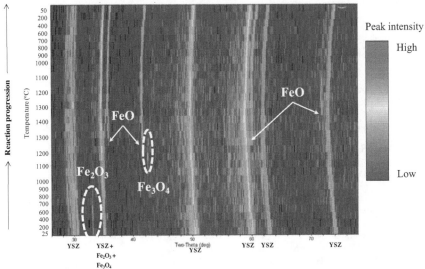

Figure 4. HT-XRD intensity plot for 10 wt.-% Fe_2O_3 / 8YSZ, 25 °C – 1400 °C – 50 °C under helium. Intensities are plotted on a log scale.

Closer examination of the variation in peak position for 8YSZ (111) versus temperature revealed that the YSZ lattice contracts slightly upon ingress of iron, and expands again when it is expelled, in agreement with the literature.[27] The contraction of the lattice on increasing iron content is a consequence of the smaller size of the Fe^{3+} cation than Y^{3+} or Zr^{4+} (ionic radii: 0.78, 1.01, and 0.84 Å, respectively).[29] Thus the plot of 8YSZ (111) lattice parameter versus temperature was not linear, nor did the forward branch (heating) coincide with the reverse branch (cooling), as shown in Figure 5. The dashed line in the figure is a guide to the eye; it lies parallel to the d-spacing versus temperature data recorded during temperature down-ramp. The decrease in slope of the forward branch for the iron-containing system around 400 – 600 °C corresponded to the migration of Fe from Fe_2O_3 into the YSZ, the slope then remained fairly constant until approximately 1100 °C, at which point the slope increased as iron oxides again became visible by HT-XRD. Once FeO had formed around 1250 °C, the lattice parameter was found to vary linearly with temperature, suggesting that Fe was no longer shuttling in and out of the YSZ lattice. Unit cell refinements carried out at room temperature before and after this experiment revealed an overall expansion of the YSZ lattice from a = 5.112 Å to a = 5.130 Å. Other prominent YSZ reflections showed similar variations in d-spacing with temperature.

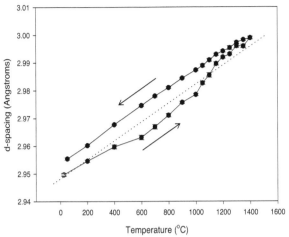

Figure 5. Variation in 8YSZ (111) peak position during heating and cooling cycles of 10 wt.-% Fe_2O_3/8YSZ under helium. The dashed line is provided as a guide for the eye, and has the same slope as the data measured during temperature down-ramp.

In a separate experiment, a freshly-prepared sample of 10 wt.-% Fe_2O_3 / 8YSZ was subjected to three consecutive thermal reduction cycles under helium. The results indicated that during the first cycle, a large unit cell expansion occurred (i.e., Fe migrated out of YSZ, in qualitative agreement with Figure 5), during the second cycle, a smaller cell expansion was observed, while the third cycle saw no overall expansion of the lattice, implying that after the second cycle, the Fe distribution had reached steady state. The variation in the 8YSZ (220) lattice parameter is shown as a function of temperature for the three cycles in Figure 6. The d-spacings for the 8YSZ (220) peak at the beginning of the three runs were 1.8064(±0.0002), 1.8132(±0.0002), and 1.8139(±0.0002) Å, the final d-spacing measured at the end of the third cycle was identical to that at the beginning of the third cycle, indicating that a steady-state had been reached with respect to Fe dissolved in YSZ. The phase evolution of Fe/YSZ species for the first of the three cycles was similar to that seen in Figure 4, however the later cycles exhibited some differences. Most notably, a gradual weakening of the reflections for iron species occurred in cycles 2 and 3; this may be caused by either a slow migration of Fe species into the <100> 9YSZ substrate, or a slow volatilization of iron species at the highest temperatures employed. Since bulk FeO has a melting point of ~ 1370 °C, some volatility at high temperatures is to be expected. Furthermore, whereas in a single cycle, FeO forms during initial TR, and persists on cooling, during the later stages of the 2nd cycle, and during the 3rd cycle FeO was observed to disappear while Fe_3O_4 and Fe metal were detected. This is probably due to a disproportionation reaction of FeO (i.e., 4FeO → Fe_3O_4 + Fe). Note that, despite this unusual behavior of the iron species in cycles 2 and 3, the 8YSZ lattice parameters did not show unusual behavior, thus the Fe species dissolved in the 8YSZ appear to be immune from the reactivity exhibited by the non-substituted iron species.

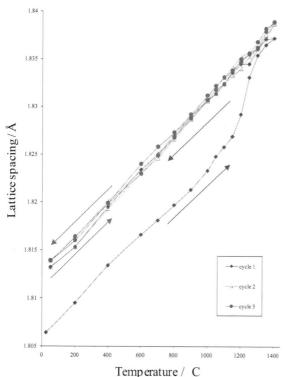

Figure 6. Variation in YSZ (220) peak position during three consecutive heating and cooling cycles of 10 wt.-% Fe_2O_3/8YSZ under helium.

Unsupported cobalt-substituted ferrite system

As-prepared, sintered cobalt-ferrite materials with the nominal composition $Co_{0.95}Fe_{2.05}O_4$ exhibited an XRD pattern characteristic of a spinel, with no detectable impurities. Upon heating to 1400 °C under He, partial reduction of the spinel (Fe^{III}/Fe^{II}) to wüstite ($Fe^{II}O$) was observed by XRD, beginning at around 1200 °C, and the wüstite persisted upon cooling to 1100 °C under He (Figure 7). On exposure to CO_2 re-oxidation of the wüstite to the spinel occurred within the first hour (each XRD pattern takes 30 minutes to scan). No other phases were detected in the course of the experiment; since separate iron- and cobalt-containing phases were not detected, it is assumed that the iron oxide and cobalt oxide constituents remained in solid solution throughout the processing.

The lattice parameters of samples after various TR/CDS protocols were evaluated, and were found to vary by at most 0.005Å between samples, indicating that while the ferrite undergoes a phase change at high temperature, the sample does not change significantly in composition.[30] Also, the spinel lattice parameter did not vary even when there was significant wüstite phase fraction (up to 4%) remaining upon cool down. This implies that the reduced phase was similar in Fe:Co ratio to the ferrite and would return to its original ferrite lattice upon re-oxidation, as corroborated by recent thermodynamic modeling.[25]

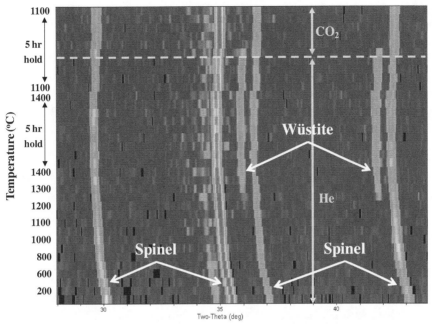

Figure 7. HT-XRD intensity plot for un-supported $Co_{0.95}Fe_{2.05}O_4$ under simulated thermochemical cycle conditions; 25 °C – 1400 °C – 1100 °C under helium, then 1100 °C under CO_2. Intensities are plotted on a log scale; see Figure 4 for scale indicator.

Thermogravimetric analysis of un-supported $Co_{0.95}Fe_{2.05}O_4$ was conducted under thermochemical cycling conditions, using argon as the inert under which thermal reduction could occur, and CO_2 to re-oxidize the reduced metal oxide. In contrast to the HT-XRD samples, the material was analyzed as a self-supporting sintered (air, 1500 °C) body of ca. 50 mg. The results shown in Figure 8 clearly show that the sample lost mass during thermal reduction due to loss of oxygen, and gained mass on exposure to CO_2 as oxygen was extracted from the CO_2 to re-oxidize the metal oxide. During initial temperature ramp under argon, the sample began losing mass (reducing) once the temperature reached 1200 °C. The initial rate of mass loss was relatively fast, which we attribute to reduction of the iron species in the surface layers of the sample. After approximately 20 minutes soak at 1400 °C the rate of mass loss decreased, indicating a possible switch from surface reduction to bulk reduction. During the two-hour isothermal hold at 1400 °C, steady state mass was not achieved, however re-oxidation of the ferrite on exposure to CO_2 was relatively rapid; essentially complete mass recovery was achieved during the two hour hold. The mass change during reduction in Figure 8 correlates with an approximate utilization of the cobalt ferrite of ~25%. Since the ferrite was not allowed to reduce to a steady-state condition, the utilization figure would be larger in an equilibrium state than that quoted. Note that the re-oxidation appears faster in the TGA than in the HT-XRD (Figure 7); this may be due in part to the large volume of the HT-XRD in-situ chamber (~850ml), leading to a relatively long lag time between switching gas streams and the environment in the chamber reaching the new composition.

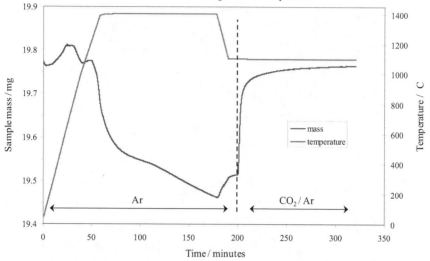

Figure 8. TGA plot for unsupported $Co_{0.95}Fe_{2.05}O_4$ measured under simulated thermochemical cycle conditions. The dashed vertical line indicates the point at which gas was switched from argon (thermal reduction at 1400 °C) to CO_2 (re-oxidation at 1100 °C).

Fusion and vaporization issues in the iron oxide / YSZ system

A simple experiment was performed to investigate the diffusion of iron oxide into 8YSZ; 4mm thick discs of the two materials were pressed, placed face-to-face (Figure 9A), and heated to various temperatures in air and held there for two hours. After heating to 1475 °C, a thin black diffusion zone could be observed in the 8YSZ where it had been in contact with the Fe_2O_3 (Figure 9B). However, after 1500 °C treatment, almost the entire 8YSZ disc had turned black (Figure 9C) indicating a substantial increase in diffusion rate of iron, suggesting the possible formation of a eutectic between the temperatures of 1475 and 1500 °C. The phase diagram for iron in zirconia reports a eutectic temperature of 1525 °C,[31] so the presence of yttria appears to depress the eutectic temperature slightly.

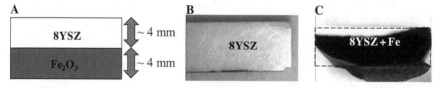

Figure 9. Illustration of iron oxide fusion during high temperature treatment. A: sample orientation during thermal treatment; B: 8YSZ disc after 2 hours at 1475 °C in air; C: 8YSZ disc after 2 hours at 1500 °C in air. Note that a section of the Fe_2O_3 disc remained bonded to the bottom of the 8YSZ disc.

In another experiment, small rods of 20 wt.-% Fe_2O_3/8YSZ, ca. 1.4 mm diameter x 20 mm, were weighed and then suspended using Pt wire such that they were centered in a ½" diameter furnace tube. The bars were then thermally treated under either flowing air at various temperatures for 16 hours, or under flowing He at 1500 °C for 16 hours. After thermal treatment, the bars were weighed again; due to the refractory nature of 8YSZ, all mass loss was attributed to loss of iron oxides. In air, the effect of increasing the temperature from 1400 to 1500 °C on sample mass was quite significant, with a large increase in mass loss between 1450 and 1500 °C (Figure 10A). This is to be expected in light of the results shown in Figure 9; once a liquid phase has formed, vaporization is enhanced. When the air atmosphere was changed to helium, a further increase in volatilization of iron was observed (Figure 10B). Even a slow flow of helium caused a large loss of ferrite from the sample. As shown by the HT-XRD results (Figure 4), under helium the ferrite has reduced to wüstite (FeO) by 1300 °C. Since wüstite has a lower melting point (ca. 1370 °C) than the other iron oxides, higher mass loss is to be expected at 1500 °C under inert atmosphere than under air. Increasing the flow rate of helium during thermal treatment induced a more pronounced loss of mass, supporting the supposition that the iron oxides are vaporizing from the sample surface and being transported away by the flowing gas.

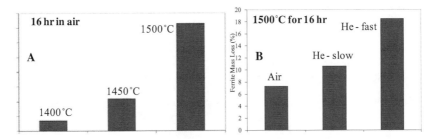

Figure 10. Illustration of iron oxide vaporization during high temperature treatment. A: effect of temperature during treatment in flowing air for 16 hours; B: effect of gas flow conditions during treatment at 1500 °C for 16 hours.

CONCLUSIONS

Sandia National Laboratories has been developing a process and prototype reactor system to enable conversion of H_2O and CO_2 into H_2 and CO (syngas) using concentrated solar power as the energy source. Preliminary work in this area has utilized iron oxide-based working materials mixed with yttria-stabilized zirconia (8YSZ) to strip oxygen atoms out of H_2O and CO_2. We have attempted to better understand the behavior of these materials at the high temperatures of the process via in-situ characterization. Iron oxide was found by HT-XRD to have a higher solubility in the 8YSZ matrix between approximately 900 and 1100 °C than at room temperature; the high-temperature solubility is estimated to be at least 14 mol-% Fe. On heating 10wt.-%Fe_2O_3/8YSZ under helium, the Fe_2O_3 reduced to Fe_3O_4 (spinel) at around 1150 °C, then to FeO (wüstite) beginning at around 1250 °C. Once FeO was formed, it remained stable during heating to 1400 °C, and cool down to ambient temperature under inert atmosphere. The 8YSZ lattice parameters were found to be sensitive indicators of the migration of Fe; decreasing lattice parameter correlated with insertion of Fe into the 8YSZ lattice, while ejection of Fe resulted in lattice expansion. At temperatures between 1400 and 1500 °C, fusion and volatility of iron species becomes an issue, particularly under inert atmosphere.

A cobalt-substituted ferrite (spinel) was observed by HT-XRD to reduce to the wüstite phase at around 1200 °C under He. Exposure to CO_2 at 1100 °C caused re-oxidation to the spinel in less than 1 hour. After several reduction-oxidation cycles, the lattice parameter of the cobalt ferrite was virtually unchanged, indicating that no gross phase separation had occurred. Thermogravimetric analysis corroborated the HT-XRD results, and a ferrite utilization of about 25% was measured during a 2 hour thermal reduction under argon for a dense, sintered body of the cobalt ferrite.

The two-step thermochemical conversion of H_2O and CO_2 to H_2 and CO can be effectively carried out using ferrite-based materials to strip oxygen from the reactants. The temperatures required to drive the reduction reaction (1st step) are well aligned with those accessible using concentrated solar power, however issues of ferrite fusion and volatility remain to be addressed.

ACKNOWLEDGMENT

This work was supported by the Laboratory Directed Research and Development program at Sandia National Laboratories, in the form of a Grand Challenge project entitled "Reimagining Liquid Transportation Fuels: Sunshine to Petrol." Sandia is a multiprogram laboratory operated by Sandia Corporation, a Lockheed Martin Company, for the United States Department of Energy's National Nuclear Security Administration under contract DE-AC04-94AL85000.

REFERENCES

1   A. Steinfeld, *Solar En.* **78**, 603-15, (2005).

2   T. Kodama, *Prog. En. Comb. Sci.* **29**, 567, (2003).

3   J.E. Miller, *Sandia Report*, **2007**, SAND2007-8012, (order at "orders@ntis.fedworld.gov").

4   T. Kodama, Y. Nakamuro, and T. Mizuno, *J. Sol. Energy Eng.-Trans. ASME*, **128**, 3-7, (2006).

5   J.E. Miller, M.D. Allendorf, R.B. Diver, L.R. Evans, N.P. Siegel, and J.N. Stuecker, *J. Mater. Sci.*, **43**, 4714-28, (2007).

6   T. Nakamura, *Solar En.* **19**, 467, (1977).

7   T. Kodama, Y. Nakamuro, and T. Mizuno, *J. Solar Energy Eng.* **128**, 3, (2006).

8   T. Kodama, et al. *Proceedings of the ASME ISES Conference*, Hawaii, **2003**.

9   T. Kodama, and N. Gokon, *Chem. Rev.* **107**, 4048, (2007).

10   R.B. Diver, J.E. Miller, M.D. Allendorf, N.P. Siegel, and R.E. Hogan, *J. Solar Energy Eng.* **130**, 041001-1, (2008).

11   S.V. Bechta, E.V. Krushinov, V.I. Almjashev, S.A. Vitol, L.P. Mezentseva, Yu. B. Petrov, D.B. Lopukh, V.B. Khabensky, M. Barrachin, S. Hellmann, K. Froment, M. Fischer, W. Tromm, D. Bottomley, F. Defoort, and V.V. Gusarov, *J. Nuclear Materials* **348**, 114–21, (2006).

12   H.P. Beck, and C. Kaliba, *Mat. Res. Bull.*, **25**, 1161-8, (1990).

13   F.J. Berry, M.H. Loretto, and M.R. Smith, *J. Solid State Chemistry* **83**, 91-9 (1989).

14   S. Davison, R. Kershaw, K. Dwight, and A. Wold, *J Solid State Chemistry* **73**, 47-51, (1988).

15   P. Ghigna, G. Spinolo, U. Anselmi-Tamburini, F. Maglia, M. Dapiaggi, G. Spina, and L. Cianchi, *J. Am. Chem. Soc.* **121**, 301-7, (1999).

16   I.B. Inwang, F. Chyad, and I.J. McColm, *J. Mater. Chem.* **5(8)**, 1209-13 (1995).

17   J.Z. Jiang, F.W. Poulsen, and S. Mørup, *J. Mater. Res.* **14(4)**, 1343-52, (1999).

18   M. Lajavardi, D.J. Kenney, and S.H. Lin, *J. Chinese Chem. Soc.*, **47**, 1065-75, (2000).

19   P. Li, and I.W. Chen, *J Am. Ceram. Soc.*, **77(1)**, 118-28, (1994).

20   G. Štefanić, B. Gržeta, and S. Musić, *Mater. Chem. Phys.*, **65**, 216–21 (2000).

21   G. Štefanić, B. Gržeta, K. Nomurab, R. Trojkoa, and S. Musić, S. *J. Alloys Compounds*, **327**, 151–60 (2001).

22   F. Wyrwalski, J.F. Lamonier, S. Siffert, E.A. Zhilinskaya, L. Gengembre, and A. Aboukaïs, *J. Mater. Sci.*, **40**, 933–42 (2005).

23   M. Hartmanova, F.W. Poulsen, F. Hanic, K. Putyera, D. Tunega, A.A. Urusovskaya, and T.V. Oreshnikova, *J. Mater. Sci.*, **29**, 2152-8 (1994).

24   T. Raming, L. Winnubst, and H. Verweij, *J. Mater. Chem.*, **12**, 3705–11 (2002).

25   M.D. Allendorf, R.B. Diver, N.P. Siegel, and J.E. Miller, Energy & Fuels **22**, 4115–24 (2008).

26   R.V. Wilhelm, and D.S. Howarth, *Ceram. Bull.*, **58**, 229 (1979).

27   M.J. Verkerk, A.J.A. Winnubst, and A.J. Burggraaf, *J. Mater. Sci.*, **17**, 3113-22 (1982).

28   Ambrosini, A.; Coker; E. N.; Rodriguez, M. A.; Livers, S.; Evans, L. R.; Miller, J. E.; Stechel, E. B.; *Advances in CO_2 Conversion and Utilization*. Y. Hu (ed.). American Chemical Society, Washington, D.C.,(2010).In Press..

29   R.D. Shannon, and C.T. Prewitt, *Acta Cryst.*, **B25**, 925 (1969).

30   E.N. Coker, M.A. Rodriguez, A. Ambrosini, and J.E. Miller, *Proceedings of SolarPACES*, Berlin, September 15-18, 2009.

31   T.S. Jones, S. Kimura, and A. Muan, *J. Am. Ceram. Soc.*, **50(3)**, 139 (1967).

ULTRASMALL ANGLE X-RAY SCATTERING (USAXS) STUDIES OF MORPHOLOGICAL CHANGES IN NaAlH₄

Shathabish NaraseGowda and Scott A. Gold
Department of Chemical Engineering
Institute for Micromanufacturing
Louisiana Tech University, Ruston, LA 71272, United States

Jan Ilavsky
Advanced Photon Source,
Argonne National Laboratory, Argonne, IL 60439, United States

Tabbetha A. Dobbins*
Department of Physics
Institute for Micromanufacturing
Louisiana Tech University, Ruston, LA 71272, United States

* Contact author

ABSTRACT

The ultra-small Angle X-Ray Scattering (USAXS) technique has been explored to study morphological changes that occur in the hydrides during various stages of hydrogen release. The power law scattering data from USAXS measurements yield the power law slope (p) which is used as a definitive parameter to study changes in surface fractal dimensions (Ds). Changes in surface area occur due to densification during desorption at elevated temperatures and the rate of those changes are mitigated by the addition of transition metal dopants. For the present study, NaAlH₄ was doped with 4 mol% TiCl₃ by high energy ball milling and subjected to USAXS measurements to determine the effect of the catalyst. USAXS measurements were also done on NaAlH₄ nano-confined within porous alumina membranes. Results showed that the power law slope from USAXS analysis increased from p=-3.6 to p=-3.46 after high energy ball milling; to p=-3.26 after catalyst addition and to p=-3 after heat treatment, indicating an increase in specific surface area accompanying hydrogen release. The particle sizes before and after heat treatment were also evaluated to quantify the extent of densification occurring due to heat treatment. The radius of gyration (R_g) for unmilled NaAlH₄ increased from 9.7nm to 26.4nm and a 5 min milled NaAlH4 particle size increased from R_g=6nm to R_g=14.5nm. USAXS on nanostructured NaAlH₄ yielded power law slopes of p=-1.41 (rod shaped particles) and p = -2.7 (disc shaped particles) which proved USAXS to be an effective tool to identify the confined particles in porous matrices by virtue of shape.

INTRODUCTION

Complex metal hydrides are undoubtedly the most promising materials for hydrogen storage. Many light metal complex hydrides such as NaAlH₄ and LiBH₄ have been studied as candidate hydrogen storage materials for many years.[1-4] NaAlH₄, among these has been an archetype for complex metal hydrides and the most widely researched material in this category so far. However, not much research has been done to quantitatively study the changes in morphological features of these materials during desorption. Many underlying challenges in developing reversible complex metal hydrides, such as hysteresis in the sorption behavior or choosing the right catalysts and mill times can adequately be addressed by examining their morphology. To study complicated morphologies, small angle scattering of light and x-rays is an invaluable tool since it covers almost seven orders of magnitude on the length scale.[5] Measuring x-ray scattering at small angles can yield information on large scale electronic inhomogeneities of the size of colloidal particles. Ultra small angle x-ray scattering finds use when there is a need to extend the length scales to micron or sub micron levels while small angle x-ray scattering is limited to the nanometer scale.[6] NaAlH$_4$ is comprised of colloidal particles with extremely complex surface fractal pore morphology and the USAXS technique is well suited for such materials.[7] The present work is intended to show some unique applications of ultra small angle x-ray scattering technique that address some of the unanswered problems in development of reversible complex metal hydrides. Catalysis, hysteresis, optimization of process parameters (eg. mill time), nanostructuring strategies and characterization for complex metal hydrides are some of the vital questions that need further investigation.

Small angle scattering yields information on colloidal attributes such as specific surface area, radius of gyration, and fractal features. Specific surface area is an important parameter for all desorbing systems and increasing the surface area enhances the catalytic activity as well as increases the desorption rates.[8] Thus, specific surface area, as a parameter, can be used to assess the efficacy of any treatment aimed at enhancing desorption properties. Dobbins et al., for the first time, showed that the most effective mill times and the most effective catalyst for NaAlH$_4$ can be singled out by tracking the surface area changes via power law slopes gained from USAXS experiments.[7] In this paper, the adeptness of each treatment performed on NaAlH$_4$ is illustrated by virtue of specific surface area changes. NaAlH$_4$ is understood to undergo densification during sintering due to agglomeration and fusing of the primary particles.[9] This actually leads to a loss in total surface area which affects the hydrogen uptake capacity of the material in subsequent cycles. Many researchers have reported the phenomena of hysteresis in NaAlH$_4$, wherein the rehydrogenation capacity of the material reduces with every subsequent sorption cycle.[10-12] NaAlH$_4$ is expected to deliver 5.6% H$_2$ by weight, but the cyclic capacity was shown to be less than 3% H$_2$ by weight.[12] This paper attempts to provide an explanation to hysteresis by quantitatively studying the particle size changes after one desorption run.

Nanostructuring of complex metal hydrides has gained tremendous interest lately due to the many advantages it offers in comparison to the bulk materials. NaAlH$_4$ [13-14] and LiBH$_4$ [15-16] have been confined in a variety of materials with a wide range of pore sizes such as carbon scaffolds [13, 15-16], MOFs [14] etc. One of the key roles of a nano-scaffold is to prevent the particles from agglomerating and sintering.[15] The particle size obtained from small angle scattering analysis can be used to decide the pore size that will effectively curb any sintering effects for that material. Various nanostructuring techniques, such as melt infiltration and solution infiltration, are adopted to load the nano-scaffolds with the hydrides but it has been a challenge to actually identify material confined within the pores from those resting on the surface. Techniques used to date for this purpose are based on measuring the changes in weight and specific surface area.[13-16] One of the recent articles showed the use of small angle neutron scattering to assess the loading of activated carbon by Mg(BH$_4$)$_2$ by showing a decrease in specific surface area.[17] An attempt has been made in the present paper to identify the confined material based on the shape acquired due to confinement.

MATERIALS AND METHODS

Sample Preparation
Commercially available sodium aluminum hydride (NaAlH$_4$) hydrogen-storage grade (Sigma-Aldrich®) and titanium (III) chloride (TiCl$_3$) hydrogen storage grade (Sigma-Aldrich®) were used in the preparation of all samples. Both chemicals were stored and handled in a glove box under a controlled dry N$_2$ environment. NaAlH$_4$ was milled with 4 mol% TiCl$_3$ in tungsten carbide vials using a Certiprep SPEX 8000M high energy ball mill for 5 minutes. Undoped (neat) NaAlH$_4$ was also milled for 5 minutes to compare with the catalyzed sample as well as the unmilled sample. For nanostructuring the NaAlH$_4$, porous alumina membranes (Whatman® Anodisc™) of 13mm diameter and 200nm pore size were used. NaAlH$_4$ was inserted into the pore wells of the alumina membranes by

the template wetting technique (discussed in the next section) using tetrahydrofuran (THF) as a precursor. For the USAXS experiment, small amounts of each sample was packed between two layers of kapton film and the edges further sealed with kapton tape to avoid any exposure to air. Two 'blank' samples were made each for the powdered and templated samples. Here 'blank' refers to the background material (that holds the actual sample) whose scattering needs to be subtracted to evaluate the scattering from the sample during the USAXS data analysis. The 'blank' for the powdered samples was just a bilayer of kapton. In the case of nanostructured NaAlH$_4$, the empty template packed in kapton was treated as the 'blank' sample.

Template Wetting

Template wetting is one of the many wet impregnation techniques that is based on forming thin uniform precursor films using a low surface energy organic solvent over substrates with high surface energy.[18] Template wetting has been widely regarded as a simple and easy way to fabricate polymer nanotubes and nanowires.[19] The procedure to fabricate functionalized nanotubes has been lucidly described in one of the earliest articles on this technique, authored by Steinhart et al. [20] The same principles were applied here, but the intention was to fill the pores completely with NaAlH$_4$ and not just wet the walls to form nanotubes. Tetrahydrofuran (THF) was used as the wetting agent forming a dilute precursor solution containing NaAlH$_4$. For the present study, 50mg of a 5 minute milled sample of NaAlH$_4$ was dissolved completely in 1ml of THF. A more concentrated solution would clog the pores and form a layer on the surface, and a more dilute solution would end up just wetting the pore walls, neither of which was desired. The alumina templates were pre-treated to remove any impurities by dipping in ethanol and acetone for 2 hours each and later heating to 200 deg C for a period of 6 hours using an autoclave oven. Later the templates were allowed to cool down to room temperature and set up on a flat Teflon plate and the NaAlH$_4$ – THF precursor was dropped on to the template using a micro-pipette.

Ultra-Small Angle X-Ray Scattering (USAXS)

USAXS data were collected at Sector 32ID, Advanced Photon Source, Argonne National Laboratory (IL) using incident photon energy of 16.9keV. The beam size employed for these experiments was 0.3mm x 0.3 mm. Powder samples were on the order of 0.2mm thick and the template samples were approximately 0.1mm in thickness, including the two layers of kapton tape. Data reduction, slit-smear correction and analysis were made using IGOR Pro (a data analysis software product of WaveMetrics, Inc.) combined with sets of macros *Indra* [21] and *Irena* [22] to analyze and fit the data. The scattering intensity was fit by the unified model (eq. 1), developed by Beaucage.[23] The unified model fits complicated SAXS data having multiple Guinier regions and power law regions without introducing any additional parameters than those used for local fits.[23]

$$I(Q) = G \exp\left(-\frac{Q^2 R_g^2}{3}\right) + B \left\{ \frac{\left[\text{erf}\left(\frac{QR_g}{\sqrt{6}}\right)\right]^3}{Q} \right\}^p$$

(1)

The unified model fit yields the power law slope (p) and the radius of gyration (R$_g$) for each structural level present. For colloidal particles, the power law slope represents the degree of surface roughness, which is defined by a parameter commonly called surface fractal dimension (D$_s$). The power law slopes are related to the surface fractal dimension by the relation [24]

$$D_s = p + 2D$$

(2)

where D = 3 for colloidal particles.

D_s varies between 2 and 3 with a corresponding variation of p between -4 and -3. D_s = 3 (or p = -3) represents the high specific surface area particles either due to agglomeration of smaller particles or due to very coarse surfaces and/or a lot of internal pores, while D_s = 2 (or p = -4) represents the smooth colloidal particle.[24] The value of the exponent (p) may be less than -4 for diffuse interfaces and more than -3 for mass fractals. The exponent p = -1 represents elongated rod like mass fractals and p = -2 represents flat disc like particles.[24] The surface fractal dimension is related to the surface area of the particle by

$$S \sim R^{D_s}$$

(3)

where R is the radius of the colloidal particle.
For spherical particles, R can be deduced from the radius of gyration (R_g) by the relation [24]

$$R = \sqrt{\frac{5}{3}}\, R_g$$

(4)

The specific surface area of the particle was evaluated by dividing the calculated surface area from the volume of an individual spherical particle (i.e $\frac{4}{3}\pi R^3$) by the radius (R) calculated from R_g.

RESULTS AND DISCUSSION

Effectiveness of the Treatments – Ball milling, Catalytic Doping and Heat treatment
Specific surface area of a material is a collective indication of the extent of porosity and surface roughness prevalent in the material.[25] Specific surface area can thus be increased by creation of internal pores by desorbing gaseous elements occupying interstitial sites, by aggregating smaller particles to make a rough surface or by breaking down a larger particle to expose new surfaces. In this study, the effectiveness of a treatment aimed at improving desorption properties is thus decided based on the extent to which it increases the particle's surface area.
Figure 1 shows the changes in power law slopes and the accompanying changes in specific surface areas of the particles due to the various treatments employed for NaAlH$_4$. The unmilled NaAlH$_4$ (as received) particles which are of the order of 12.5 nm in effective radius yield an absolute power law slope value of 3.6 corresponding to a calculated specific surface area of 0.05 nm^{-1}. High energy ball milling reduces the particle size thus creating new surfaces due to which the absolute slope reduces to 3.46 indicating an increase in specific surface area. Addition of the catalyst (4 mol% TiCl$_3$) and ball milling further reduces the absolute value of the exponent to 3.26. This decrease in absolute value of the slope can be attributed to the loss of interstitial hydrogen due to the combined effect of catalyst and heat generation during ball milling. Catalyzed NaAlH$_4$ desorbs at 180°C and USAXS data for heat-treated samples at this temperature indicates particles with a maximally increased surface fractal dimension (p = -3) and a significantly increased specific surface area (that is evidence to aggregation of particles). This work confirms the proposition that USAXS is a potentially robust technique to arrive at effective processing treatments (eg. ball milling) and their parameters (eg. mill time) by studying the morphological changes induced during the course of desorption.

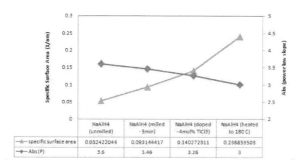

Figure1. Specific surface area and power law slope changes for NaAlH$_4$ due to various treatments

Sintering, Hysteresis and Effective Pore Size

Prior reports of USAXS data have shown that a 5 min milled sample creates more surface area than a 25 min milled sample or a 1 min milled sample.[7] Figure 2 shows a unified fit for a 5min milled sample of NaAlH$_4$. The curve is characterized by two power law slopes corresponding to a large population (at low Q) and a smaller population of particles (at high Q) separated by an exponential shoulder that represents the Guinier scattering regime. This Guinier region scattering characterizes the small population of particles (referred to as primary colloids) by a parameter called radius of gyration. The radius of gyration for the primary colloidal NaAlH$_4$ particles is found to be approx. 6.5nm, which translates to a particle radius of approx. 8nm using eq(4).

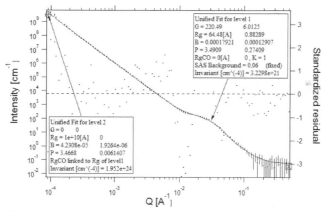

Figure 2. Unified fit of the USAXS intensity as a function of wave vector Q for 5 min milled NaAlH$_4$.

To address the hysteresis phenomena (that reduces the cyclic capacity), the morphology of the particles before and after heat treatment has been evaluated. Figure 3 illustrates the changes in particle radius due to heat treatment for the unmilled NaAlH$_4$ and the 5 min milled NaAlH$_4$. It can be seen that there is more than a two-fold increase in particle size after heat treatment, indicating agglomeration of particles due to heat treatment, regardless of whether the particles were milled or not. Agglomeration of particles leads to densification which reduces the overall surface area available for re-adsorption, thus reducing the hydrogen up-take capacity with every subsequent cycle. This is a feasible explanation to the hysteresis phenomena in NaAlH$_4$. In order to avoid the agglomeration and the associated loss in total surface area, nano-confinement of NaAlH$_4$ is suggested. Particle size measured by USAXS can also be used as a reference when choosing the pore-size for nano-confinement applications. While it is a widely accepted idea that, 'the smaller the pore size, the better the desorption properties', it should also be noted that nano-confinement becomes increasingly difficult as one goes down in size. Therefore in reality, one should be choosing an optimum pore size that is both reasonably easy to load and also enhances the (re)hydrogenation properties. This idea can be generalized for all of the metal hydride systems, as a tool to decide on the best possible size of the nanostructures.

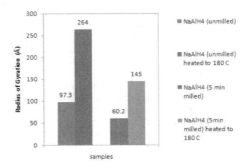

Figure 3. The radius of gyration before and after heat treatment for the unmilled and 5min milled NaAlH$_4$

Identifying Confined NaAlH$_4$ by Particle Shape

As pointed out earlier, the background chosen for the nanostructured NaAlH$_4$ was the combination of a bilayer of kapton and an empty template, whose scattering was subtracted to identify confined NaAlH$_4$. The USAXS data for nanostructured samples revealed that the primary particles on the porous template had absolute power law slope values less than 3 indicating mass fractal morphology. The template wetting technique involved dissolution of the NaAlH$_4$ particles in the volatile THF solution and re-crystallizing NaAlH$_4$ back by allowing the THF to evaporate. The NaAlH$_4$ re-crystallizes into tightly packed particles (negligible internal porosity), thus possessing significant material (mass) within a shape acquired by the terrain of the template. This explains the mass fractal morphology detected by USAXS. The data was collected from six different positions on the template containing NaAlH$_4$. Some positions revealed slopes of p = -1.41 [Figure 4] indicating rod like particles and other positions revealed slopes of p = -2.72 [Figure 5] indicating flat disc like particles. It can be inferred that the particles acquire the shape of an elongated rod due to their confinement inside the pore well. The particles that do not enter the pore wells due to clogging of the pores rest in a flat disc-like configuration atop of the pore wall.

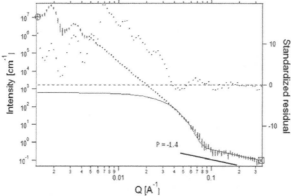

Figure 4. Unified fit (level 1 fit - high Q region) showing slope p = -1.4 indicating rod-like mass fractals of NaAlH$_4$ confined within the pores of alumina template

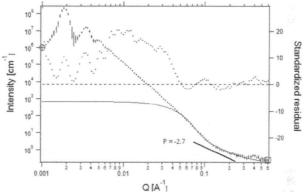

Figure 5. Unified fit (level 1 fit - high Q region) showing slope p = -2.7 indicating disc like mass fractals of NaAlH$_4$ on the surface of the alumina template

CONCLUSION

USAXS has been used to quantitatively study the morphology (SSA and particle size) of NaAlH$_4$. These changes in morphological features during desorption illustrated the effectiveness of the various treatments (such as ball milling and heat treatment) in aiding hydrogen desorption from NaAlH$_4$ by virtue of an increase in the specific surface area. USAXS data also served as a tool to accurately identify confined and unconfined NaAlH$_4$ based on particle shapes (rod like particles vs disc like particles). This paper also gives an insight into the role of morphology of the material in phenomena such as hysteresis and suggests nanostructuring strategies such as pore size selection to curb it.

ACKNOWLEDGEMENTS
Funding for this project was provided by US Department of Energy (DOE), Office of Basic Energy Sciences under contract no. DE-FG02-05ER46246 (AY 2008-09) and National Science Foundation (NSF) CAREER Award, Division of Materials Research, Ceramics Program under contract no. DMR-0847464 (AY 2009-10). Use of the Advanced Photon Source was supported by the U. S. Department of Energy, Office of Science, Office of Basic Energy Sciences, under contract no.DE-AC02-06CH11357.

REFERENCES
1. B. Bogdanovic, M. Schwickardi, Ti-doped Alkali Metal Aluminum Hydrides as Potential Novel Reversible Hydrogen Storage Materials, *J. Alloys Comp*., **253-254**, pp 1-9 (1997)
2. D. L. Anton, Hydrogen Desorption Kinetics in Transition Metal Modified NaAlH$_4$, *J. Alloys Comp*., **356-367**, pp 400-404 (2003)
3. Q. Ge, Structure and Energetics of LiBH$_4$ and Its Surfaces: A First-Principles Study, *J. Phys. Chem. A*., **108(41)**, 8682-8690 (2004)
4. P. Mauron, F. Buchter, O. Friedrichs, A. Remhof, M. Bielmann, C. N. Zwicky, A. Zuttel, Stability and Reversibility of LiBH$_4$, *J. Phys. Chem. B*., **112**, 906-910, (2008)
5. P. Rajan, Understanding Aggregate Morphology in Colloidal Systems through Small Angle Scattering and Reverse Monte Carlo (RMC) Simulations, *(M.S. Thesis)*, retrieved from http://etd.ohiolink.edu/view.cgi?acc_num=ucin1218832905 *last accessed on 03/15/2010*.
6. J. Ilavsky, P. R. Jemian, A. J. Allen, F. Zhang, L. E. Levine, G. G. Long, Ultra-small-angle X-ray Scattering At the Advanced Photon Source, *J. Appl. Cryst*., **42**, 469-479 (2009)
7. T. A. Dobbins, E. B. Bruster, E. U. Oteri, J. Ilavsky, Ultrasmall Angle X-ray Scattering (USAXS) Studies of Morphological Trends in High Energy Milled NaAlH$_4$ Powders, *J. Alloys Comp*., **446-447**, 248-254 (2007)
8. D. Lupu, G. Blanita, I. Misan, O. Ardelean, I. Coldea, G. Popeneciu and A. R. Biris, Hydrogen desorption from NaAlH$_4$ catalyzed by ball-milling with carbon nanofibers, **182**, *J. of Physics: Conference Series*, pp 012050 (2009)
9. D. A. Mosher, S. Arsenault, X. Tang, D. L. Anton, Design, fabrication and testing of NaAlH$_4$ based hydrogen storage systems, *J. Alloys Comp*, **446-447**, 707-712 (2007)
10. H. Yang, A. Ojo, P. Ogaro, A. J. Goudy, Hydriding and Dehydriding Kinetics of Sodium Alanate at Constant Pressure Thermodynamic Driving Forces, *J. Phys. Chem. C*., **113(32)**, 14512-14517 (2009)
11. D. Chandra, J. J. Reilly, R. Chellappa, Metal Hydrides for Vehicular Applications: The State of The Art, *JOM*, **58(2)**, 26-32 (2007)
12. G. Sandrock, K. Gross, G. Thomas, C. Jensen, D. Meeker, S. Takara, Engineering Considerations in the Use of Catalyzed Sodium Alanates for Hydrogen Storage, *J. Alloys Comp*., **330-332**, pp 696-701 (2002)
13. J. Gao, P. Adelhelm, M. H. W. Verkuijlen, C. Rongeat, M. Herrich, P. J. M. Van Bentum, O. Gutfleisch, A. P. M. Kentgens, K. P. de Jong, P. E. de Jongh, Confinement of NaAlH$_4$ in Nanoporous Carbon: Impact on H$_2$ Release, Reversibility and Thermodynamics, *J. Phys. Chem. C*., **114(10)**, 4675-4682 (2010)
14. R. K. Bhakta, J. L. Herberg, B. Jacobs, A. Highley, R. Behrens, Jr., N. W. Ockwig, J. A. Greathouse, M. D. Allendorf, Metal-Organic Frameworks As Templates for Nanoscale NaAlH$_4$, *J. Am. Chem. Soc*., , **131 (37)**, pp 13198–13199 (2009)
15. A. F. Gross, J. J. Vajo, S. L. Van Atta, G. L. Olson, Enhanced Hydrogen Storage Kinetics of LiBH$_4$ in Nanoporous Carbon Scaffolds, *J. Phys. Chem. C*., **112(14)**, 5651-5657 (2009)
16. S. Cahen, J. B. Eymery, R. Janot, J. M. Tarascon, Improvement of the LiBH$_4$ Desorption by Inclusion Into Mesoporous Carbon, *J. Power Sources*, **189**, 902-908 (2009)

17. S. Sartori, K. D. Knudsen, Z. Z. Karger, E. G. Bardaij, M. Fichtner, B. C. Hauback, Small-Angle Scattering Investigations of Mg-Borohydride Infiltrated in Activated Carbon, *Nanotechnology*, **20**, pp 7 (2009)

18. M. Steinhart, R. B. Wehrspohn, U. Gosele, J. H. Wendorff, Nanotubes by Template Wetting: A Modular Assembly System, *Angew. Chem. Int. Ed*, **43**, 1334-1344 (2004)

19. J. P. Cannon, S. D. Bearden, F. M. Khatkhatey, J. Cook, S. Z. Selmic, S. A. Gold, Confinement-Induced Enhancement of Hole Mobility in MEH-PPV, *Synth. Met.*, **159(17-18)**, 1786-1791 (2009)

20. M. Steinhart, J. H. Wendorff, R. B. Wehrspohn, Nanotubes a la Carte: Wetting of Porous Templates, *Chemphyschem*, **4**, 1171-1176 (2003)

21. J. Ilavsky, A. J. Allen, G. G. Long, P. R. Jemian, Effective Pinhole-Collimated Ultra-small Angle X-ray Scattering Instrument for Measuring Anisotropic Microstructures, *Rev. Sci. Instrum.*, **73(2)**, 1660-1662 (2002)

22. J. Ilavsky, P. R. Jemian, *Irena*: Tool Suite for Modeling and Analysis of Small Angle Scattering, *J. Appl. Cryst.*, **42**, 347-353 (2009)

23. G. Beaucage, Approximations Leading to a Unified Exponential/Power-Law Approach to Small Angle Scattering, *J. Appl. Cryst.*, **28**, 717-728 (1995)

24. Ryong-Joon Roe, Methods of X-Ray and Neutron Scattering in Polymer Science, *Oxford University Press, NY*, (2000)

25. D. M. Cox, High Surface Area Materials, *Nanostructure Science and Technology: A worldwide Study*, *WA*, 49-66 (1999) *full article can be accessed online at* http://www.wtec.org/loyola/nano/IWGN.Worldwide.Study/ch4.pdf, *last accessed 03/16/2010.*

CARBON BUILDING MATERIALS FROM COAL CHAR: DURABLE MATERIALS FOR SOLID CARBON SEQUESTRATION TO ENABLE HYDROGEN PRODUCTION BY COAL PYROLYSIS

John W. Halloran and Zuimdie Guerra
Materials Science and Engineering Department, University of Michigan,
Ann Arbor, Michigan USA.

ABSTRACT

The pyrolysis of coal can produce hydrogen-rich gases for clean fuel. It also produces coke-like coal char. Rather than using the coal char as a fuel, it could provide the basis for carbon building materials that sequester the carbon in the solid state while performing an essential function. Data is presented on the properties of prototype of coal-char based solid carbons, which is stronger and lighter than conventional fired clay or concrete masonry. A technoeconomic analysis shows co-production of hydrogen fuels and carbon building materials from coal can be cost competitive with steam reforming of natural gas. The environmental benefits are analyzed in terms of carbon dioxide emissions and land spoilage for substituting carbon building materials for Portland cement concrete.

INTRODUCTION

Coal is an abundant resource and, with a hydrogen/carbon atomic ratio in coal about H/C~0.7, a very large reservoir of reduced hydrogen. However, hydrogen is typically produced from coal by reacting carbon with steam to reduce the water to H_2 while oxidizing the carbon to CO_2. Thus the conventional manufacture of hydrogen is accompanied by a large production of carbon dioxide. The coal-steam reaction increases the yield of hydrogen, which is important if hydrogen is the only valuable product and the carbon dioxide can be vented to the atmosphere. But in the context of producing hydrogen as a "green" energy carrier, a co-production of CO_2 greenhouse gas is undesirable. But note that coal is already a rich source of reduced hydrogen, and the H_2 can be effectively separated from the carbon without oxidation[1]. Coal pyrolysis is a *direct method* of hydrogen separation. Pyrolysis produces a small weight fraction of hydrogen-rich gases and liquids, which carry a significant fraction of the fuel value, and a large fraction of solid carbon char or coke. Since the carbon is not oxidized during pyrolysis, *no carbon dioxide is produced*, except for CO_2 from the oxygen naturally present in the coal. If the solid carbon char co-product could be used in a valuable application, pyrolysis could be an effective method of hydrogen production. This paper explores solid carbon used as a building material, *not as a fuel*. Solid carbon used as a *durable material* could be used in place of conventional masonry. Since manufacture of conventional masonry from concrete has a large CO_2 emission, solid carbon building materials can also reduce CO_2 emissions from cement manufacture. Coal can serve as a "carbon ore" for solid carbons and a "hydrogen ore" for fuel.

Coal pyrolysis was widely used during the gaslight era of the 19th Century in gasworks to manufacture "town gas" as a domestic fuel and illuminant, with the co-product coke sold as a "smokeless" solid fuel and metallurgical reducing agent. Coal pyrolysis produced other valuable co-products, including gas, liquid, and solid. It was the original source of feedstock at the beginning of the organic chemical industry, and a major source of liquid fuels (benzene and "coal oil"). By the middle of the 20th Century, petroleum had replaced coal liquids, natural gas had replaced town gas, and electric lights had replaced gaslights. Now coal pyrolysis is mostly used to manufacture metallurgical coke at a global rate[2] around 0.5 Gt/yr, with the coke oven gas and liquids as minor co-products.

The production of hydrogen by coal pyrolysis might be considered to be impractical and uneconomic because of the relatively small yield of hydrogen, and because the solid carbon char is considered to be a low-value waste. Consequently, the conventional approach to acquire gases or liquids from coal uses the carbon as an indirect fuel, by reacting the carbon with water to produce hydrogen-rich products and carbon dioxide[3]. But it might be possible to return to direct pyrolysis if solid carbon co-products could be developed using the char. In the sense that char is now a "product without a market"[4] the challenge is to create a valuable market for very large quantities of solid coal char. Previously, we have suggested[5] that building materials, manufactured from coal char, could become a valuable marketable product. The market for building materials (masonry and concrete) is large enough (several Gt/yr) and valuable enough ($20-200/ton) to serve as a co-product market for the large-scale production of hydrogen from coal pyrolysis. This concept, coal-derived hydrogen energy and carbon materials (HECAM), requires the development of a set of carbon building materials (CBM) to be used in place of conventional masonry, such as concrete blocks or fired clay bricks. Note that carbon masonry is not new. This is essentially an ambient temperature application of carbon refractory masonry, made from coke and pitch, which has been in use for decades[6].

This paper addresses three issues: 1) the economics of HECAM, as compared with the standard hydrogen production process, steam reforming of methane (SRM); 2) the materials technology of CBM from coal char, and their properties compared to clay and concrete masonry; and 3) the environmental issues related to carbon dioxide emissions and land spoilage. To address the economics of HECAM-hydrogen compared to SRM-hydrogen, we present some results from Guerra[7]. To address the properties, we contrast literature values for the strength and density of conventional concrete and clay bricks with preliminary results for coal char-based materials[8]. To address atmospheric spoilage, we compare the carbon dioxide production (per GJ hydrogen energy) for HECAM and SRM, and the carbon dioxide production avoided by substituting carbon dioxide-producing Portland cement with carbon-sequestering CBM. To address land spoilage, we compare the total volume of material that must be mined or quarried to produce a particular amount of building materials.

TECHNOECONOMIC ANALYSIS OF CO-PRODUCTION OF HYDROGEN AND CARBON MATERIALS

Guerra[7] has done extensive technoeconomic analyses of hydrogen produced from coal pyrolysis. Costs and yields are based on a model for a metallurgical coke oven[9]. Here we present results for a particular case of HECAM where the methane in the pyrolysis gas was decomposed by the hot char to increase the hydrogen yield. This was compared to the standard process for hydrogen produced by steam reforming of methane, where reliable cost models are available[10]. The relative production costs for hydrogen depends on natural gas price, coal price, and the carbon co-product value. The cost for the natural gas is presented as 2007 dollars per GJ of the hydrogen product, based on the high heating value. The line in Figure 1 is a *cost-equivalence line*, showing the particular combinations of natural gas prices (for SRM) and carbon co-product values (for HECAM) where the hydrogen production costs are the same for both processes. For all points below the cost-equivalence line, hydrogen can be made by SRM at lower cost; for points above the line, HECAM produces lower cost hydrogen. Note that this particular analysis neglects the contribution of other pyrolysis co-products (ammonium compounds, oils, pitch, etc.), which are common byproducts of metallurgical coke production. These contribute about $18 of byproduct value per ton of coal[7], so by neglecting them Figure 1 is conservative.

Figure 1A is for a baseline case using the average 2007 spot price for Illinois bituminous coal ($47.91/metric ton). It assumes that each ton of coal produces about 0.7 tons of CBM products (which is the approximate yield of coke in a coke oven). For the natural gas at the 2007 average city gate price ($7.24/GJ hydrogen), the cost for coal-pyrolysis hydrogen would be the same as hydrogen from steam reforming of methane if the carbon co-product were worth about $88/ton. Notice that this cost-equivalence co-product value is larger than the value of conventional concrete, which is about[11] $45/ton, but less than the value of standard clay face brick, which is about[12] $95/ton (EPA 1997, adjusted to 2007 market price.) Thus it appears that CBM marketed at typical construction material values make HECAM economic with SRM, using 2007 coal and gas prices. Figure 1 has several annotations on the graph. These show the recent range of natural gas prices, from the 1998 low price to the July 2008 peak price, to illustrate recent cost ranges of this commodity. For the July 2008 peak gas price, cost equivalence requires only $69 of CBM value per ton, while at the 1998 gas cost, cost equivalence requires a market value of about $115/ton for CBM. These CBM values *per ton* are higher than concrete masonry but lower than clay brick.

However, building materials are rarely sold by weight. Rather they are sold by volume (such as a cubic yard for concrete, or per piece for concrete blocks, or per 1000 for bricks). The large difference in density of CBM ($\sim$0.8-1.2 gm/cm^3) and OPC ($\sim$2.4 gm/cm^3) makes the gravimetric value ($/ton) quite different than the volumetric value ($/m^3$. Figure 1B presents the same analysis in terms of value of carbon co-products per unit volume. It shows equivalence lines for CBM over a range of densities. Notice that the benchmark value of concrete moves up to $105/m^3$, and the clay brick value is $230/m^3$. For the 2007 gas price, the hydrogen from coal by HECAM is the same cost as hydrogen from natural gas by SRM if the CBM can be sold for about $90/m^3$, which is less than the cost of concrete. Thus it seems that H_2 and CBM from coal pyrolysis can compete well with conventional sources. Now let us consider if carbon has good enough mechanical properties.

MATERIALS PROPERTIES OF BUILDING MATERIALS

Figure 2 shows the microstructure of two of the prototype CBM we consider. Figure 2A is the structure of a formed coke[14], which is typical of metallurgical coke. It is a cellular solid, with relatively dense solid struts between large (20-100 micrometer) pores. These large pores were created while the hot coal was in the plastic stage as pyrolysis gases evolved. For metallurgical cokes, no effort is made to prevent the porosity, as this material is not intended for use as a structural material. Figure 2B shows a prototype CBM made by bonding pulverized coke fines with 30 wt% coal tar pitch[8], after firing to 1200°C. It has a microstructure typical of bonded aggregates, where in this case the aggregate is 2-5 micrometer particles of coke. These pulverized coke particles are smaller than the typical coke pore size, so are dense carbon, rather than a cellular solid. We will compare the mechanical properties of these materials with conventional concrete and fired clay masonry.

Construction materials must possess a wide range of properties, including mechanical properties, which are characterized by a variety of test methods and standards[14]. For the present purpose, we must over-simplify the issue by considering only one mechanical characteristic, the tensile strength as determined by the diametral compression[15] or the "indirect tensile" or "splitting" test used for concrete. Typical values[16] for the indirect tensile strength for ordinary type III concrete with a density of 2.4 gm/cm^3 range from 2 MPa to 5.8 MPa. Very high strength concrete has an indirect tensile strength up to 7.7 MPa. These values appear in Figure 3 as indirect tensile strength vs. density for concrete. Structural clay products are not typically evaluated by the indirect tensile strength, but rather by compressive crushing strength[17] or flexural strength[18]. Using relations commonly used to estimate indirect tensile strength from compressive or flexural strength for concrete[16], we show for bricks on Figure 3 a strength range of 5-7 MPa and a density range of 1.7-1.9 gm/cm^3.

These strength-density values for common building materials can be compared with prototype CBM made by pyrolyizing mixtures of pulverized coal char and coal tar pitch. The strength depends upon the pyrolysis temperature and relative amount of pitch binder[8]. Figure 3 includes two compositions: char+13 wt% pitch, with a density of 1.14 gm/cm^3 and an indirect tensile strength of 8.4- 11 MPa; and char+23 wt% pitch with a density of 1.17 gm/cm^3 and strength range 29-41 MPa. Note that these prototype CBM materials are significantly stronger than concrete or clay bricks, but have a much lower density.

Coke itself is a very porous cellular solid with appreciable strength, although it is not considered a structural material. Figure 2 has some data for strength of specimens core-drilled from a continuously processed formed coke[13]. These specimens were not intended for structural use, but simply harvested from foundry-grade formed coke. This formed coke strength is similar to literature data[19] for blast furnace-grade metallurgical coke, which ranges from 4-5 MPa, with a density of 0.8-0.9

gm/cm^3. But coke has an indirect tensile strength comparable with concrete. From the standpoint of strength, it appears that CBM from coal char could readily substitute for bricks or concrete masonry.

ENVIRONMENTAL IMPACTS: CARBON DIOXIDE EMISSIONS FROM SRM AND HYDROGEN-FUELED HECAM HYDROGEN PRODUCTION

Guerra also calculated the amount of carbon dioxide produced by the manufacture of hydrogen[7]. This is expressed below as the mass of carbon dioxide per unit energy from hydrogen combustion, using the high heating value of hydrogen (141.9 MJ/kg of hydrogen). The standard process of steam reforming of methane (SRM) produces carbon dioxide at a rate of 47 $kg_{CO2}/GJ_{hydrogen}$, using natural gas as the feedstock. The coal-fueled HECAM process, with thermal decomposition of the methane in the pyrolysis gas, produces hydrogen *from coal* with 22.6 $kg_{CO2}/GJ_{hydrogen}$. Note that this coal-fed process can produce hydrogen with about *half* the carbon dioxide emissions as natural gas reforming. Of course a great deal more of the fossil resource is needed. SRM uses only 14.1 kg of natural gas to produce 1 GJ of hydrogen, while producing 1 GJ of hydrogen energy uses 318 kg of coal is for HECAM. To yield the same amount of hydrogen, the mass of coal consumed is more than 20 times the mass of natural gas. This is because coal is a much leaner "hydrogen ore" than methane, and also because the SRM process oxidizes the carbon in the methane to produce more hydrogen by the water gas shift reaction. Carbon is not oxidized by HECAM. Rather, most of that mass of coal is converted to solid carbon materials, so the carbon is used in the *materials sector* rather than the *energy sector*.

MATERIALS AND ENERGY: COMPARING LAND AND ATMOSPHERE IMPACTS OF CONCRETE AND COAL COMBUSTION VS. CBM AND HYDROGEN COMBUSTION

HECAM implies co-products entering the economy in both the *energy sector* and the *materials sector*. Metrics must be introduced that combine energy sector and materials sector inputs. Substitution of conventional concrete with carbon building materials also implies assessing benefits and impacts of coal mining and quarrying for limestone and aggregate. Thus one should consider impacts on atmosphere, water, and land. This is a complex issue. This paper will consider only two simple metrics. For atmospheric impacts, it considers carbon dioxide emissions for both energy and materials production. For land impacts, it considers only the volume of extracted raw materials. The raw material for HECAM is coal, and the products are hydrogen for heat and carbon for building materials. Consider the volume of coal that must be mined to produce 1000 GJ of heat from hydrogen (on HHV basis) and a certain volume of CBM. The conventional alternative is to use concrete for materials and obtain energy by burning coal. Consider the amount of coal that must be mined to produce 1000 GJ of heat, and the amount of limestone and aggregate that must be quarried to produce the same volume as the CBM. Clearly this is a simplification for land spoilage, since it only compares the size of the hole in the ground, and not the other impacts of coal mining and stone quarrying.

Cement manufacture produces massive amounts of anthropogenic carbon dioxide amounting to 3.8% of global emissions[20]. Consider ordinary Portland cement concrete (OPC), which is manufactured by heating limestone, to remove the CO_2. Producing a ton of OPC (at 12 wt% cement) produces about 0.13 tons CO_2, consumes 0.3 tons of quarried limestone, and about 0.88 tons of quarried aggregate. A ton of OPC (with a density about 2.3 tons/m^3) has a volume of 0.43 m^3. Producing a ton of OPC requires the extraction of about 0.54 m^3 of materials, including about 0.17 m^3 removed from the limestone quarry and 0.37 m^3 removed from the aggregate quarry (which might be sand and gravel or crushed stone).

Consider now the process of hydrogen-fueled HECAM to produce 1000 GJ of heat, based on the Guerra's technoeconomic analysis. To produce enough hydrogen for 1000 GJ of heat (at HHV), 22.4 tons of carbon dioxide is produced. The volume of coal consumed is 236 m^3 or 318 tons of coal. The co-production of carbon building materials is 185 m^3 or 222 kilograms, with the 70% mass yield, which amounts to a volume of 185 m^3 (if the density is 1.2 gm/cm^3). The atmospheric impact is 22.4 tons of carbon dioxide emissions and the land spoilage impact is a volume of 236 m^3 extracted from coal mines. This appears in Table 1 and Table 2.

Assume that carbon building materials can substitute for concrete at an *equal volume basis*. This should be conservative, since CBM can be significantly stronger than OPC (as shown in Figure 3). Consider the situation if 185 m^3 of carbon building materials were used to displace 185 m^3 of OPC. This volume corresponds to about 425 tons of concrete. About 55 tons of carbon dioxide is produced in the manufacture of 45 tons of OPC. The volume of 230 m^3 must be quarried to produce this OPC, which includes 72 m^3 extracted from the limestone quarry and 157 m^3 extracted from the aggregate quarry. These figures appear in Table 2. Considering only the *materials sector*, the atmospheric impact of the OPC is 55 tons of carbon dioxide emissions and the land spoilage impact is 230 m^3 extracted from stone quarries.

Now consider the *energy sector*. Burning a ton of bituminous coal produces 38 GJ of heat on HHV basis, and 3.6 tons carbon dioxide. A volume of 0.56 m^3 is extracted from coal mines per ton of coal. To produce the 1000 GJ of heat from coal combustion, 26 tons of coal is consumed and 94.7 tons of carbon dioxide is produced. The volume of coal extracted from the coal mine is small: only 15 m^3 of coal. This appears in Table 1.

Thus to acquire 185 m^3 of building materials from OPC and 100 GJ of heat from coal burning, the combined atmospheric impact is the emission of about 102 tons of carbon dioxide, and the combined land spoilage is about 245 m^3 of volume extracted from limestone quarries, aggregate quarries, and coal mines. The combined land and atmospheric spoilage for the HECAM scenario and the coal burning with concrete scenario are combined in Table 3 and Figure 3. Also shown is an intermediate case where materials are OPC, but the heat is obtained by burning hydrogen manufactured from natural gas by SRM. For the HECAM-H$_2$/CBM scenario, the atmospheric impact is smaller by a factor of 3 compared to the Coal burning/OPC case, while the land impact is slightly smaller. Compared with the SRM-H$_2$/OPC scenario, HECAM reduces the atmospheric impact by about a factor of 2, with a small increase in the land impact.

Thus if CBM can substitute for OPC on an equal-volume basis, hydrogen extracted by coal pyrolysis has a *reduction* of 68% of the carbon dioxide released into the atmosphere, and 4% *reduction* in volumetric land spoilage (for coal burning) or a 3% *increase* in volumetric land spoilage (for SRM-hydrogen burning). Most of the reduction comes from the elimination of the cement manufacture and the stone quarrying. The major reduction is the carbon dioxide emissions, where the benefits come from *not manufacturing cement* from limestone and *not burning coal*. Even though much *more* coal mining is required to obtain the hydrogen from HECAM, much *less* stone quarrying is required. The land spoilage impact is essentially unchanged, considering only the volume of material extracted from the mines and quarries. This simple analysis presumes that stone quarries and coal mines are equivalent. Of course they differ in the nature of their environmental impacts, and often in their location. Many stone quarries are in urban areas, while most coal mines are in rural areas.

GENERAL DISCUSSION

Carbon building materials, as a co-product with hydrogen fuel from coal pyrolysis, can enable hydrogen to be made *without* oxidizing the carbon, and thereby avoid the production of carbon dioxide. Coal can then be used as an economic "hydrogen ore" for clean energy, using only the H present in the coal, which is about 40 atomic percent hydrogen (for a nominal coal at CH$_{0.7}$). The remaining ~ 60 atomic percent carbon is not used as a fuel (to avoid CO$_2$), but rather in the materials sector as a durable building material. The CBM sequesters carbon while putting it to productive use as a durable material. If CBM *displaces* conventional materials, such as Portland cement concrete, it also eliminates the CO$_2$ emissions associated with cement manufacture. The use of carbon char as a solid sequesterant implies that CBM will not later be burned. So at end-of-life, the rubble from carbon building materials should be land filled, as is commonly done now with building debris.

Of course, building-grade carbons do not presently exist, so the materials and processes must be invented. They could be made by direct pyrolysis of coal as a materials-grade variety of formed coke, if a well-controlled pyrolysis process were developed which could reduce the porosity. Or they could be made as a carbon-bonded carbon ceramic, using the pyrolysis product coal tar pitch as a binder for pulverized char. Preliminary indications of the mechanical strength achievable from both pitch-bonded char ceramics and formed metallurgical coke are quite promising. The strength matches or exceeds the strength of common masonry products, even though the density is significantly lower. New manufacturing methods must be devised to realize these products. The production of CBM involves temperatures on the order of 1000°C, which correspond also to the temperatures where volatilization is complete. These temperatures are high, but are comparable to what is common for structural clay products, architectural glass, or the production of cement for concrete. Our building materials are all derived from high temperature manufacturing, so large-scale production of CBM at 1000°C compares favorably with current practice.

Clearly many other issues must be defined for the practical application of CBM. Some of these are durability, degradation mechanism, freeze thaw stability, leaching of species from the minerals in the coal, etc. New standards and methods will have to be developed. Carbon, of course, burns, while concrete or structural clay products do not. However, many building products (lumber, polymers, asphalt) also burn, so the risk of fire is commonly accepted. Note that these bonded carbons have been thoroughly de-volatilized (as the volatiles are extracted for the hydrogen product), so the ignition temperature of CBM would be more like coke, which is much higher than the ignition temperature than coal[21] or wood[22]. Fire resistance and other durability issues must be addressed for CBM. These application-related issues, as well as the manufacturing issues, must be resolved to make carbon building materials a practical reality, and realize the environmental benefits of extracting clean energy with low-CO_2 emissions from our abundant coal resources.

CONCLUSIONS

A technoeconomic analyses show that manufacturing hydrogen by coal pyrolysis can be cost competitive with steam reforming of methane for certain ranges of natural gas cost, and certain ranges of solid carbon co-product value. For 2007 natural gas prices, a volumetric carbon co-product value similar to ordinary concrete and less than structural clay products suffices for cost equivalence of hydrogen from coal pyrolysis and steam reforming of methane. A coal pyrolysis route to hydrogen can be economic. Hydrogen from coal pyrolysis has less than half the CO_2 emissions as hydrogen from steam reforming of methane.

The mechanical strength of prototype carbon building materials, such as coal tar pitch-bonded char and metallurgical coke is comparable to the strength of fired clay brick and concrete. Carbon building materials are strong enough to be used in place of conventional building material.

Compared with coal burning and Portland cement manufacturing, the combination of energy from coal-pyrolysis derived hydrogen, and carbon building materials, can reduce carbon dioxide emissions by 68% with little impact on land spoilage, assessed by volume extracted from mines and quarries. The additional coal mining necessitated by not using the carbon in coal for energy is compensated by the reduced stone quarrying for the limestone and aggregate supplanted by using the carbon as building materials.

ACKNOWLEDGEMENTS
Steven J. Norton did the microscopy and strength measurements for the formed coke.

REFERENCES

[1] J.W. Halloran. Carbon-Neutral Economy with Fossil Fuel-Based Hydrogen Energy and Carbon Materials. *Energy Policy* **35** 4839-46 (2007)

[2] V. I. Rudyka and V.P. Malina. Recent trends in hot metal and coke production,*Coke and Chemistry* **51** (1) 7-9 (2008)

[3] M. Ball and M. Wietschel.*The Hydrogen Economy, Opportunities and Challenges* Cambridge University Press (2009)

[4] J. Kronenberg and R. Winkler.Wasted waste: An evolutionary perspective on industrial by-products. *Ecological Economics* **68** 3026-33 (2009)

[5] J. W. Halloran. Extraction of Hydrogen from Fossil Resources with Production of Solid Carbon Materials. *International Journal of Hydrogen Energy* **33** 2218-24 (2008)

[6] F.H.Norton. *Refractories, 3rd Ed.*, McGraw-Hill Book Co., New York. p.196 (1949)

[7] Z. Guerra. *Technoeconomic Analysis of the Co-Production of Hydrogen Energy and Carbon Materials*. Ph.D. Thesis, University of Michigan Ann Arbor, MI (2010)

[8] A. Wiratmoko and J.W. Halloran. Fabricated Carbon from Minimally-Processed Coke and Coal Tar Pitch as Carbon-Sequestering Construction Materials. *J. Materials Science* **44** [8] 2097-00 (2009)

[9] M.E. Ertem and A. Ozdaback. Energy balance application for Erdemir coke planet with thermal camera measurements. *Applied Thermal Engineering* **31** 426-33 (2005)

[10] L.Basye. S. Swaminathan. Hydrogen production costs – a survey. Department of Energy Report DOE/G0/10170-T-18 December 4 (1997)

[11] M. Pastilli. "The Cost of Doing Business with Concrete". *Concrete Construction Magazine*. Article ID 231054, November 1 (2005)

[12] United States Environmental Protection Agency Final Report "Brick and Structural Clay Product Manufacturing" (August 1997) www.epa.gov/ttn/chief/ap42/ch11/final/c11s03.pdf

[13] R. A. Wolfe. "A Continuous Carbonite™ Process for Making Foundry-Sized Formed Coke". *Ductile Iron News*. Ductile Iron Society (2009)

[14] J.M. Illston and P.J. L. Domone, editors, *Construction Materials: Their Nature and Behavior*, Spon Press, London (2001)

[15] B. Bhatia and R.K. Aggarawal. A Comparative Study of Baked Carbon Mixes Employing Different Filler Materials. *J. Materials Science*.**14** 1103-10 (1979)

[16] N. Arioglu, Z. C. Girgin, and E. Arioglu. "Evaluation of Ratio between Splitting Tensile Strength and Compressive Strength for Concretes up to 120 MPa and it Application in Strength Criterion" *ACI Materials Journal*, Technical Paper No. 103-M03, January/February 18-28 (2006_

[17] D.V. Oliveira, P.B. Lourenco, and P. Rosa. "Cyclic Behavior of Stone and Brick in Compressive Loading", *Materials and Structures* **39**, pp. 247-57 (2006)

[18] G.L. Robinson. "Structural Clay Products" pp. 924-95 in *Engineering Materials Handbook, Vol 4 Ceramics and Glass*, S.J. Schneider, editor, ASM International, Pittsburgh PA (1991)

[19] H. Sato, J. W. Patrick, and A.Walker. "Effect of coal properties and porous structure on the tensile strength of metallurgical coke". *Fuel* **77** (11) pp. 1203-08 (1998)

[20] J.G. Cansell, C. Le Quere, M.R. Raupach, C.B. Field, E.T Buitenhuis, F. Ciais, T.J. Conway, N/P. Gillett, R.S. Houghton, and G. Marland. Contributions to accelerating atmospheric CO_2 growth from economic activity, carbon intensity, and efficiency of natural sinks. *Proceedings of the National Academy of Sciences*, **104**. 47. 18866-70 (2007)

[21] L. Jia, E.J. Anthony, I. Lau, and J. Wang. Study of Coal and Coke Ignition in Fluidized Beds. *Fuel* **85**,(5-6) 635-42 (2006)

[22] A. B. Fuertes, E. Hampartsoumian, and A. Williams. Direct Measurement of Ignition Temperatures of Pulverized Coal Particles. *Fuel* **72** (9) 1287-91 (1993)

Table I: Resource Consumption and Products for 1000 GJ of Heat from Hydrogen

Scenario	Coal Consumed (tons)	Coal Mine Volume (m³)	CO₂ Emissions (tons)	CBM Produced (tons)	CBM Volume (m³)
Coal combustion	26	15	94.7	----	----
SRM-H₂ combustion	----	----	47	----	----
HECAM-H₂ combustion	318	236	22.4	222	185

Table II: Consumption of Resources and Production of Products for 185 m³ of Building Materials

Scenario	Mass Consumption (tons)	Volume extracted from mine or quarry (m³)	CO₂ Emissions Produced (tons)	Building Materials Volume (m³)
CBM from HECAM	318 coal	236 coal	----	185
435 tons OPC concrete	51 limestone 374 aggregate	72 limestone 157 aggregate	55	----
Total for 425 tons OPC	425	230	55	185

Table III: Total Spoilage of Land and Atmosphere to Produce 1000 GJ of Heat from Hydrogen for HHV and 185 m³ of Building Materials

Scenario	Total Mine or Quarry Volume (m³)	Total CO₂ Emissions (tons)
Coal combustion-Build with OPC concrete	245	150
SRM-H₂ Combustion-Build with OPC concrete	230	102
HECAM-H₂ Combustion-Build with CBM carbon	236	47

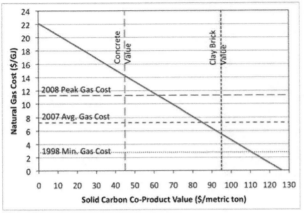

Figure 1A

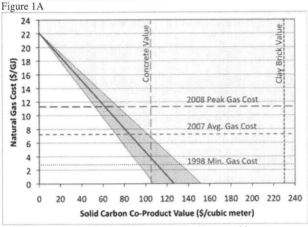

Figure 1 B Volumetric – Carbon co-product value per cubic meter

Figure 1 Natural Gas prices and solid carbon co-product values for equivalent hydrogen production cost, from steam reforming of methane (SRM) using natural gas and coal pyrolysis (HECAM) using Illinois bituminous coal, using baseline cost parameters by Guerra (2010)

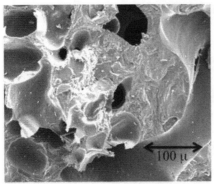

Figure 2A

Figure 2B

Figure 2 Typical microstructures of coal-derived carbons
Figure 2A Porous cellular structure of formed coke (from Carbonite Corporation)
Figure 2B Pulverized coke fines bonded with 30 wt% coal tar pitch fired to 1200°C
Note the magnification marker is 100 μm for Figure 2A but 10 μm for Figure 2B.

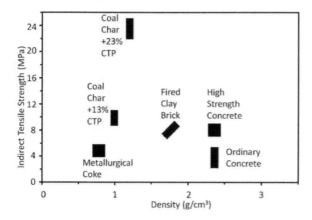

Figure 3 Indirect tensile strength vs. density of carbon materials based on coke and coal tar pitch-bonded coal char compared with typical values for clay masonry and concrete.

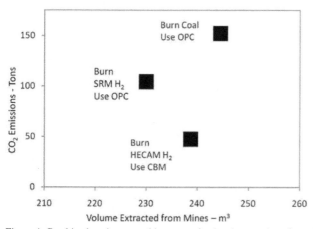

Figure 4 Combined environmental impact on land and atmosphere from the production of 1000GJ of heat energy and 185 cubic meters of building materials, assuming carbon building materials (CBM) can substitute for an equivalent volume of ordinary Portland cement concrete (OPC). The heat is produced by burning coal, or burning hydrogen produced from natural gas by steam reforming (SRM) or from coal by pyrolysis (HECAM)

THERMAL DECOMPOSITION OF T-BUTYLAMINE BORANE STUDIED BY IN SITU SOLID STATE NMR

Jordan Feigerle, Norm Smyrl, Jonathan Morrell, Ashley C. Stowe*

Y-12 National Security Complex, P. O. Box 2009, MS 8097, Oak Ridge, TN 37831
*corresponding author: stoweac@y12.doe.gov; 865-241-0675

ABSTRACT

Amine boranes are a class of hydrogen storage materials which release significant yields of hydrogen gas upon heating to temperatures relevant for automotive applications. The thermodynamic stability of this class of amine boranes is correlated with the extensive di-hydrogen bonding between hydrides bound to boron and amine protons. Further, the hydrogen evolution follows a bimolecular pathway via the di-hydrogen bonding network. The thermal decomposition of t-butylamine borane (tBuAB), $(CH_3)_3CNH_2BH_3$, has been studied in order to understand the reaction pathway of hydrogen sorption and the impact of the t-butyl substitution dehydrogenation thermodynamics. ^{1}H, ^{11}B, and ^{13}C solid state nuclear magnetic resonance (NMR) spectroscopy has revealed that heating initiates two separate reaction pathways: isomerization and hydrocarbon abstraction resulting in varying yields of isobutane and hydrogen. It is also possible that tBuAB dissociates about the N-B bond giving rise to borane stretching modes in the gas FTIR. Trapped t-butylamine (tBuA) which slowly diffuses from the tBuAB solid in ^{13}C NMR studies appears to be present; however, this spectral region is convoluted by other decomposition products. ^{11}B NMR indicates that the major reaction pathway results in hydrogen evolution with isobutane formation being present in smaller yields. The t-butyl substitution lowers the thermodynamic stability—compared to NH_3BH_3—but results in impure hydrogen gas stream and lowered capacity due to isobutene evolution.

INTRODUCTION

Development of suitable materials to store hydrogen for automotive use has received pointed attention over the past decade [1]. Significant progress has been made with the discovery of novel chemical hydrides, complex metal hydrides, and adsorption substrates which continue to optimize both thermodynamics and kinetics of hydrogen sorption [2]. Chemical hydrides typically offer the largest theoretical gravimetric capacities. Autrey *et al.* [3] have recently shown that mechanical milling of alkali metal hydrides with ammonia borane can further lower the decomposition temperature. In all cases, however, many challenges remain in order to meet the current US DOE performance targets [4].

Amine boranes are being considered for hydrogen storage materials since they contain significant quantities of hydrogen which potentially can be released at low temperatures (80-150 $^{\circ}$C) via chemical reactions. Ammonia borane, NH_3BH_3, is one of the most promising in this class as it decomposes to release greater than two moles of pure hydrogen gas (14 wt %) below 160 $^{\circ}$C [5-19]. Although isoelectronic to ethane, NH_3BH_3 is a solid at room temperature due to the di-hydrogen bonding network formed between the amine protons and boron hydrides in the solid state lattice. Further, it has been shown that the hydrogen release mechanism involves transformation and isomerization to an ionic dimer where a hydride migrates from one boron to the adjacent boron in the dimer [20]. The greatest challenge to the use of ammonia borane as a hydrogen fuel is the regeneration path from spent fuel to ammonia borane again. The proposed chemical synthesis involves complicated organometallic reactions to form boron hydrogen bonds from the thermodynamically stable polyimidoborane products $(BNH)_n$ [21].

Recent theoretical calculations suggested that incorporation of carbon atoms into the $(BNH)_n$ product would be less thermodynamically stable. These $(CBNH)_n$ compounds are potentially less energy intensive making regeneration of the amine borane fuel more feasible [22]. In the present study, *tert*-butylamine borane is investigated by heteronuclear *in situ* solid state NMR to understand hydrogen release from a hydrocarbon containing amine borane. t-butylamine borane has similar physical properties to amine borane with a melting point of 96 °C. A single proton has been replaced with a t-butylamine group resulting in a weakening of the di-hydrogen bonding framework. t-butylamine borane has a theoretical gravimetric hydrogen density of 15.1%; however, isobutane can also be evolved rather than hydrogen. If decomposition yields one mole isobutane and two moles hydrogen, 4.5 wt% H_2 gas will be evolved. More importantly for the present work, the resulting spent fuel should be comprised of both $(BNH)_n$ and $(CBNH)_n$ polyimidoboranes.

EXPERIMENTAL

tert-butylamine borane, $(C_4H_9)NH_{2*}BH_3$, hereafter tBuAB, is a white solid obtained from Sigma Aldrich and used as received without further purification. Solid state magic angle spinning NMR experiments were conducted on a Bruker Avance III 300 MHz spectrometer with a 4mm CP/MAS probe and forced air heating of the sample. A powder sample of t-butylamine borane was placed in the MAS-NMR sample rotor and with a spinning rate of 8 kHz. Isothermal *in situ* decomposition experiments were conducted at 80, 85, 90, and 95 °C for as long as 10 hr to study the decomposition pathway. A maximum temperature of 95 °C was chosen as to prevent t-butylamine borane from melting at the literature value of 96-100 °C. ^{11}B, ^1H, and ^{13}C experiments were conducted by collecting a spectrum every minute.

Evolution of volatile decomposition products was studied by coupling thermogravimetric isotherms with FTIR and mass spectrometry analysis. 5 mg of tBuAB was placed on a platinum sample holder in a TA instruments TGA Q500 Thermogravimetric Analyzer for isothermal experiments at 80, 85, 90, and 95 °C. Evolved gas passed through a heated quartz capillary (250 °C) to a Thermo Scientific FTIR 2-meter gas cell. FTIR spectra were collected at 30 second intervals. A Netzsche STA-449 simultaneous thermal analyzer coupled to a residual gas analyzer was used to obtain mass spectra during decomposition.

RESULTS

Thermal decomposition of tBuAB resulted in the evolution of hydrogen gas as well as a variety of non-hydrogen volatiles. Figure 1 presents the stacked ^{11}B MAS-NMR spectra at 95 °C where the reaction begins almost immediately after heating. The starting material (tBuAB, δ = -33 ppm) becomes more mobile resulting in resolution of the quartet pattern representative of a boron nuclei being coupled to three equivalent hydrogen atoms. A slight downfield shift to -27 ppm is observed for the mobile phase. The ^{11}B-^1H J-coupling constant is 128 Hz. The expected quartet pattern resulting from splitting of the ^{11}B resonance by three equivalent protons (-BH_3) for both the starting material and new phase indicate that no reaction has taken place. This fact is confirmed by FTIR and MS evolved gas analysis in which no volatile species are observed. The origin of the new phase of tBuAB has previously been suggested for ammonia borane to be related to a disruption in the solid state di-hydrogen bonding lattice which promotes greater molecular motion [20]. Shaw *et al.* further investigated the nature of the mobile phase of ammonia borane [23].

A quintet resonance pattern at -40 ppm appears after 2 minutes of heating. The five line feature arises from hydrogen coupling hydrides bonded to the boron atom (-BH$_4$). A corresponding resonance in the –BH$_2$ region of the spectrum is also observed growing in at -8 ppm. This feature is split into the expected triplet pattern. The correlated appearance of a –BH$_4$ and –BH$_2$ resonance upon heating an amine borane suggests that isomerization of a t-butylamine borane dimer (**I**) has occurred in a similar fashion to ammonia borane [20]. The isomeric dimer — the diammoniate of diborane — is the nucleation species upon which hydrogen release occurs.

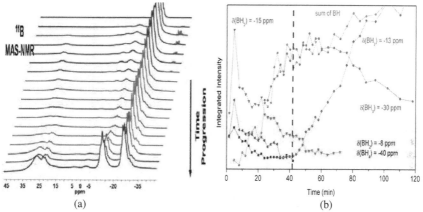

Figure 1. (a) ^{11}B MAS-NMR of t-butylamine borane heated isothermally to 95 °C. Isothermal decomposition products are observed as a function of elapsed time. (b) Plot of the ^{11}B MAS-NMR integrated intensities for each resonance as a function of elapsed isothermal heating time at 95 °C. δ(BH$_2$) = -13 ppm represents the total intensity of **B, II,** and **III**. Note that the starting material is not shown.

A second –BH$_2$ resonance appears with the ionic dimer. In order for this feature to be identified, FTIR evolved gas analysis was conducted. Aliphatic hydrocarbon peaks appear in the FTIR spectrum which correspond to the appearance of this second –BH$_2$ feature at -15 ppm. Further, no new spectral feature was observed in the ^{13}C MAS-NMR at the identical decomposition time suggesting that the boron hydride reacts with the t-butyl ligand to release isobutane gas in competition with hydrogen gas formation. Once t-butylamine borane evolves isobutane, unsubstituted aminoborane (NH$_2$BH$_2$; **A**) remains and continues to decompose separately from t-butylamine borane dehydrogenation products according to the reaction pathway of ammonia borane described by Stowe *et al.* [20]. One exception to this amino borane pathway is observed. Loss of isobutane appears to favor formation of the aminoborane monomer rather than the dimer reported in ref. 20.

As the reaction progresses, both hydrogen and isobutane continue to be released as additional –BH$_2$ and multiple –BH resonances appear. To better understand the reaction products, the integrated intensity of each spectral feature was calculated. The intensities were plotted as a function of elapsed time (Figure 1b) showing which features were correlated and which reaction intermediates were consumed to form additional spectral features. Aminoborane (δ = -15 ppm) can cyclize to form **B** (δ = -15 ppm) and as it evolves hydrogen, it forms a common cyclic –BH species called borazine ($B_3N_3H_6$; **C**; δ = 27 ppm) which polymerizes to form polyborazyline (**D**; δ = 31 ppm).

Decomposition of the t-butylamine borane ionic dimer creates the possibility of two similar BN linear chains. If hydrocarbon abstraction occurs, only one terminal t-butyl group remains where as hydrogen loss yields **II** containing two t-butyl groups. A –BH_3 (δ = -30 ppm) resonance and –BH_2 resonance appear as the substituted diammoniate of diborane (**I**) resonances begin to be consumed. Observation of the δ = -13 ppm resonance is initially masked by the aminoborane resonance. Proton decoupled $^{11}B\{^1H\}$ experiments indicate two distinct resonances which are masked due to the proton splitting. These two structures cannot be distinguished under the current ^{11}B experimental conditions. If **III** reacts to remove another isobutane molecule, aminoborane products are formed. This reaction does not appear to be favored since a significant increase in the aminoborane peak does not occur. As the reaction of these two species continues, both hydrogen and isobutane continue to be released (FTIR and MS data) resulting in formation of extended, unsaturated BN chain compounds. These species have –BH groups which have lost two hydrides from the original stoichiometry. The –BH resonance is observed at 23 ppm. This peak—which shows a doublet splitting—transfers intensity to a new peak at 17 ppm corresponding to polymerization of a $[CBNH]_x$ (**V**) species. The complete decomposition mechanism is shown in Scheme 1.

^{13}C MAS-NMR was also conducted at multiple temperatures to further understand the chemistry. In particular, ^{13}C NMR provides a direct probe of the reaction product after isomerization. Both hydrogen and isobutane release are observed in evolved gas analysis resulting in a product differing by one t-butyl group only. The ^{13}C resonance should differ between these compounds. Figure 2 compares the ^{13}C spectrum when heating is initiated to the spectrum of decomposed product after 30 minutes. For the t-butyl group, there are two distinct regions of the ^{13}C spectrum. The central carbon which has three methyl groups bound to it appears near δ = 50 ppm while the methyl carbons are centered in the 25-35 ppm region. The methyl carbon nuclei have three protons which create a quartet splitting pattern. As t-butylamine borane decomposes, isobutane release removes all carbon atoms from the solid such that no changes in the ^{13}C MAS-NMR spectrum are observed.

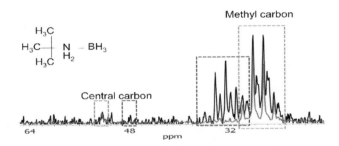

Figure 2. ^{13}C MAS-NMR spectrum of t-butylamine borane at t = 0 (*red*) and t = 30 minutes (*black*) of heating at 95 °C.

Dehydrogenation, however, results in reaction products containing t-butyl groups. The carbon atoms are multiple atoms away from the dehydrogenation such that changes to the spectrum are small. This can be observed in Figure 2 (*black trace*) where many features are observed. The proton coupling convolutes the spectrum and a series of decoupling experiments was conducted to simplify the result.

In Figure 3a, the proton decoupling ^{13}C MAS-NMR experiment is stacked as a function of elapsed heating time. The starting materials (δ = 28 ppm) is transformed to a mobile phase (δ = 29 ppm) as observed in ^{11}B experiments. Isomerization to the ionic dimer (**I**) will not result in a significant shift in the ^{13}C signal. Subsequent decomposition from the ionic dimer, however, results in the appearance of new resonances. The terminal t-butyl group of **II** and **III** are the same and appears at δ = 34 ppm. The second t-butyl group in **II** (product of dehydrogenation) has a distinct ^{13}C resonance at δ = 28.3 ppm. A multinuclear approach allowed a more complete understanding of the decomposition chemistry because the two species resulting from dehydrogenation and hydrocarbon abstraction could not be separated by ^{11}B NMR. Comparison of the ^{13}C resonances also revealed that hydrogen release was the favored reaction over hydrocarbon abstraction as shown in Figure 3. A similar preference to dehydrogenation was seen for methylamine borane [23].

One surprising result was the observation of t-butylamine gas in the solid state ^{13}C NMR spectrum. t-Butylamine forms via dissociation of the boron-nitrogen bond and is apparently trapped within the solid lattice. FTIR analysis of evolved gas showed a boron hydride in the spectrum at comparable reaction time, however, no amine vibrations were observed. This is consistent with slow diffusion of bulky t-butylamine from the *sticky* solid lattice. The methyl carbon resonance is centered at δ = 33 ppm as expected for t-butylamine and the concentration of this resonance remains consistent and minor (as shown in Figure 3b).

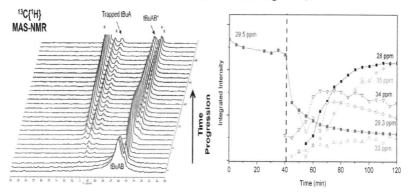

Figure 3. (a) ^{13}C{1H} MAS-NMR of t-butylamine borane at 95 °C. t-butylamine gas was observed to be trapped in the solid material. (b). Plot of the ^{13}C{^{1}H} MAS-NMR integrated intensities for each resonance as a function of elapsed isothermal heating time at 95 °C. No new resonance is observed for the ionic dimer as described in the text.

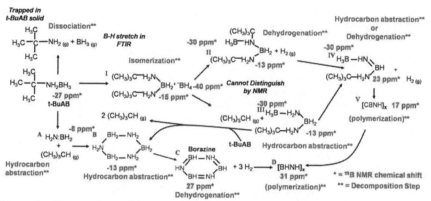

Scheme 1. The complete solid state reaction pathway of t-butylamine borane decomposition under isothermal conditions. Hydrocarbon abstraction, dehydrogenation and isomerization all occur yielding $[BNH]_x$ and $[CBNH]_x$ polymers with both H_2 and C_4H_{10} evolution. ^{11}B chemical shift values are included in red.

DISCUSSION

The overall reaction mechanism consisting of three distinct reaction pathways is shown in Scheme 1. Dehydrogenation and hydrocarbon abstraction both occur with significant hydrogen and isobutane evolution. The hydrogen reaction is favored based on ^{13}C NMR results. To a minor extent, dissociation across the boron-nitrogen bond occurs as well. Carbon insertion into the dehydrogenation product occurs with the formation of carbon substituted polyimidoborane polymers $(CBNH)_n$. The overall reaction mechanism to achieve the carbon insertion is similar to that of ammonia borane such that efficient, low temperature hydrogen release is still possible. In the present case, the carbon atoms originate from hydrocarbons on the amine which also results in hydrocarbon release which is detrimental to the purity of the hydrogen gas stream and the overall fuel capacity. While this is undesirable, a mixed amine borane system where a small ratio of a substituted amine borane is combined with ammonia borane could result in sufficient quantities of $(CBNH)_n$ formation. Dixon et al. [22] have shown significant reductions to the melting point of mixed methylamine borane/ammonia borane systems. The resultant material likely must still undergo chemical processing to regenerate the original hydrogen fuel; however, this chemical processing potentially will be less energy intensive.

ACKNOWLEDGEMENTS

This work was funded through the Y-12 Plant Directed Research and Development program under contract no. DE-AC05-00OR22800.

REFERENCES

1. S. Satyapal, J. Petrovic, C. Read, G. Thomas, G. Ordaz. *Catalysis Today* 2007, **120**, 246-256
2. S. Orimo, Y. Nakamori, J. Eliseo, A. Zuttel and C. Jensen, *Chem. Rev.*, 2007, **107**, 4111–4132

3. Z. Xiong, C. Yong, G. Wu, P. Chen, W. Shaw, A. Karkamkar, T. Autrey, M. Jones, S. Johnson, P. Edwards, B. David. *Nature Materials* 2008, **7**, 138-141

4. Recent information can be found at: http://www.eere.energy.gov/hydrogenandfuelcells.

5. W. T. Klooster, K. T. F., P. E. M. Siegbahn, T. B. Richardson, R. H. Crabtree, *J. Am. Chem. Soc.*, 1999, **121**, 6337–6343

6. D. J. Heldebrant, A. Karkamkar, N. J. Hess, M. Bowden, S. Rassat, F. Zheng, K. Rappe and T. Autrey, *Chem. Mater.*, 2008, **20**, 5332–5336

7. N. J. Hess, M. E. Bowden, V. M. Parvanov, C. Mundy, S. M. Kathmann, G. K. Schenter and T. Autrey, *J. Chem. Phys.*, 2008, **128**, 034508

8. A. Paolone, O. Palumbo, P. Rispoli, R. Cantelli and T. Autrey, *J. Phys. Chem. C*, 2009, **113**, 5872–5878

9. H. Cho, W. J. Shaw, V. Parvanov, G. K. Schenter, A. Karkamkar, N. J. Hess, C. Mundy, S. Kathmann, J. Sears, A. S. Lipton, P. D. Ellis and S. T. Autrey, *J. Phys. Chem. A*, 2008, **112**, 4277–4283

10. O. Gunaydin-Sen, R. Achey, N. S. Dalal, A. Stowe and T. Autrey, *J. Phys. Chem. B*, 2007, **111**, 677–681

11. N. J. Hess, M. R. Hartman, C. M. Brown, E. Mamontov, A. Karkamkar, D. J. Heldebrant, L. L. Daemen and T. Autrey, *Chem. Phys. Lett.*, 2008, **459**, 85–88

12. N. J. Hess, G. K. Schenter, M. R. Hartman, L. L. Daemen, T. Proffen, S. M. Kathmann, C. J. Mundy, M. Hartl, D. J. Heldebrant, A. C. Stowe and T. Autrey, *J. Phys. Chem. A*, 2009, **113**, 5723–5735

13. S. M. Kathmann, V. Parvanov, G. K. Schenter, A. C. Stowe, L. L. Daemen, M. Hartl, J. Linehan, N. J. Hess, A. Karkamkar and T. Autrey, *J. Chem. Phys.*, 2009, **130**, 024507

14. V. M. Parvanov, G. K. Schenter, N. J. Hess, L. L. Daemen, M. Hartl, A. C. Stowe, D. M. Camaioni and T. Autrey, *Dalton Trans.*, 2008, 4514–4522

15. R. Raja, S. Chellappa, M. Somayazulu, V. V. Struzhkin, T. Autrey and R. J. Hemley, *J. Chem. Phys.*, 2009, **131**, 224515

16. R. Custelcean and Z. A. Dreger, *J. Phys. Chem. B*, 2003, **107**, 9231–9235

17. Y. Lin, W. L. Mao, V. Drozd, J. Chen and L. L. Daemen, *J. Chem. Phys.*, 2008, **129**, 234509

18. S. Trudel and D. F. R. Gilson, *Inorg. Chem.*, 2003, **42**, 2814

19. G. Wolf, J. Baumann, F. Baitalow and F. P. Hoffmann, *Thermochim. Acta*, 2000, **343**, 19–25

20. A. C. Stowe, W. J. Shaw, J. C. Linehan, B. Schmid and T. Autrey, *Phys. Chem. Chem. Phys.*, 2007, **9**, 1831–1836

21. In http://www.hydrogen.energy.gov/annual_review09_storage.html

22. D. Grant, M Matus, K. Anderson, D. Camaioni, S. Neufeldt, C. Lane, D. Dixon, *J. Phys. Chem. A*, 2009, **113**, 6121-6132

23. W. Shaw, M. Bowden, A. Karkamkar, C. Howard, D. Heldebrant, N. Hess, J. Linehan, T. Autrey, *Energy Environ. Sci.* 2010, doi: 10.1039/b914338f

24. M. Bowden, I. Brown, G. Gainsford, H. Wong, *Inorg. Chim. Acta*, 2008, **361**, 2147.

THE PERFORMANCES OF CERAMIC BASED MEMBRANES FOR FUEL CELLS

Uma Thanganathan[1,*] and Masayuki Nogami[2]

[1]RCIS (Research Core Interdisciplinary Science), Okayama University, Tsushima-Naka, Okayama, 700-8530, Japan
[2]*Department of Materials Science and Engineering, Nagoya Institute of Technology, Showa, Nagoya, 466-8555, Japan*

ABSTRACT

A new class of proton-conducting ceramic composite membranes was developed for low-temperature fuel cells. According to characterizations, these membranes demonstrated good mechanical, structural, thermal and textural properties. Moreover, they yielded a maximum proton conductivity of 10^{-2} S cm^{-1} at room temperature. An electrode consisting of a heteropolyacid dispersed in a catalyst layer was also prepared and characterized, and the cell performances when using composite membranes of PWA/PMA doped with ZrO_2/TiO_2 together with an electrode contain heterpolyacid were extremely high as compared to counterparts with a phosphosilicate electrolyte at room temperature. The cell evaluation was performed on PWA/PMA-TiO_2-P_2O_5-SiO_2 and PWA/PMA-ZrO_2-P_2O_5-SiO_2 composite membranes with hydrogen and oxygen. These membranes presented the best properties.

1. INTRODUCTION

At the present time, environmental issues due to increasing emissions of air pollutants and greenhouse gases are spurring the development of technologies for the delivery of clean energy, such as fuel cells. Low-temperature proton exchange membrane fuel cells (PEMFCs) use hydrogen as fuel and their only emission is water. Over the last decade, the interest for utilizing ceramic materials as possible electrolyte membranes for fuel cell applications has grown.[1-4] Ceramics are good candidates for electrolyte materials for PEMFCs because of their thermal, chemical, and mechanical stability, as well as their lower material costs.

The proton conduction is highly dependent on the coverage of adsorbed water on the surface of the inorganic material. While significant advances have been made in recent years, a major limitation of the current technology is the cost and material restrictions of the proton conduction membrane. For transport applications, fuel cell companies require more durable, cost effective membrane technologies that are capable of delivering enhanced properties at low temperatures (from room temperature up to 80 °C). As a result, research is being driven towards a wide range of novel organic and inorganic materials showing a potential of being good proton conductors and forming coherent membranes. However, Nafion® and similar membranes suffer serious disadvantages such as high cost, poor hydrophilicity, fuel crossover, low proton conductivity at low humidity and high temperature.[5]

Attempts have been made to produce cheaper proton-conducting membranes that can equal the electrochemical performance of perflourinated ionomers as well as their mechanical, thermal and electrochemical stability. Membranes fabricated from the heteropolyacid (HPA) family are currently the best suited for proton conduction at both low and high temperatures. This paper describes the synthesis and structural properties of novel proton-conducting ceramic composite membranes based on heteropolyacids. Such materials are both cheaper and more thermally stable than Nafion®. Recent studies have already reported on the preparation of highly proton- conducting glasses containing HPAs which have provided high power densities at room temperature.[6, 7]

Since increasing or changing the operating properties of a fuel cell is not an energetically efficient solution, it would be preferable to modify the properties of the proton conducting membrane. For that purpose, heteropolyacids have been incorporated into phosphosilicate ceramic membrane and ZrO_2 and TiO_2 added in order to enhance the properties even further. Heteropolyacids (HPAs) are some of the most attractive inorganic modifiers due to their crystalline form. It is in fact the crystallinity that causes them to be highly conductive and thermally stable.

HPAs have several hydrated structures depending on the environment.[8] In the dehydrated phase in polar solvents, the primary structure is called a Keggin unit. The structure, conductivity, thermal and chemical stability of numerous protonated materials have been investigated, and studies on many more will follow. We have already proposed glass ceramic composite membranes with fuel cell performances for H_2/O_2 at low temperatures.[6, 7] Synthesis and characterization have been performed of inorganic composite-based structures that have been made highly proton-conductive through doping with a mixture of PWA and PMA. Such membranes are expected to have proton conductivities exceeding 10^{-1} S cm^{-1} at room temperature and superior thermal, mechanical, chemical and electrochemical stabilities.[9, 10]

This work describes the design and development of a new class of proton-conducting composite membranes. Our main target was to increase the power efficiency at low temperature with these materials. A detailed report is given on the development of PWA/PMA-TiO_2-P_2O_5-SiO_2 and PWA/PMA-TiO_2-P_2O_5-SiO_2 ceramic composite systems and their properties for fuel cell application are clarified.

II EXPERIMENTAL SECTION

Ceramic composite membranes can be fabricated by a sol-gel process, according to a technique largely utilized in our lab.[6,7, 11, 12] All the materials were used as received. The catalyst layers consisted of a porous mixture of the ionomer and platinum-loaded carbon particles, thus allowing contact between the solid, -ionomer and the gas phases. The catalyst ink was prepared with a Pt/C powder mixed with Nafion and a PTFE solution, and this was followed by blending the heteropolyacid (PWA) and solvents (water and ethanol) under ultrasonic stirring at 50 °C for 1 day. Subsequently, the ink was uniformly sprayed on the surface of a carbon sheet, polished with a brush and dried at 80 °C. The catalyst layer was then attached to two sides of a membrane using catalyst ink, manually fabricating a membrane-electrode-assembly (MEA) at room atmosphere. The active area of the electrode was 0.7 cm^2 corresponding to the membrane surface during fuel cell tests. The amount of Pt/C was 0.05 mg/cm^2.

TG/DTA measurements were carried out with a model SSC-5200 (SEIKO Instrument) under a nitrogen atmosphere at a heating rate of 10 °C/min. Fourier-transform infrared (FTIR) spectra of the glass composite samples were collected with an FTIR spectrometer (JASCO–FTIR–460) in the range 600 – 4000 cm^{-1}. The average pore diameter was measured from gas adsorption analysis (NOVA-1000, Quantochrome). The pore size distribution was established using the BJH method[13] and specific surface areas were determined by the BET method.[14] Nuclear magnetic resonance (NMR) spectroscopy was carried out on Varian (model unity 400) NMR spectrometer, operating at a spinning speed of 5 kHz. The electrochemical characterization of H_2/O_2 fuel cells was performed by measuring a polarization curve and carrying out impedance measurements on a Solartron 1287 electrochemical interface and a Solartron 1260 frequency response analyzer. The cell was operated at room temperature (25 °C) with a constant humidity of 30 %. The cell conditions were controlled by an NF- Fuel cell evaluation system (Japan) and the gas flow rates were 20 and 100 mL/min to the anode and cathode sides, respectively.

PWA (phosphotungstic acid, Aldrich), PMA (phosphomolybdic acid, Aldrich), Ti $(OC_4H_9)_4$, (titanium (IV) isopropoxide, 99.0 %, Kishida chemical), $Si(OC_2H_5)_4$ (TEOS, tetraethoxysilane, 99.9 % Colcote) and $HCON(CH_3)_2$ (N, N- Dimethylformamide, Kishida chemical) were all used as received. Di-isopropyl phosphite was obtained from Wako chemicals.

III RESULTS AND DISCUSSION

Fig. 1 a and b show the results form thermal gravimetric analysis (TGA) and differential thermal analysis (DTA), respectively. Two weight loss processes can be observed: one around 200 °C and a second around 400 °C, which were assigned to water. The first transition peak was the sum of two components indicating the evaporation of weakly and more strongly bound water. The second transition corresponded to the degradation of the membrane. The locations of the weight loss stages observed here were very similar to previous results reported by our group.[6, 12] The weight losses in the temperature range 200-400 °C, resulted from the desorption of chemisorbed water and hydroxyl groups via hydrogen bonding in the structure. The TGA curves from room temperature to 200 °C were assigned to the desorption of physically adsorbed water due to the presence of hydroxyl groups from Si-OH groups capable of attracting water molecules. The last weight loss occurred at a temperature of 400 °C and was attributed to the final decomposition of the ceramic matrix network. At temperatures much higher than 400 °C, the ceramic composite membrane started to degrade. The assignment of these weight loss variations was supported by the results of the DTA.

In Fig. 2 a and b, broad and small exothermic peaks could be observed at approximately 250 °C. In these DTA curves, there occurred no phase changes up to 600 °C indicating that the ceramic membranes were thermally stable at high temperatures because of the temperature-tolerant inorganic SiO_2 framework in the composite matrix.

Fig. 3 a and b show the FTIR absorption spectra of the PWA/PMA-TiO_2/ZrO_2-P_2O_5-SiO_2 composite membranes. It has been speculated that heteropolyacids (PWA and PMA) can be incorporated and dispersed in a TiO_2/ZrO_2-P_2O_5-SiO_2 ceramic matrix by a sol-gel route.[15] Antisymmetric stretching vibrations of the O-P and O-Si bonds in the P-O-Si linkage were seen at 1050 cm^{-1} and the O-P bands in the P-O-P linkage was present at 1280 cm^{-1}. These bands were all assigned to the OH stretching modes of the Si-OH bonds. The broad bands induced around 3700 cm^{-1}, were attributed to the free and hydrogen-bonded molecular water. In the stretching mode (P), the O-H was strongly hydrogen-bonded with the non-bridging oxygen of an inorganic matrix. This band was visible at around 2350 cm^{-1}.[16]

The average pore size and pore volume were measured by nitrogen adsorption for the PWA/PMA-TiO_2/ZrO_2-P_2O_5-SiO_2 composite membranes, and the obtained nitrogen adsorption/desorption isotherms for both tested materials showed a type-I hysteresis loop [Fig. 4 (a and b)]. A total pore volume of 0.28 cm^3/g and 0.16 cm^3/g in addition to average pore sizes of 2.4 and 2.1 nm were determined for the PWA/PMA-TiO_2/ZrO_2-P_2O_5-SiO_2 composite membranes. This result was similar to those presented in previous reports [12]. Materials with a macroporous structure are less able to retain the physisorbed water on the metal oxide surface, which is essential for proton transfer. On the other hand, materials with a microporous structure will also hinder proton transfer.[17] The proton conductivity of the membranes was measured at room temperature, and was found to be on order of 10^{-2} S/cm.

The ^{31}P NMR spectra of the PWA/PMA-TiO_2/ZrO_2-P_2O_5-SiO_2 composite membranes are displayed in Fig. 5 a, b and c. A significant intensity of the chemical shift was observed at 6.9 ppm for the PWA/PMA-P_2O_5-SiO_2 composite membrane, and this intensity decreased with the further addition of TiO_2/ZrO_2 mixed with PWA/PMA-P_2O_5-SiO_2 composite membrane. Moreover, significant line broadening for the peaks at 5.5 ppm and 4.7 ppm took place [Fig. 5 (b and c)]. These changes in intensity and chemical shifts of the proton signals in the complexes supported the idea of the formation of composite membranes between heteropolyacids and phosphosilicate. ^{31}P NMR analysis revealed the protonic transport in the PWA/PMA-TiO$_2$/ZrO$_2$-P$_2$O$_5$-SiO$_2$ composite membrane. The proton conduction favored proton hopping between hydroxyl bonds and water molecules. The dissociated protons moved to a neighboring water molecule, leading to the formation of an activated H$_2$O:H$^+$ state. The second stage of the proton hopping involved the dissociation of the proton from the activated H$_2$O:H$^+$. Thus, the protons hopped between the adsorbed water molecules.

Fuel cell measurements performed on the PWA/PMA-TiO$_2$/ZrO$_2$-P$_2$O$_5$-SiO$_2$ composite membranes demonstrated that H$_2$/O$_2$ fuel cell performances at room temperature, using humidified H$_2$ and O$_2$ at a pressure of 1 atm, were enhanced as, compared to the original PWA/PMA glass membrane.[18] Similar results were found for PWA doped with TiO$_2$/ZrO$_2$-P$_2$O$_5$-SiO$_2$ and PMA doped with TiO$_2$/ZrO$_2$-P$_2$O$_5$-SiO$_2$.[19, 20] At 24 °C, the maximum current density was 200 and 135 mA/cm^2 and the maximum power density was 50.5 and 33.5 mW/cm^2 for the PWA/PMA-TiO$_2$/ZrO$_2$-P$_2$O$_5$-SiO$_2$ composite membranes [Fig. 6 a and b] under a hydrogen and oxygen gas feed. The fuel cell test confirmed that the PWA/PMA-TiO$_2$/ZrO$_2$-P$_2$O$_5$-SiO$_2$ composite membrane had a superior performance as opposed to its PWA/PMA-P$_2$O$_5$-SiO$_2$ counterpart when utilized in hydrogen and oxygen fuel cells. However, in order for the cell performances of ceramic composite membrane to be satisfactory at low temperature, they need to be operated above room temperature as this will enable them to be effective with Pt/C- based electrodes.

IV SUMMARY

A low-temperature proton-conducting ceramic membrane consisting of PWA/PMA-P$_2$O$_5$-SiO$_2$ mixed with TiO$_2$ and ZrO$_2$ was synthesized through a sol-gel process. Novel ceramic composite membranes for use in high-performance fuel cells could thus be developed. These new nano-engineered ceramic proton-conducting membranes presented an excellent thermal stability up to 250 °C, and were also mechanically stable. The thermal properties of the composites were investigated by TGA and DTA under air. The TG/DTA measurements indicated that both samples were stable up to 250 °C. ^{31}P NMR analysis revealed the protonic transport in the PWA/PMA-TiO$_2$/ZrO$_2$-P$_2$O$_5$-SiO$_2$ composite membrane. Moreover, the composites had large pore sizes and surface areas, as determined by the textural properties. A maximum current density of 200 mA/cm^2 was obtained for the PWA/PMA-TiO$_2$/ZrO$_2$-P$_2$O$_5$-SiO$_2$ composite membrane at 25 C under a relative humidity of 30 %, and further improvements should be able to be achieved by fabrication of novel catalysts for these ceramic composite materials in low-temperature fuel cells.

ACKNOWLEDGEMENTS

The authors wish to thank the Special Coordination Funds for Promoting Sciences and Technology of MEXT (Ministry of education, sport, culture, science and technology) of Japan & the Japan Society for Promotion of Science (JSPS) program.

REFERENCES

[1] F. M. Vichi, M. T. Colomer, M. A. Anderson, "Nanopore ceramic membranes as novel electrolytes for proton exchange membranes," *Electrochem. Solid St.* **2**, 313-316 (1999).

[2] F. M. Vichi, M. I. Tejedor-Tejedor, M. A. Anderson, "Effect of pore-wall chemistry on proton conductivity in mesoporous titanium dioxide," *Chem.Mater.* **12**, 1762-1770 (2000).

[3] S.-P. Tung, B.-J. Hwang, "High proton conductive glass electrolyte synthesized by an accelerated sol-gel process with water/vapor management," *J. Membr. Sci.* **241**, 315-323 (2004).

[4] M. Yamada, D.L. Li, I. Honma, H.S. Zhou, "A self-ordered, crystalline glass, mesoporous nanocomposite with high proton conductivity of 2 x 10(-2) S cm(-1) at intermediate temperature," *J. Am. Chem. Soc.* **127**, 13092-13093 (2005).

[5] S. R. Samms, S. Wasmus, R. F. Savinell, "Thermal stability of Nafion(R) in simulated fuel cell environments," *J. Electrochem. Soc.,* **143**, 1498-1504 (1996).

[6] T. Uma, M. Nogami, "Influence of TiO_2 on proton conductivity in fuel cell electrolytes based sol-gel derived P_2O_5-SiO_2 glasses," *J. Non-Cryst Solids*, **351**, 3325-3333 (2005).

[7] T. Uma, M. Nogami, "High performance of H_2/O_2 fuel cells using Pt/C electrodes and P_2O_5-SiO_2-PMA glasses as an electrolyte in low temperature," *J. Ceramic Society of Japan*, **114**, 748-753 (2006).

[8] Y.S. Kim, F. Wang, M. Hickner, T.A. Zawodzinski, J.E. McGrath, "Fabrication and characterization of heteropolyacid ($H_3PW_{12}O_{40}$)/directly polymerized sulfonated poly(arylene ether sulfone) copolymer composite membranes for higher temperature fuel cell applications," *J. Membr. Sci.* **212**, 263-282 (2003).

[9] O. Nakamura, T. Kodama, I. Ogino, Y. Miyake, "High-conductivity solid proton conductors-dodecamollybdophosphoric acid and dodecatungstophosphoric acid crystals," *Chem. Lett.* **1**, 17-18 (1979).

[10] O. Nakamura, I. Ogino, "Electrical conductivities of some hydrates of dodecamolybdophosphoric acid and their mixed-crystals," *Mat. Res. Bull.,* **17**, 231-234 (1982).

[11] T. Uma, M. Nogami, "Synthesis and characterization of P_2O_5-SiO_2-X (X = phosphotungstic acid) glasses as proton conducting electrolyte for H_2/O_2 fuel cell in low temperature," *J. Membr Sci*, **280**, 744-751 (2006).

[12]T. Uma, M. Nogami, "Development of new glass composite membranes and their properties for low temperature H_2/O_2 fel cells," *ChemPhyChem*. **8**, 2227- 2234 (2007).

[13]S. Brunauer, P. H. Emmett, E. Teller, "Adsorption of gases in multimolecular layers," *J. Am. Chem. Soc*. **62**, 309-319 (1938).

[14]E. P. Barret, L. G. Joyner, P. H. Halenda, "The Adsorption of Water Vapor on Silica Surfaces, by Direct Weighing," *J. Am. Chem. Soc*. **62**, 2839-2844 (1940).

[15]T. Uma, M. Nogami, "Structural and textural properties of mixed PMA/PWA glass membranes for H_2/O_2 fuel cells," *Chem. Mater*., **19,** 3604-3610 (2007).

[16]H. Scholze, Glass: Nature, Structure and Properties, Springer, NY, 1991.

[17]M. T. Colomer, M. A. Anderson, "High porosity silica xerogels prepared by a particulate sol-gel route: pore structure and proton conductivity," J. *Non-cryst. Solids*, **290**, 93-104 (2001).

[18]T. Uma, M. Nogami, "A highly proton conducting novel glass electrolyte," *Anal Chem,* **80**, 506-508 (2008).

[19]T. Uma, M. Nogami, "Properties of PWA/ZrO_2-doped phosphosilicate glass composite membranes for low- temperature H_2/O_2 fuel cell applications," *J. Membr Sci*. **323**, 11-16 (2008).

[20]T. Uma, M. Nogami, "PMA/ZrO_2–P_2O_5–SiO_2 glass composite membranes: H_2/O_2 fuel cells," *J. Membr. Sci*. **334,** 123-128 (2009).

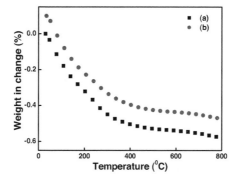

Fig. 1. TGA plots of (a) the PWA/PMA-TiO$_2$-P$_2$O$_5$-SiO$_2$ and (b) the PWA/PMA-ZrO$_2$-P$_2$O$_5$-SiO$_2$ composite membranes. All TGA curves were recorded on powder samples in scans from room temperature to 800 °C, under a nitrogen atmosphere and at a heating rate of 10 C/min.

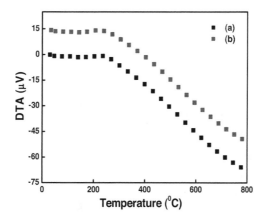

Fig. 2. DTA plots of (a) the PWA/PMA-TiO$_2$-P$_2$O$_5$-SiO$_2$ and (b) the PWA/PMA-ZrO$_2$-P$_2$O$_5$-SiO$_2$ composite membranes. The DTA curves were recorded on powder samples in scans from room temperature to 800 °C under a nitrogen atmosphere and at a heating rate of 10 C/min.

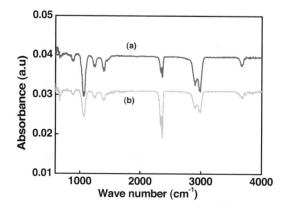

Fig. 3. FTIR spectra of (a) the PWA/PMA-TiO$_2$-P$_2$O$_5$-SiO$_2$ and (b) the PWA/PMA-ZrO$_2$-P$_2$O$_5$-SiO$_2$ composite membranes. The spectra were recorded in absorbance mode on powder samples in the wave number range 4000-600 cm^{-1}.

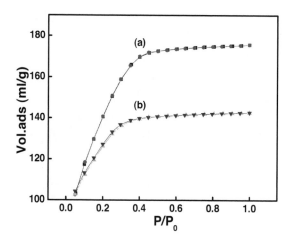

Fig. 4. N$_2$ adsorption-desorption isotherms of the PWA/PMA-TiO$_2$-P$_2$O$_5$-SiO$_2$ and PWA/PMA-ZrO$_2$-P$_2$O$_5$-SiO$_2$ composite membranes. The samples were degassed at 250 C for 5 h prior to the measurements.

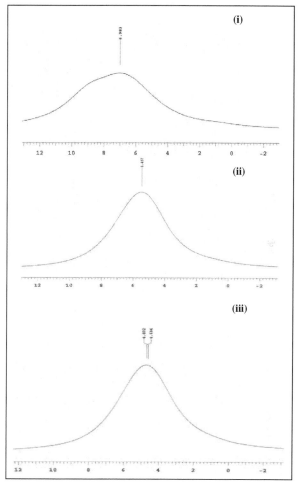

Fig. 5. ^{31}P NMR spectra of (i) PWA/PMA, (ii) PWA/PMA-TiO$_2$-P$_2$O$_5$-SiO$_2$ and (iii) PWA/PMA-ZrO$_2$-P$_2$O$_5$-SiO$_2$ composite membranes.

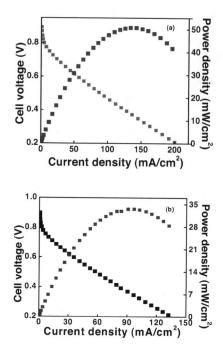

Fig. 6 Polarization curves of (a) the PWA/PMA-TiO_2-P_2O_5-SiO_2 and (b) the PWA/PMA-ZrO_2-P_2O_5-SiO composite membranes when employed as electrolytes for H_2/O_2 fuel cells.

MICROCRACK RESISTANT POLYMERS ENABLING LIGHTWEIGHT COMPOSITE HYDROGEN STORAGE VESSELS

Kaushik Mallick, John Cronin, Paul Fabian and Mike Tupper
Composite Technology Development, Inc.
Lafayette, Colorado, 80026

ABSTRACT

Robust, lightweight, storage vessels are needed for on-board storage of hydrogen. Challenges with currently used composite overwrapped vessels include weight due to the non-load bearing liner, performance reliability resulting from separation of the liner, and costs of extra manufacturing steps to fabricate the liner. Linerless composite vessels, where the composite shell serves both as a permeation barrier and a structure, can provide for the lightest weight vessels for a given set of requirements. Preliminary designs show up to 25% weight savings allowing reduced storage system mass and more internal volume. These tanks are targeted to attain hydrogen mass storage efficiency of 15-18% as compared to 3-4% from lined vessels. Manufacturing cost, operational risks and maintenance costs can be reduced due to inherently simple construction. Engineering methods that define material performance requirements, such as polymer strain requirements in a lamina have been used to guide the development of microcrack resistant polymers. Performance of linerless composite tanks has been demonstrated and qualification is on-going.

INTRODUCTION

Hydrogen is an ideal energy carrier that can help increase our energy diversity and security by reducing our dependence on hydrocarbon-based fuels. Hydrogen can be produced from domestic resources that are clean, diverse, and abundant; fuel cells provide a technology to use this energy in a highly efficient way, in numerous applications, with only water and heat as byproducts. The US Department of Energy's initiative and push has already put fuel cell buses on the road, and may soon put new fuel-cell powered vehicles on the nation's rails and waterways.

Among the obstacles to commercializing hydrogen-powered vehicles—besides production and infrastructure—is the need for storage systems that can contain sufficient hydrogen onboard a car to compete with the range and performance of gasoline-powered autos. Robust, lightweight, high-strength pressure vessels that can store gaseous hydrogen under high pressure still provide the most commercially viable approach to driving fuel cell cars (Figure 1). The driving range of fuel cell vehicles with compressed hydrogen tanks depends, of course, on vehicle type, design, and the amount and pressure of stored hydrogen. By increasing the amount and pressure of

Figure 1: Hydrogen fuel cell cars

hydrogen, a greater driving range can be achieved but at the expense of cost and valuable space within the vehicle.[1] Volumetric capacity, high pressure, and cost are thus key challenges for compressed hydrogen tanks. Currently, pressure vessels for ambient high-pressure storage are fabricated with metal lined (Type III) or polymer lined (Type IV) composite overwrapped pressure vessels. Tanks and pressure vessels used to store hydrogen at cryogenic temperatures are fabricated from metals (aluminum or titanium). Challenges associated with the use of Type III and Type IV vessels for storage of ambient high pressure gases include additional weight due to the liner (which is generally not load bearing), performance reliability resulting from separation of the liner from the composite, additional cost associated with extra manufacturing steps for liner fabrication (of particular concern with Type IV vessels), and the high cost of tooling modification for tank geometry changes in Type III and IV vessels for different applications.

OVERVIEW OF KIBOKO® TECHNOLOGY

To address the above-mentioned issues, CTD has developed KIBOKO® lightweight all-composite pressure vessels (Figure 2), which have been designated as Type V pressure vessels. Due to the lack of a metallic or polymeric liner, Type V vessels can provide the lightest possible weight vessels for a given set of requirements. Preliminary designs have shown an approximate 50% weight savings over all metal (Type I), 25% over Type III, and a 10% weight savings over Type IV vessels, allowing reduced total storage system mass. In addition, by eliminating the liner, KIBOKO® Type V vessels can provide more internal volume, which is critical for storage of low energy density gaseous fuel like hydrogen. Due to the combined effect of reduced vessel mass and increased internal volume, KIBOKO® vessels are

Figure 2: CTD has fabricated many KIBOKO® all-composite pressure vessels in various sizes and configurations

expected to attain gaseous hydrogen mass storage efficiency of 10-15%. This is a significant performance jump over the current state of the art of 3-4% in commercially available pressure vessels. If properly designed, KIBOKO® vessels can also reduce the manufacturing cost, operational risks, and maintenance costs over their lifetime due to their inherently simple construction.

The improved storage efficiency of a KIBOKO® Type V vessel is easily appreciated by comparing different classes of composite pressure vessels in terms of a common metric. One such industry-accepted yardstick is *Pressure Vessel Efficiency*, commonly defined as $\eta = pV/W$, where p is the design burst pressure, V is tank volume, and W is tank mass. To improve efficiency, it is imperative that the composite structure be optimized to provide the highest burst strength, while reducing tank weight and maximizing storage volume. The bar chart in Figure 3 shows the efficiency advantage of linerless designs, compared to other types of composite pressure vessels commonly used in the aerospace industry.[2]

Previous attempts at linerless composite vessels by others, showed that these vessels prematurely leaked and structurally degraded. CTD has overcome these inadequacies by developing engineering methods that define specific material performance requirements to prevent this premature leakage and structural failure. Furthermore, CTD has developed and demonstrated materials that provide the performance dictated by these engineering models. Key to the successful performance of these materials is that they do not microcrack within the operating range

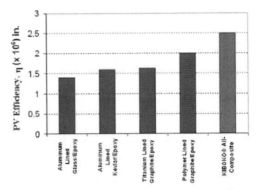

Figure 3: Pressure vessel efficiencies compared for different types of pressure vessels

of the tanks. Understanding the strains at which microcracking initiates in a composite material, and how this inhibits permeation and leakage of fluids is a primary criterion for optimizing the design of these lightweight pressure vessels. The substantial advancements to the technology of all-composite pressure vessels made by CTD can be attributable to an integrated systematic approach that looks concurrently at the totality of critical issues, including material capabilities and tailoring, fabrication process optimization, and structural design optimization (Figure 4).[3]

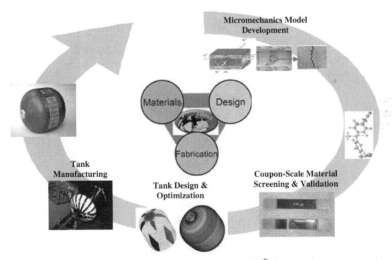

Figure 4: CTD's integrated systematic approach to developing KIBOKO® all-composite pressure vessels

DEVELOPMENT OF MICROCRACK RESISTANT MATERIALS FOR KIBOKO® PRESSURE VESSELS

KIBOKO® vessels, which do not utilize a polymer or metallic liner use the composite shell to serve both as a permeation barrier as well as to provide the structure necessary to carrying all pressure and environmental loads. CTD has developed engineering models and methods that define specific material performance requirements, such as polymer matrix strain requirements for a particular lamina layer within the composite, that prevent microcracking of the resin and thus premature leakage and structural failure. Key to the successful performance of these materials is that they do not microcrack within the operating range of the vessels. From the structural perspective, the reinforcing fiber of the composite material defines the stress and strain limits of the composite material. To eliminate the potential for leakage and permeation, the matrix, or resin, material used to bind the fibers together into the structural composite, must not microcrack under the tank's operating conditions (Figure 5). Furthermore, the composite structure is constructed of multiple layers of fibers and resins, called lamina. It is also important that the matrix be strong enough to prevent lamina from delaminating until the fiber performance limits are reached.

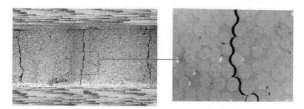

Figure 5: Microcrack in a composite laminate

Key to this effort is the development and validation of new matrix materials that meet these requirements. CTD has developed novel toughened epoxy systems such as CTD-7.1 and CTD-9HX that meet the requirements defined above (Figure 6). Furthermore, to complement the design of composite materials suitable for pressure vessels, CTD has developed micromechanics based material test methods based on microcrack fracture toughness to evaluate and rank their performance.[5] These test methods enable the determination of the strain level at which the matrix starts to degrade. CTD has found that the inclusion of nano-reinforcements in the matrix improves resin modulus and significantly increases the inter-laminar shear strength at the ply interfaces, thus enabling the composite material to fail only when the limits of the fiber performance have been reached. CTD has also found that the addition of suitable nano-reinforcements within composite plies can significantly reduce the permeability of the structure to low-molecular gaseous contents like hydrogen.

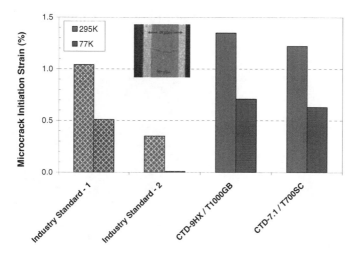

Figure 6: CTD's toughened polymers enable improved microcrack resistance over industry standard materials

DEMONSTRATION OF KIBOKO® TECHNOLOGY

KIBOKO® Pressure Vessels for Satellites

CTD designed, built, qualified, and supplied two Space flight qualified Argon gas storage vessels and two spares to the University of Texas FASTRAC (the Formation Autonomy Spacecraft with Thrust, Relay, Attitude and Crosslink) nanosatellite program. These vessels were 6" dia. x 6.875" long to fit within the available envelope of the satellite, a size and design not previously built by CTD. A total of 21 pressure vessels were fabricated by CTD, at the Composite Laboratory of AFRL, Kirtland AFB, under a Cooperative Research and Development Agreement (CRADA) between CTD and AFRL, to ensure a sufficient number of pressure vessels to test and deliver. The vessel's 1.8 liter capacity was 15% greater and its 0.35 kg mass was 25% lighter than the previously baselined aluminum pressure vessels that did not pass the qualification tests (Figure 7). However, the KIBOKO® pressure vessels for FASTRAC were far from optimized for minimal weight. With only six weeks to deliver, these pressure vessels were designed for a burst pressure

Figure 7: KIBOKO® linerless composite tanks for Air Force sponsored nanosatellite program

of 3,000 psi, pressure tested to 2,000 psi prior to leak and proof tested to 500 psi to ensure a safe operating pressure of 100 psi with negligible permeation of gaseous content.

Permeation Testing of 14-liter KIBOKO® Pressure Vessel

As part of the US Air Forces' Fully Reusable Access to Space Technology (FAST) program, CTD fabricated two 10 in. diameter x 18 in. long, 14L KIBOKO® pressure vessels and tested them at ambient conditions (295 K). The test procedure consisted of three acceptance tests: pre-hydrostatic helium leak test (Figure 8), hydrostatic pressure test, and post-hydrostatic helium leak test. Both pressure vessels exhibited maximum leak rates of 1.4 x 10^{-5} scc/sec, which compare favorably with the FE-model-predicted leak rate of 3.4 x 10^{-5} scc/sec. This result indicates that there was no significant microcrack or void damage in the pressure vessels after fabrication.

Figure 8: CTD's test bench to measure permeation rate in KIBOKO® tanks using a leak chamber and helium mass spectrometer

Following the permeation testing, the pressure vessels were subjected to hydrostatic pressure test. The limiting pressure that the vessel can experience without causing any leakage was determined to be 2,800 psi, which results in a Maximum hoop strain of 1.2%. To test this design values, the vessels were first pressurized with water to 1,900 psi and subsequently to 2,800 psi. Both pressure vessels successfully completed the test with no leakage reported during holds of ten minute at 1,900 psi and 2,800 psi.

One of the pressure vessels was instrumented with multiple strain gages to measure the strain on the external surface of the cylindrical section of the vessel during the pressure test (Figure 9). This included 2 strain gages aligned along the circumferential (hoop) direction and 2 bellyband gages to measure the hoop strain located about 2" away from the hoop strain gages. Figure 10 shows a photo of the strain gages and bellybands. shows the plots of pressure vs. strain, comparing the finite element model predictions and experimental results. The agreement between the FE analysis and test results is very good.

Figure 9: KIBOKO® 14L vessel with strain gages prior to hydrostatic pressure test

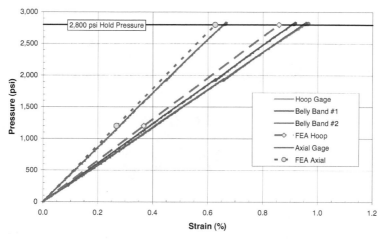

Figure 10: Measured and predicted strain in 14L KIBOKO® pressure vessel

Following the hydrostatic pressurization test up to 2,800 psi, the pressure vessels were tested again for leaks using the helium-leak test. The first pressure vessel exhibited a maximum leak rate of 2.0×10^{-4} scc/sec and the second pressure vessel's maximum leak rate was recorded at 1.0×10^{-5} scc/sec. Prior studies of helium leakage through highly microcracked pressure vessels typically yielded leak rates on the order of 10^{-3} scc/sec. Therefore, the lower measured leak rate in the pressure vessels seems to be indicative of molecular permeation, rather than viscous flow (i.e., true "leakage"). The changes in measured permeation rates of both pressure vessels after hydrostatic testing were within the experimental error of measuring the helium permeation rate.

Cyclic and Permeation Testing

One of the challenges of the pressure vessels intended for H2 storage is to limit the permeation rate of gaseous hydrogen through the vessel wall at the vessel's operating pressure and throughout its life consisting of numerous fill and drain cycles. Under a program funded by General Motors (GM), CTD performed a preliminary evaluation of the structural performance of the KIBOKO® pressure vessels under hydraulic and pneumatic pressure and measured their permeation rate before and after subjecting them to a moderate number of pressure cycles. Seven pressure vessels were fabricated for the program, some of which were tested to failure and the remainder used for measuring helium permeation rates. The permeation rate was measured at the vessel's working pressure of 15.0 MPa (2,175 psi) and measured over an extended time period (see Figure 8). Test results show that the helium permeation rate through the vessel wall has good correlation with the predicted value at the vessel's design operating pressure. Additionally, one pressure vessel was tested for permeation rate both before and after pressure cycles. Results from this test indicate that the permeation rate in these vessels remain unaffected by the hydraulic cycles. The average permeation rate of CTD's KIBOKO® composite pressure vessels was measured as 1.4 scc/liter/hr.

Burst Pressure Testing of 6.8L KIBOKO® Pressure Vessels

As part of technology demonstration efforts, CTD wound three (3) KIBOKO® Type V cylinders in collaboration with Luxfer Gas Cylinders (see Figure 11). The design MEOP for these KIBOKO® Type V cylinders was 3,600 psi and the target first failure pressure for these cylinders was 8,100 psi (2.25 x 3,600 psi MEOP) conforming to NGV-2 specifications. The average weight of the CTD KIBOKO® cylinders was 5.75 lbs.

Figure 11: KIBOKO® 6.8L tank with MEOP of 3,600 psi

Two cylinders that were used for pattern development achieved a first failure (leak) of 7,200 and 7,500 psi respectively during pressure testing. Failure in the form of leakage was observed in the knuckle (cylinder-to-dome transition) area. It was surmised that the hoop transitions in the wind pattern were not adequate and the hoop patterns were revised. A subsequent cylinder with a refined wind sequence layup achieved a failure (burst) pressure of 9,000 psi and successfully met the design goals for first failure pressure.

SUMMARY

The technical feasibility of KIBOKO® Type V all-composite pressure vessels has been demonstrated through an integrated systematic approach to analytical modeling, material development and testing of prototype vessels of various sizes. These vessels promise to provide excellent mass storage efficiency for gaseous hydrogen fuels compared to traditional metal or polymer lined composite overwrapped pressure vessels. The testing of prototype pressure vessels has illustrated the need for a repeatable manufacturing process. CTD hopes that continued work in this area will lead to a robust product that can provide storage efficiencies not yet achievable with other storage vessels for hydrogen storage.

REFERENCES

1. http://www1.eere.energy.gov/hydrogenandfuelcells/storage/hydrogen_storage.html
2. *Handbook of Composites*, G. Lubin, ed., Van Nostrand Reinhold Co., 1982.
3. K. Mallick, J. Cronin, S. Arzberger, N. Munshi, C. Paul and J. Welsh, "An Integrated Systematic Approach to Linerless Composite Tank Development," to be presented at the AIAA Structures, Structural Dynamics and Materials Conference, Austin, TX, Spring 2005.
4. Nairn, John A., "Matrix Microcracking in Composites," *Polymer Matrix Composites*, Elsevier Science, R. Talreja and J-A, Manson eds., Chapter 13, 2001.
5. Nairn, John A., "The Strain Energy Release Rate of Composite Microcracking: A Variational Approach", *Journal of Composite Materials.*, Vol. 23, pp 1106-1129, 1989.

A Study of the Thermodynamic Destabilization of Sodium Aluminum Hydride (NaAlH₄) with Titanium Nitride (TiN) using X-ray Diffraction and Residual Gas Analysis

Whitney Fisher Ukpai
Institute for Micromanufacturing
Louisiana Tech University, Ruston, LA 71272, United States

Tabbetha A. Dobbins*
Institute for Micromanufacturing
Louisiana Tech University, Ruston, LA 71272, United States

* Contact author

ABSTRACT
This project is designed to extend the limits of hydrogen storage technology for practical purposes. Currently, there is a need to develop systems which release hydrogen at lower temperatures. The addition of destabilizers are believed to lower the H_2 gas desorption temperatures by forming a stable product phase comprised of the hydrogen coordinated cation and the destabilizer phase cation. In this case, TiN is added to NaAlH₄ in order to destabilize the AlH_4^- complex by forming a stable Ti-Al alloy. Although the bond energy in the nitride phase is high, x-ray diffraction shows the product intermetallic phase TiAl to form. The sodium alanate powders were mixed using titanium nitride (TiN) and the mixture was high energy ball milled. The samples had varying concentrations of TiN (e.g. 25 mol%, 50 mol%, and 75 mol%). X-ray relative peak intensity analysis shows that the content of TiAl formed increases with increasing TiN added to the NaAlH₄ system. Moreover, residual gas analysis of the 25 mol % TiN in NaAlH₄ sample shows that the onset of desorption occurs at 60-70°C, with the peak temperature of hydrogen evolution from this stoichiometry occurring at 110°C.

INTRODUCTION

The primary source of power for most of the automobiles that are currently in production is gasoline. Due to reduced (and fluctuations in) availability of gasoline, the Department of Energy has begun looking to hydrogen as an alternative power source. There are several challenges involved with the utilization of hydrogen in automobiles; one such challenge is the safe on-board storage of hydrogen. To overcome this challenge, the energy industry needs to find storage medium that meets stringent constraints of cost, safety, efficiency, weight, and volume.[1] The problem with many metal hydrides is they cannot reversibly store hydrogen at moderate temperature and pressures.[1]

Of the metal hydrides that are under study, sodium aluminum hydride (NaAlH₄) has been the most investigated compound because of its reasonable hydrogen content and reversibility. The release of hydrogen from the NaAlH₄ occurs in a three-step process:[2-5]

$$NaAlH_4 \leftrightarrow \frac{1}{3}Na_3AlH_6 + \frac{2}{3}Al + H_2 \tag{1}$$

$$\frac{1}{3}Na_3AlH_6 \leftrightarrow NaH + \frac{1}{3}Al + \frac{1}{2}H_2 \tag{2}$$

$$NaH \leftrightarrow Na + \frac{1}{2}H_2 \tag{3}$$

The first two reactions give a theoretical reversible storage capacity of 5.6 wt %, while the third reaction occurs at a temperature that is too high for practical use.[2] By using transition metals or rare earth metals as catalyst, the H$_2$ desorption reaction in NaAlH$_4$ and other complex hydrides can be made reversible.[3-5]

Thermodynamic destabilization occurs because of the emergence of a stable alloy comprised of cations in the destabilizer and the hydrogen coordinated cation on the hydride. The modified system is then able to cycle between the hydride and the stable alloy instead of the hydride product phase. Formation of the alloy drives the overall dehydrogenation reaction to a lower enthalpy, which, in turn, increases the equilibrium hydrogen pressure for the reaction.[6,7] According to measurements made by Vajo et al.[6] on MgH$_2$ destabilized LiBH$_4$ systems, the possibility of hydrogen to be detached from both the simple metal hydride and the complex metal hydride at lower temperatures by an alternative reaction pathway involving the formation of the stable alloy MgB$_2$ is feasible. Several more studies have examined the destabilization reactions occurring between LiBH$_4$ and an assortment of metals, metal hydrides, and carbon.[8,9] This research seeks to determine whether a pathway for desorption from NaAlH$_4$ occurring at lower temperatures via stable alloy formation also exists. In this case, Al^{3+} is tetrahedrally coordinated to H$^-$ ions in the tetragonal unit cell of NaAlH$_4$ and the destabilizer used is TiN (added by high energy ball milled with NaAlH$_4$).

MATERIALS AND METHODS

Nanoparticle titanium nitride (TiN) and hydrogen storage grade sodium aluminum hydride (NaAlH$_4$) were obtained commercially from Sigma Aldrich® and stored in a dry nitrogen atmosphere glove box to reduce exposure to moisture and oxygen. In preparing the TiN destabilized NaAlH$_4$ powders, the powders were measured to 25 mol%, 50 mol%, and 75 mol% TiN in NaAlH$_4$ mixtures and ball milled using tungsten carbide (WC) mill jar and milling media within an SPEX Certiprep 8000M mixer mill at 1080 cycles/minute for 10 minutes. The samples were then sealed within Kapton tape (x-ray transparent) and transported to the Rigaku Miniflex X-ray Diffractometer at Grambling State University (Grambling, LA) for analysis of crystal structures (2θ varied between 28° and 100°). Residual gas analysis (RGA) using the Stanford Research Systems QMS 200 Gas Analyzer was also performed on the 25 mol % sample. Gas analysis scans were collected at fixed temperatures (gained via oil immersion bath) ranging from 60°C to 180°C (in 10°C increments) with evolved gas flow directly into the RGA instrument.

RESULTS AND DISCUSSION

X-ray Diffraction, Chemical Analysis, and Morphologies of TiN and NaAlH$_4$ Mixtures

Figure 1 shows scanning electron microscopy (SEM) and energy dispersive spectroscopy (EDS) elemental analysis of 25 mol% TiN in NaAlH$_4$ powders after milling for ten minutes. Although some isolated regions rich in TiN were found (Figure 1c-d), elemental analysis of the NaAlH$_4$-rich region shows TiN dispersed within it (Figure 1a-b). Similar morphologies were found for the 50mol % and 75 mol % TiN in NaAlH$_4$ powders.

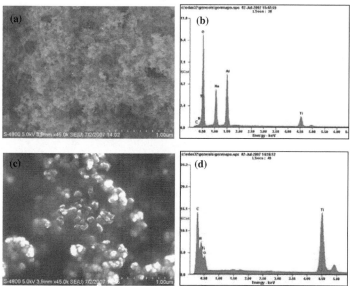

Figure 1. Scanning Electron Microscopy (SEM) with Energy Dispersive Spectroscopy (EDS) elemental analysis of 25 mol % TiN in NaAlH4. (a) Morphology of regions rich in NaAlH4, (b) Elemental analysis of the region shown in (a), (c) Morphology of regions rich in TiN, and (d) Elemental analysis of the region shown in (c).

Figure 2 shows x-ray diffraction analysis of samples prepared with 0 mol%, 25 mol%, 50 mol%, and 75 mol% TiN in NaAlH$_4$. All samples were found to comprise x-ray reflections attributable to NaAlH$_4$ phase. Four TiN peaks are clearly distinguished in the 25 mol%, 50 mol%, and 75 mol% TiN x-ray diffraction data at 2θ=36.9°, 42.4°, 61.9° and 74.1°. The peak at 61.9° is overlayed with the NaAlH$_4$ peak found at nearly the same 2θ value. However, the intensity of this peak in samples prepared at 50 mol% and 75 mol% TiN relative to the neighboring NaAlH$_4$ (only) crystalline peak at 60.5° clearly shows its increased intensity (indicative of increasing amounts of TiN). Two additional peaks at 2θ=39.5° and 44.2° are found in the samples containing TiN (and are absent from the 0 mol % TiN control sample). Comparison with crystallographic databases show that the peaks at 2θ=39.5° and 44.2° may be attributed to the titanium aluminide intermetallic phase TiAl. Both of the peaks attributed to TiAl grow in intensity as the mol% of TiN is increased and Figure 3 shows the relative peak intensities of TiAl (at 2θ=39.5°) and NaAlH$_4$ (at 2θ=29.9°). Although the bond strength in TiN is high, still, some of the titanium is available to react with Al^{3+} in NaAlH$_4$ yielding the intermetallic alloy TiAl as product phase. One of the criterion for the destabilization of complex metal hydrides to occur is that the thermodynamic driving force to form an alloy between the destabilizer cation (in this case Ti^{3+}) and the hydride cation which is coordinated to the H$^-$ anion (in this case Al^{3+}) is high. These x-ray diffraction data show the formation of TiAl product phase—and hence, show that this criterion is met by TiN introduced into the complex metal hydride NaAlH$_4$.

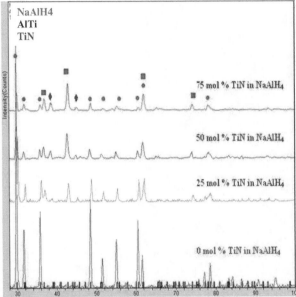

Figure 2. X-ray diffraction of 0 mol %, 25 mol %, 50 mol %, and 75 mol % TiN in NaAlH₄. Search match was used for peak assignment. Dots are used to label NaAlH₄ peaks, Diamonds are used to label TiAl peaks, and Squares are used to label TiN peaks.

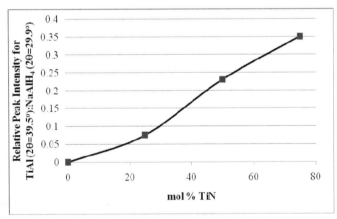

Figure. 3. Ratio of peak intensities for TiAl (at 2θ=39.5°) product phase and NaAlH₄ (at 2θ=29.9°) reactant phase for 0, 25, 50, and 75 mol % TiN in NaAlH₄ samples. The relative peak intensities increase as TiN content is increased.

Residual Gas Analysis (RGA) by Mass Spectrometry of Gases Liberated from 25 mol% TiN in NaAlH$_4$ Mixtures

A second criterion for the destabilization of complex metal hydrides is that the desorption pathway cycles between the hydride and the intermetallic alloy at temperatures lower than the hydrogen release temperature for the pure hydride. For NaAlH$_4$, the hydrogen release temperature by reaction equation (1) is ~183°C for pure uncatalyzed NaAlH$_4$ (and ~125°C for 4 mol % TiCl$_3$ catalyzed NaAlH$_4$). To explore release temperature, isothermal holds ranging between 60°C and 180°C were employed with gases liberated from the sample entering directly into a mass spectrometer residual gas analyzer. Figures 4(a) and (b) show measurable hydrogen gas pressures (in Torr), revealing that hydrogen emerges from these samples at temperatures as low as 60-70°C. The hydrogen pressure is 52 Torr at 60°C and 110 Torr at 70°C. For these temperatures, the peak corresponding to nitrogen pressure (at 28 atomic mass units) dominates the spectrum (at ~560 Torr). This nitrogen is believed to come from atmosphere control within the sample chamber. The hydrogen pressure increases to a maximum value of 3500 Torr at 110°C—while the nitrogen pressure remains at ~560 Torr. The occurrences of peaks which may be attributed to H$_2$O pressure are believed to be caused by moisture in the gas handling system rather than moisture in the sample chamber itself. Were the H$_2$O present in the sample chamber where it may react with NaAlH$_4$, no crystalline product phase associated with NaAlH$_4$ after exposure to TiN would remain. The magnitudes of those peaks were always lower than 110 Torr pressure. Figure 5 shows hydrogen pressure as a function of isothermal hold temperature. The emergence of hydrogen gas at temperatures as low as 60-70°C demonstrates that the ability for the 25 mol% TiN in NaAlH$_4$ system to form the TiAl intermetallic phase may have also led to a reduction the H$_2$ desorption onset temperature. The absolute desorption temperature of 110°C determined by these measurements is much lower than the desorption temperature from pure NaAlH$_4$.

CONCLUSION

The scope of this paper was to demonstrate the destabilization reaction to yield desorption of H$_2$ at reduced temperatures occurs in NaAlH$_4$ when exposed to TiN. Two criterion for the success of a destabilization scheme are that (1) the destabilizer cation form an alloy with the H$^-$ coordinated complex metal hydride cation and (2) during desorption, the system cycle from the hydride to the alloy phase at temperatures lower than the desorption temperature in the pure hydride. Using x-ray diffraction and residual gas analysis, this work shows that both of these criteria are met for the TiN-destabilized NaAlH$_4$ system. The predicted reaction between NaAlH$_4$ and TiN is:

$$NaAlH_4 + TiN \leftrightarrow Na + \frac{1}{2}N_2 + TiAl + 2H_2. \qquad (4)$$

X-ray diffraction analysis confirms the presence of NaAlH$_4$ and TiN reactant phases along with TiAl product phase. Furthermore, residual gas analysis confirms the presence of H$_2$ product phase. It is unclear from the x-ray diffraction analysis and residual gas analysis (performed in this research) what phase is formed by Na and N—and so the exact reaction pathway remains speculative. Additionally, the TiAl alloy formed is believed to persist—and not cycle to deliver Al for the reformation of NaAlH$_4$. The destabilization of the recyclable hydride, NaAlH$_4$, shows here a key feature of the reduced the H$_2$ desorption onset temperature to 60°C-70°C. It remains for future study to assess the amount of TiN needed to induce the lowered desorption temperature—and the relative amount of recyclable NaAlH$_4$ compound present.

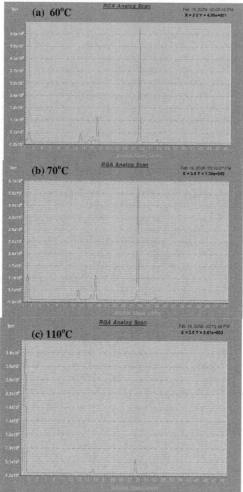

Figure 4. Residual Gas Analysis of 25 mol % TiN in NaAlH$_4$ at (a) 60°C, (b) 70°C, and (c) 110°C. H$_2$ gas liberation (peak at 1.5 atomic mass units) is onset at 60°C-70°C and is maximum at 110°C. Other peaks are attributed to N$_2$ (for atmosphere control) and to H$_2$O (due to contamination of the carrier lines). It H2O were present in the sample chamber inducing reaction with NaAlH4, amorphous, not crystalline product phase associated with NaAlH4 would remain after exposures to TiN.

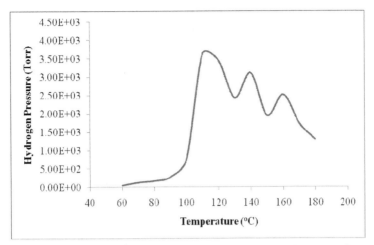

Figure 5. Residual gas analysis showing H_2 pressure (Torr) vs. temperature for a series of 25 mol % TiN in $NaAlH_4$ samples held isothermally at $60°C$ to $180°C$ (at $10°C$ increments).

ACKNOWLEDGEMENTS
Funding for this project was provided by Louisiana Board of Regents Pilot Funding for New Research (PFUND) (AY 2009-10) contract no. NSF(2009)-PFUND-146 and National Science Foundation (NSF) CAREER Award, Division of Materials Research, Ceramics Program under contract no. DMR-0847464 (AY 2009-10). We wish to thank Mr. Shathabish NaraseGowda for kind and helpful assistance in residual gas analysis measurements.

REFERENCES
1. C.M. Andrei, J.C. Walmsley, H.W. Brinks, R. Holmestad, D. Blanchard, B.C. Hauback, and G.A. Botton, Analytical Electron Microscopy Studies of Lithium Aluminum Hydrides with Ti- and V- Based Additives, *Journal of Physics and Chemistry*, **109[10]**, pp 4350-4356 (2005).
2. C.M. Andrie, J.C. Walmsley, H.W. Brinks, R. Holmestad, D. Blanchard, S.S. Srinivasan, C.M. Jensen, and B.C. Hauback, Electron-Microscopy Studies of NaAlH4 with TiF3 Additive: Hydrogen-Cycling Effects, *Applied Physics A*, **80**, pp 709-715 (2005).
3. B. Bogdanovic, M. Schwickardi, Ti-doped Alkali Metal Aluminum Hydrides as Potential Novel Reversible Hydrogen Storage Materials, *J. Alloys Comp.*, **253-254**, pp 1-9 (1997).
4. D. L. Anton, Hydrogen Desorption Kinetics in Transition Metal Modified NaAlH4, *J. Alloys Comp.*, **356-367**, pp 400-404 (2003)
5. B. Bogdanovic, M. Schwickardi, A. Marjanovic, J. Tolle, Metal-doped sodium aluminum hydrides as potential new hydrogen storage materials, *Journal of Alloys and Compounds* **302** pp 36–58 (2002).
6. J.J. Vajo, G. L., Olson, Hydrogen Storage in Destabilized Chemical Systems, *Scripta Materialia* **56** pp 829-834 (2007).

7. F.E. Pinkerton, M.S. Meyer, G.P. Meisner, M.P. Balogh, J.J. Vajo, Phase Boundaries and Reversibility of LiBH4/MgH2 Hydrogen Storage Materials, *Journal of Physical Chemistry C*, **111** pp 12881-12885 (2007).

8. J. Yang, A. Sudik, C. Wolverton, Destabilizing LiBH4 with a Metal (M=Mg, Al, Ti, V, Cr, or Sc) or Metal Hydride (MH2 = MgH2, TiH2, or CaH2), *Journal of Physical Chemistry C*, **111[51]** pp 19134-19140 (2007).

9. X.B. Yu, Z. Wu, Q.R. Chen, Z.L. Li, B.C. Weng, T.S. Huang, Improved Hydrogen Storage Properties of LiBH4 Destabilized by Carbon, *Applied Physics Letters* **90[3]** pp 034106 (2007).

Batteries and Energy Storage Materials

RAPID SYNTHESIS OF ELECTRODE MATERIALS ($Li_4Ti_5O_{12}$ and $LiFePO_4$) FOR LITHIUM ION BATTERIES THROUGH MICROWAVE ENHANCED PROCESSING TECHNIQUES

K. Cherian[1], M. Kirksey[1], A. Kasik[2], M. Armenta[2], X. Sun[2] & S.K. Dey[2]
[1] Spheric Technologies, Phoenix, AZ, USA
[2] Arizona State University, AZ, USA

ABSTRACT
        The role of new batteries and energy storage materials are key factors in the new energy economy; certain new ceramic phases and rapid methods for their synthesis are, in turn, key to more efficient and cost-effective (especially Li ion) batteries. Such key phases can be prepared more rapidly and effectively through microwave enhanced processing wherein direct microwave heating as well as indirect and anisothermal microwave heating effects could occur; separated E & H field processing configurations could also offer tremendous advantages in this regard.
        Two examples of electrode materials which have been confirmed to form rapidly through special microwave processing approaches are $Li_4Ti_5O_{12}$ and $LiFePO_4$. Lithium carbonate and titania were the major starting reactants for microwave enhanced synthesis of the former, while the major starting reactants for the latter included lithium carbonate, ammonium dihydrogen phosphate and iron(II) oxalate. XRD analyses have confirmed the formation of the required phases. This paper discusses some preliminary experiments, results and subsequent investigations focused on developing faster, cheaper and greener synthesis routes for Li ion battery electrode materials and advanced microwave processing furnaces for scale-up.

KEYWORDS: Li ion battery, electrode material, microwave synthesis, $Li_4Ti_5O_{12}$, $LiFePO_4$.

INTRODUCTION
        The role of new batteries and energy storage materials are key factors in the new energy economy; certain new ceramic phases and rapid methods for their synthesis are, in turn, key to more efficient and cost-effective (especially Lithium ion) batteries. Factors such as low cost, environmental compatibility, theoretical specific capacity and perfect thermal cycling stability has led to $LiFePO_4$ growing in use as an electrode material for lithium batteries. This material is in use, following thorough investigation, for lithium-ion battery application in full hybrid electric vehicles[1] and other large energy storage applications. Though many synthesis routes for $LiFePO_4$ have been explored it has been recognized that there is still a need for rapid and commercially viable process(es) to produce the material at low cost. This paper outlines advanced microwave processing equipment and some preliminary experiments with them focused on developing faster, cheaper and greener synthesis routes for Li ion battery electrode materials.

CONVENTIONAL VS. MICROWAVE ASSISTED/ENHANCED CHEMICAL SYNTHESIS
        Many routes to synthesize $LiFePO_4$ have been investigated by various labs worldwide to obtain the required form with the desired electrochemical properties. These include solid state reaction[2-4], sol-gel[5, 6], co-precipitation in aqueous medium[7, 8], hydrothermal[9, 10], and, more recently, microwave (MW) enhanced routes[11-14]. These routes, except those using MW, involve a very large heating step running into several hours (up to 15 hours) in inert or reducing atmospheres. Recent patent applications on battery materials synthesis too apparently involve long process times[15, 16]. MW synthesis routes, on the other hand, have been shown to enable much shorter process times – minutes in the place of hours of other conventional synthesis routes. Shorter process times also appear to result in smaller

particles/grains in the synthesized product, and this would be another advantage of a microwave assisted/enhanced chemical synthesis route.

ADVANCED MICROWAVE SYSTEMS FOR CHEMICAL SYNTHESIS AND PROCESSING

Almost all the microwave assisted/enhanced LiFePO$_4$ synthesis successes reported so far have made use of small laboratory scale microwave systems. While the results thus obtained serve to experimentally demonstrate the advantages of microwave processing, the limitations in controlling process parameters together with limited process volumes have, until now, placed severe constraints in scaling up the process to pilot scale and onward to commercially viable production levels. The availability now of advanced larger scale batch and continuous 2.45 GHz advanced microwave furnace systems removes those limitations and changes the scenario for the better; Figures 1 & 2 represent the advanced batch and continuous microwave furnaces for high temperature materials processing.

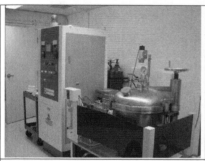

| Fig 1: Advanced Microwave Batch Furnace, Model HamiLab-V6 | Fig 2: Advanced Microwave Continuous Pusher Furnace, Model AMPS-9 |

The batch system (HamiLab-V6) has an effective uniform heating volume of ~200mm ID x 200mm HT within an applicator of dimensions ~ 500mm ID x 500mm HT; the continuous system (AMPS-9) is a pusher type with rectangular crucibles ~ 150mm L X 100mm B X 60mm H and lateral system speed of 15mm – 600mm /hour. These Spheric-SynoTherm microwave furnace systems, with MW power output of 0.6 - 6KW variable for the batch system and 0.9 – 9KW (and above, according to customer needs and specifications) for the continuous system, have inbuilt computer capabilities for temperature programming and control in the 450-2250 C range. Various temperature ramp-up rates to predetermined temperature hold stage(s) (up to ~1550C) and hold time(s) are possible. These provide the systems the versatility required to develop processing regimes and profiles of relevance to various materials processing applications; these are of specific relevance to microwave assisted chemical synthesis of inorganic materials and battery electrode materials. The process development strategy will be to develop the best processing profile for the required product in the batch system and then adapt this knowledge to developing the continuous processing profile for larger scale production in the continuous MW furnace – that is, transfer the 'processing profile in time' to 'processing profile in space'.

MICROWAVE SYNTHESIS OF Li-ION BATTERY ELECTRODE MATERIALS WITH ADVANCED MICROWAVE SYSTEMS AT SPHERIC TECHNOLOGIES INC.

Various battery electrode materials synthesis trials were carried out in a HamiLab-V6. Two examples of electrode materials which have been confirmed to form rapidly through special microwave processing approaches are $Li_4Ti_5O_{12}$ and $LiFePO_4$. Lithium carbonate and titania were the major starting reactants for microwave enhanced synthesis of the former, while the major starting reactants for the latter included lithium carbonate, ammonium dihydrogen phosphate and iron(II) oxalate. XRD analyses were done to confirm the formation of the required phases; varying process profile and parameters were found to be helpful in minimizing or eliminating undesirable phases. Additional analytical techniques including SEM have been employed to characterize the phases formed.

$LiFePO_4$ Synthesis
Two processing approaches were investigated, as described below:

1) Processing with a specific microwave power and exposure time (temperature ramp up followed by rapid cool down, unprogrammed).
In these runs, microwave exposure time and microwave power were the controllable variables; these runs were similar to some of the earlier reported work employing modified kitchen microwave and small laboratory microwave systems which did not have temperature programming capabilities. The experiments and results are summarized in Table 1 and Figs 1a-c respectively.

Sample ID	MW Exposure Time	Atm	MW Power	T max observed
#3	2 min	Air	3 kW	<450C
#2	2 min	Air	3 kW	665C
#4	3 min	Air	3 kW	930C
#1	5 min	Air	3 kW	1162C
#5	3 min	90N10H	3 kW	771C
#6	4 min	90N10H	3 kW	1042C
#7	5 min	90N10H	3 kW	1176C
#8	4 min	90N10H	2 kW	551C
#9	4 min	90N10H	2.5 kW	951C

Table 1: 'Temp ramp up and down' trials data for different microwave exposure times and power levels, in two different processing atmospheres

The corresponding XRD analytical results of the reaction products are summarized below:

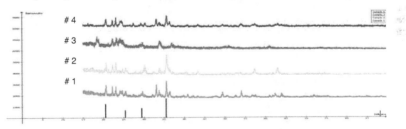

Fig. 1a: XRD spectra for samples #1 - #4 with some major $LiFePO_4$ peaks outlined. Also present: Fe_2O_3, Fe_3O_4 (and LiP in #1?)

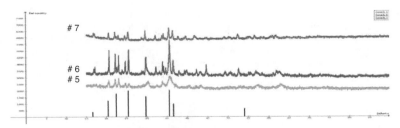

Fig. 1b: XRD spectra for samples #5 - #7 with LiFePO$_4$ peaks outlined
Also present: Fe$_2$O$_3$ (lesser) Fe$_3$O$_4$ (increased) Best: #6

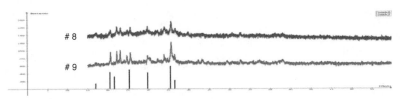

Fig. 1c: XRD spectra for samples #8 & #9 with LiFePO$_4$ peaks outlined
Also present: Fe$_2$O$_3$ (lesser), Fe$_3$O$_4$ (lesser). Best: #9

These XRD analysis results indicate the following:
a) The desired LiFePO$_4$ phase has formed, but with substantial undesired iron oxide phases as well
b) With change of the microwave power and exposure times process variables, some degree of decrease of the undesired phases may be observed, though the desired phase purity level of LiFePO$_4$ was apparently not achieved.

2. Processing with a programmed temperature-time profile with specific high temperature hold times

Anisothermal heating of the bulk due to the rapid heat and cool down process could lead to non uniform or incomplete reactions occurring at different regions. Therefore a temperature-time programmed approach to microwave processing was adopted, which enabled controlled temperature ramp up rates and holds at predetermined temperatures for predetermined periods for more uniform bulk heating. The advanced programming and process control features of the HamiLab-V6 enabled carrying out temperature–time programmed microwave processing through controlled power input. The ramp up rates, hold temperatures and hold times thus became programmable process variables together with other variables such as starting materials and their stoichiometries, processing atmospheres and the sample container–insulation configurations.

Based on the best results achieved in the first processing approach ('temp ramp up – cool down') and prior experience in various materials processing runs in the HamiLab-V6 batch furnace, a judicious selection of the variables were made and initial programmed runs carried out. The processing profile had a temperature hold of 15 min at 800C.

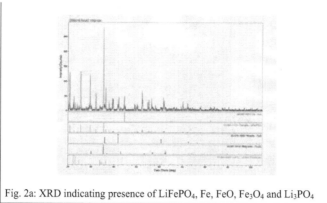

Fig. 2a: XRD indicating presence of LiFePO$_4$, Fe, FeO, Fe$_3$O$_4$ and Li$_3$PO$_4$

The XRD analysis of the product revealed the formation of the desired phase together with other oxide phases. Subsequent runs were carried out with change in the processing parameters, including rapid temperature ramp rates and processing atmospheres. Some initial results have been obtained which show that suitable process modifications could apparently help reduce the undesired oxide phases, as evidenced by the typical XRD of Fig. 2b.

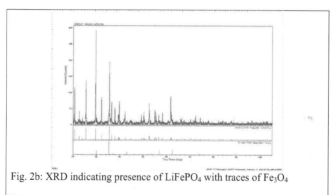

Fig. 2b: XRD indicating presence of LiFePO$_4$ with traces of Fe$_3$O$_4$

It may be noted that together with an overwhelming presence of the desired LiFePO$_4$ phase, there is probably only a relatively weak indication of the presence of an undesired Fe$_3$O$_4$ phase. Further refinement of the processing parameters therefore should enable achievement of maximum phase purity of LiFePO$_4$. Process modifications for reducing undesired oxide phases for different reactant volumes are being investigated as well, and these will be published after reconfirmation.

Li$_4$Ti$_5$O$_{12}$ Synthesis

A similar approach was adopted for microwave solid state synthesis of Li$_4$Ti$_5$O$_{12}$ with lithium carbonate and titania being the major starting reactants for the microwave enhanced synthesis process.

The resulting product was subjected to XRD analysis for phase identification. The XRD analysis result of the synthesized product is given below (Fig 3).

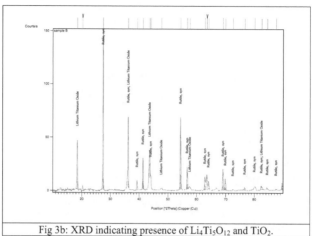

Fig 3b: XRD indicating presence of Li$_4$Ti$_5$O$_{12}$ and TiO$_2$.

The experimental results indicate the following:
a) The desired Li$_4$Ti$_5$O$_{12}$ phase has formed, but the XRD shows the presence of TiO$_2$ as well.
b) Adopting suitable process modification strategies, better yield and purity of the desired phase should be attainable as was the case for LiFePO$_4$ synthesis.

This paper deals with only some of the preliminary results in utilizing advanced microwave processing furnace system facilities for developing faster cheaper and greener synthesis routes for Li ion battery materials. Further results may be reported later at appropriate forums.

CONCLUSIONS
1. Microwave assisted processing offers a more rapid and efficient synthesis route for Li ion battery electrode materials LiFePO$_4$ and Li$_4$Ti$_5$O$_{12}$.
2. Advanced microwave furnace systems, batch and continuous, with larger processing volumes /throughput and temperature-time-power programming facilities are now available; these offer the opportunity of scaling up small lab scale experimental results to pilot scale and commercially viable production levels.
3. Li-ion battery materials Li$_4$Ti$_5$O$_{12}$ and LiFePO$_4$ have been successfully synthesized in the larger advanced microwave batch system; suitable modification of process profiles to minimize undesired phases and increase yield of required phase were successfully implemented for LiFePO$_4$ through the temperature-time-power programming facility of the advanced MW furnace system.

REFERENCES
[1]M. Anderman, Status and prospects technology for hybrid electric vehicles, including plug-in hybrid vehicles, Briefing to the U.S. Senate Committee on Energy and Natural Resources by the President of Advanced Automotive Batteries, 26 January 2007

(http://energy.senate.gov/public/_files/andermantestimony.pdf)
[2]A.K. Padhi, K.S. Nanjundaswamy, J.B. Goodenough, J. Electrochem. Soc. **144**, 1188 (1997).
[3]P.P. Prosini, D. Zane, M. Pasquali, Electrochim. Acta **46**, 3517 (2001).
[4]M. Takahashi, S. Tobishima, K. Takei, Y. Sakurai, J. Power Sources **97/98**, 508 (2001).
[5]K.F. Hsu, S.Y. Tsay, B.J. Hwang, J. Mater. Chem. **14**, 2690 (2004).
[6]R. Dominko,M. Bele, M. Gaberscek,M. Remskar, D. Hanzel, S. Pejovnik, J. Jamnik, J. Electrochem. Soc. **152**, A607 (2005).
[7]J.F. Ni, H.H. Zhou, J.T. Chen, X.X. Zhang, Mater. Lett. **59**, 2361 (2005)
[8]M.R. Yang, W.H. Ke, S.H. Wu, J. Power Sources **146**, 539 (2005) .
[9]S. Yang, P.Y. Zavalij, M.S. Whittingham, Electrochem, Commun., **3**, 505 (2001)
[10]Ch. Xu, J. Lee, A.S. Teja, J. Supercrit. Fluids, **44,** 92 (2008).
[11]M. Higuchi, K. Katayama, Y. Azuma, M. Yukawa, M. Suhara, J. Power Sources **119–121**, 258 (2003).
[12]S. Beninati, L. Damen, M. Mastragostino,. J. Power Sources **180,** 875 (2008).
[13]Xiang-Feng Guo, Hui Zhan, Yun-Hong Zhou, Solid State Ionics, **180**, 386 (2009)
[14]Hongli Zou, Guanghui Zhang, Pei Kang Shen , Mater. Res. Bull., **45,** 149 (2010)
[15]Y-M. Chiang, A.S. Godz, M.W. Payne, United States Patent Application 20070190418
[16]M.M. Saidi, H. Huang, United States Patent 7060238

*Contact for further info.: kcherian@spherictech.com and mkirksey@spherictech.com

LITHIUM STORAGE CHARACTERISTICS IN NANO-GRAPHENE PLATELETS

S. L. Cheekati[a], Y. Xing[b], Y. Zhuang[b], and H. Huang[a]

a. Department of Mechanical and Materials Engineering, Wright State University
b. Department of Electrical Engineering, Wright State University
Dayton, Ohio, USA

ABSTRACT

Improvements of the anode performances in Li-ions batteries are in demand to satisfy applications in transportation. The emerging new class of materials – nano graphene platelets (NGPs) and their composites - are promising alternative anodes. In this paper, a brief review was presented focusing on the high-capacity lithium storage characteristics in pyrolyzed carbons and carbon nanotubes as well as the computational results on lithium – graphene interactions and diffusion. These results led to the research towards both high lithium storage capacity and rapid lithium kinetics in NGPs. Afterwards, experimental results on lithium storage characteristics in three different kinds of NGPs were presented. High reversible capacities and rate capabilities were achieved in the NGPs. Some comments on the necessity and feasibility of mechanistic studies on lithium storage in graphene-based nanomaterials were also addressed.

1. INTRODUCTION

High power density energy production and storage systems, such as batteries, fuel cells, electrochemical capacitors, and solar cells, are currently under intense research and development. Lithium-ion batteries, which are superior to other conventional batteries in the aspects of energy density and cycle life etc., are considered as a primary battery power source in hybrid or plug-in electric vehicles to curtain greenhouse gas emission. However, Li-ion batteries have yet met the targets of low-cost and high-performance in such applications due to the lack of advanced cathode and anode materials [1-5]. The state-of-the-art anodes suffer from one or more of the following problems: limited Li storage capacity, large irreversible capacity loss, low charge/discharge rate capability, and poor capacity retention upon charge/discharge cycling.

Since it was experimentally isolated in 2004 [6-8], graphene, the unique 2-dimensional atomically-thick honey-comb structured carbon, has revealed various novel properties. Graphene has the highest intrinsic mechanical strength (1060 GPa), the highest thermal conductivity ($3000 Wm^{-1}K^{-1}$), a high surface area (2630 m^2/g), and a high electronic mobility (10000 $cm^2/V/s$). However, mass-production of monolayer graphene sheet is still challenging to date. This leads to the emerging of a new class of material - nano-graphene platelets (NGPs) [9] which render an alternative route to harness graphene's outstanding properties for immediate applications. NGPs collectively refer to platelets, sheets, or ribbons of monolayered and/or multiplelayered graphene from "pristine graphene" (containing insignificant amount of oxygen) to "graphene-oxide" (containing substantial oxygen contents). NGPs, usually with x and y dimensions less than hundreds of nanometers and z dimension less than ten of nanometers, have exhibited similar outstanding properties to the pristine monolayer graphene sheets as well as carbon nanotubes (CNT). Distinguished from graphene and CNTs, NGPs can be achieved via exfoliation from graphite at significantly low cost [8, 9].

NGPs and their nanocomposites have triggered a "gold rush" in exploring their potential applications in electronics, sensors, medicines, and alternative energy technologies [10-21]. The functionalized graphene nanosheets exhibited high specific capacitances in electrochemical double layer capacitors [15, 16]. Graphene films, with a high conductivity of 550 S/cm and a transparency of more than 70% over a wavelength range of 1000 to 3000 nm, were demonstrated as an alternative to metal oxide window electrodes for solid-state dye-sensitized solar cells [17]. NGPs also found their applications in fuel cells as bipolar plates and/or active electrode catalyst supporters [18-21]. Recently, graphene-based anodes for Li-ion batteries are under extensive investigations. As a result, improved lithium storage capacity, cycle life, and rate capability were achieved in graphene-based nanomaterials [22-27]. In this paper, a brief review will be emphasized on 1) the high-capacity lithium storage characteristics in pyrolyzed carbons and carbon nanotubes; and 2) the computational results on lithium – graphene interactions and kinetics. Afterwards, experimental results will be presented focusing on lithium storage characteristics in three different kinds of NGPs. Some thoughts on the necessity of systematically mechanistic studies on lithium storage in graphene-based nanomaterials will also be addressed.

2. BACKGROUND

2.1 Lithium storage in low-temperature pyrolyzed hard carbons

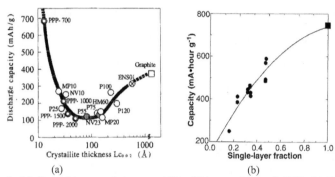

(a)   (b)

Figure 1. (a) Reversible capacity vs. crystallite thickness, from ref. [37]; (b) Reversible capacity vs. single layer fraction in hard carbons, from ref. [28]

Pyrolyzed carbonaceous materials are generally classified into three basic categories, i.e. well-graphitized carbon, disordered soft (graphitzable) carbon, and disordered hard (nongraphtizable) carbon, based on the final structures determined by the precursors and thermal processing temperatures. In the 1990s hard carbons, fabricated via low-temperature pyrolysis (500 – 1000 °C) from various organics and polymers, were extensively investigated [28-39]. Reversible capacities in the range of 400 – 1000 mAh/g, higher than theoretical value of 372 mAh/g (corresponding to the formation of LiC_6) in the graphitic carbons, were constantly achieved in the potential range of 0 – 2.5 V vs. metallic Li electrode.

Endo et al. [37] studied the reversible capacities of various hard carbons as a function of crystallite thicknesses L_{c002} which was estimated from the Small-Angle X-ray Scattering (SAXS) Spectra. It was discovered that lithium storage capacity increased dramatically upon decreasing the carbon crystallite thickness, as shown in Figure 1(a). As the carbon thickness was reduced to about 1 nm (equivalent to the tri-layered graphene sheets), the capacity increased to 700 mAh/g. Endo submitted that a different lithium storage process from intercalation into graphite might have occurred as the crystallite thickness (L_{c002}) became less than 10 nm resulting in the high lithium storage capacities. Meanwhile, based on the hypothesis that Li could completely absorb on both sides of each monolayer graphene sheet (corresponding to formation of Li_2C_6), Dahn *[28]* discovered a relationship between lithium storage reversible capacity and the mono-layered graphene fraction in various hard carbons. As can be seen in Figure 1(b), the trend of the lithium storage capacity increasing with the fraction of monolayered graphene corroborated very well with the guiding line towards the maximum value of 740 mAh/g. Dahn and Zheng *[34-36]* later proposed that hard carbons consisted predominantly of monolayered, bi-layered, and tri-layered graphene sheets arranged like a "house of cards". Lithium could be adsorbed on both sides of monolayered graphene sheets and possibly onto the internal surfaces of nanopores formed by the carbon nanosheets. The two storage mechanisms were attributed to the high lithium capacities. There were other interpretations on the high Li storage capacities in hard carbons, such as Sato's Li_2 covalent molecules intercalation model [38] and Yazami's multilayer (2-3 layers) Li adsorption model [39].

2.2 Lithium storage in carbon nanotubes

Li storage capacities as high as $Li_{1.6}C_6$ were experimentally demonstrated in single-wall carbon nanotubes (SWCNTs) and further increased up to $Li_{2.7}C_6$ after the ball-milling process [40-42]. The reversible capacity (over 1000mAh/g) of the chemically-treated SWCNTs was about three times greater than that of graphite. Shimoda [42] systematically investigated lithium storage behaviors in the closed and opened SWCNTs of different lengths. It was observed that 1) closed long SWCNTs could store less than one Li per six carbons; but 2) opened SWCNTs with short lengths could uptake over two Li per six carbons. Figure 2 presented the discharge and charge profile of lithium in and out of short open SWCNTs.

Zhao and Meinier et al [43, 44], after a series of first principle computational studies, suggested that: 1) the cohesion between Li and carbon nanotubes mainly ionic resulting by the strong electron transfer, which might be related to the potential hysteresis observed in the voltage-capacity profile; 2) the energy of Li inside the tube was comparable to that outside the tube, implying lithium could be stored in both nanotube exteriors and interiors. The simultaneous lithium insertion/adsorption on the sides of SWCNTs were ascribed to the high Li storage densities; and 3) Li motion through perfect sidewalls was forbidden and Li-ions could only enter tubes through topological defects containing at least nine-sided rings or through the ends of open-ended nanotubes. As Li-ions had to enter through the open ends, diffusion length along the nanotubes determined the kinetics.

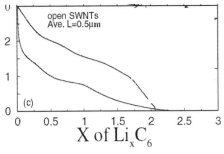

Figure 2. Discharge/charge profile of short open SWCNTs, from ref. [42]

2.3 Computational results on lithium interaction with graphene

Graphene is the building block of all kind of carbonaceous materials. A monolayered graphene sheet can be spirally wrapped into a SWCNT. Multiple graphene sheets can be stacked into graphite. Multi-layered graphene sheets can be clustered into hard carbons or wrapped into mutil-wall carbon nanotubes (MWCNTs). Therefore, graphene structure provides an ideal platform for fundamental understandings of Li-C interactions in the carbonaceous anodes. In accordance, computations based on density function theory have been performed to elucidate the potential energy for the interaction between Li and graphene and the nature of electron transfer in Li-graphene structure [45-50].

It was consistently reported that Li preferred the hexagonal center (H-site) to the atom top (T-site) and the bridge center (B-site). The Li-C binding distance varied from 1.60 nm to 2.10 nm and binding energies were in a broad range of 0.5 eV to 1.7 eV, depending on detailed computational approaches. The interaction between Li and graphene sheet, similar to that of the Li-nanotube, was found both ionic and van der Waals characteristic. Valencia [47] revealed that there was a strong interaction between the resulting Li ion and the cloud π electrons in the graphene because a significant polarization of the π -cloud for Li adatoms existed in the cation - π complexes. Recently, Wang [26] computed two Li atoms on the each side of a graphene monolayer structure. It was found that the most stable configuration, after relaxation, was Li at symmetric H-sites with the graphehe sheet as a mirror, resulting in the formation of a stable LiC_3 structure. Additionally, Fermi level of graphene with the two-side Li configuration was higher than that with the one-side Li adsorption, which resulted about 0.36 charge transfer from the two Li atoms to every six carbon atoms.

Li-ion diffusion processes on a model surface of graphene sheet terminated with hydrogen atoms ($C_{96}H_{24}$) was investigated by means of the direct molecular orbital dynamics method. Tachikawa et al [50] showed that the Li-ion (Li^+) started to move on the surface above 250K. The diffusion coefficient of Li^+ in the model graphene cluster was 9* 10^{-11} m^2/s at 300K, which was several orders magnitude higher than on graphite ($10^{-15} - 10^{-12}$ m^2/s).

3. EXPERIMENTAL PROCEDURE AND RESULTS - LITHIUM STORAGE IN NGPS

Previous experimental investigations on hard carbons and carbon nanotubes combined with the computational results on the simplified lithium-graphene system led to a reasonable hypothesis of both high lithium storage capacity and rapid lithium kinetics in NGPs. Moreover, NGPs are advantageous over both hard carbons and carbon nanotubes because NGPs circumvent cross-linkage in hard carbons and eliminate slow diffusion in the long tube interior. Therefore, graphene-based materials are promising anode candidates for lithium-ion batteries.

A large reversible specific capacity up to 784 mAh/g of graphene nanosheets incorporated with carbon nanotube and C_{60} were reported by Yoo et al [22]. Lithium storage characteristics in NGP-based composites, such graphene-Sn, graphene-SnO_2, and graphene-Si etc., were also investigated and reported [23-27]. Both high lithium storage capacities and improved cycleability were observed.

Up to date, various experimental approaches of fabricating graphene have been demonstrated including wet-exfoliation, mechanical exfoliation, dry printing transfer, and thermal decomposition on atomic flat SiC substrate, etc [51-58]. Being used as the anode materials in Li-ion batteries, graphene nanosheets were typically fabricated from natural graphite powders via the low cost oxidation-exfoliation-reduction process.

Recently we have studied lithium storage characteristics in two different types of NGPs – pristine NGPs (less than 0.1 at% of oxygen) and oxide NGPs (containing over 1 at% of oxygen). The pristine NGPs were fabricated via low-cost wet intercalation followed by thermal shock, provided by

Angstron Materials [9]. The extent of intercalation with a liquid agent was well-controlled to obtain large-batch NGPs with a narrow thickness distribution. A controlled-atmosphere furnace was used for thermal shock exfoliation and reduction. The physical properties of the pristine NGPs such as particle distribution and BET surface area were quantized. Two pristine NGP specimens, i.e. NGP001 and NGP010, with different average thicknesses were compared in this work. NGP001 had an in-plane dimension of up to 300 nm and vertical dimension of 1-2 nm. The NGP010 specimen had an in-plane dimension of up to 600 nm and vertical dimension of average 10 nm. Oxide NGPs (herein abbreviated as GO) were synthesized by the traditional Hummers' approach without the hydrazine reduction step. The oxidation intercalation agents included $KMnO_4$, H_2O_2, $NaNO_3$, and concentrated H_2SO_4 and the thermal exfoliation was 250°C.

Figure 3 exhibited Atomic Force Microscope (AFM) image of NGP001, prepared by thermal shock followed by high-temperature hydrogen reduction. One folding monolayer graphene sheet was clearly visualized. The step height was measured approximately 0.39 nm. The slightly larger step height than the theoretically monolayer thickness 0.335 nm might have resulted from the gap between the top and bottom folded area.

Figure 4 exhibited AFM topography-mode and friction-mode images obtained on NGP010. The thickness measured along the marking line in the figure was measured around 2 nm corresponding to 6-layered graphene. The lateral dimension was 200-300 nm for this specific NGP sheet. Better contrast was obtained in the friction mode because graphene nanosheet had less friction against the AFM probe relative to the mica substrate.

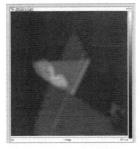

Figure 3. AFM image of the specimen NGP001

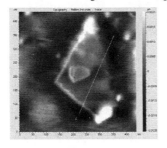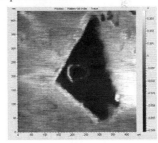

Figure 4. AFM topography-mode (left) and friction-mode (right) images of the specimen NGP010.

To assess lithium storage electrochemical characteristics, the pristine and oxide NGPs powders, i.e. NGP001, NGP010, and GO specimens, were mixed with 6 wt% polyvinyliene fluoride (PVDF) in N-Methyl-2-pyrrolidone (NMP), respectively. The mixed paste was then coated on a copper foil and then dried at 120 °C under vacuum for 24 hours. The Swagelok prototype testing cells were assembled in a glove box with moisture and oxygen contents less than 1 ppm. The NGP electrode was used as the working electrode, $LiPF_6$ - EC (ethylene carbonate) – EMC (ethyl methyl carbonate) as the electrolyte, and Li foil as the counter and reference electrode. The lithium storage characteristics, including potential profile, irreversible loss, and reversible capacity, were accurately recorded on a battery testing station at the galvanostatic or intermittent mode.

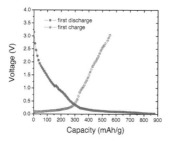

Figure 5. The first discharge/charge profile from the NGP001 specimen, prepared by thermal shock followed by HT hydrogen reduction.

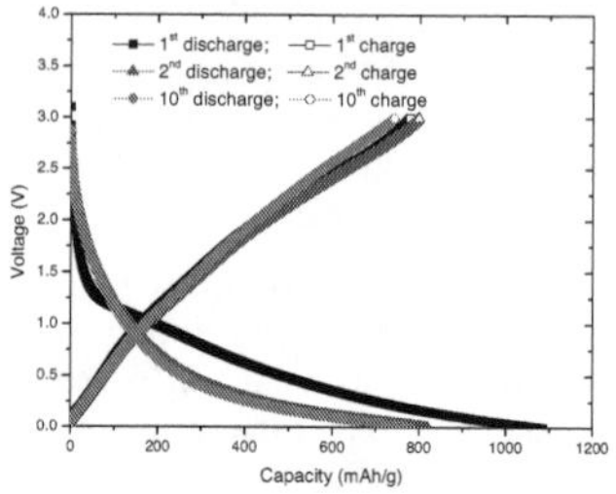

Figure 6. The discharge/charge profiles of the GO specimen prepared by Hummer's approach without hydrazine reduction.

Figure 5 showed the first discharge and charge (d/c) profile obtained on the pristine NGP001 at a galvanostatic intermittent discharge/charge mode. The current density was 50 $\mu A/cm^2$ and the current loading as well as intermittent rest time was 10 minutes. The first discharge capacity was 890 mAh/g and the charge capacity 560 mAh/g in average. Hence, the first charge/discharge columbic efficiency was 62%. Two distinguished potential regions were observable in the d/c profile. The high potential region above 0.2V showed similar characteristics of lithium storage in the short single-wall nanotube (compared with Figure 2). The low potential region below 0.2 V showed the typical characteristics of lithium intercalation in graphite. The electrochemical characteristic profile, together with AFM observations, indicating that NGP001 specimen was made up of monolayered and multilayered graphene nanosheets. The NGP001 specimen was estimated to contain approximately 40 - 50% monolayer graphene based on the reversible capacity data, as well as the assumptions that lithium storage capacities in monolayered and multilayered graphene were 700 - 740 mAh/g and 320-360 mAh/g, respectively.

Figure 6 showed the first, second, and tenth d/c profiles from the GO specimen recorded at a constant current density of 100 $\mu A/cm^2$. The GO's potential-current profile differed significantly from the pristine NGP001. A large potential hysteresis between discharge and charge was apparent. No low potential plateau was observed in the GO profile. The charge potential increased rapidly to 1V and then gradually to the cut-off voltage 3.0 V. The characteristics resembled to the reported results on graphene nanosheets prepared by Hummer's approach with hydrazine reduction [22, 23]. It is noteworthy that the present results were achieved on the GO specimen with no hydrazine reduction procedure. The resemblance of the lithium insertion/removal characteristics between GOs with and without hydrazine reduction suggested that the hydrazine might incompletely reduce the graphene oxide. A high discharge and charge capacities of 1086mAh/g and 780mAh/g were achieved during the first discharge/charge cycle, and hence the first c/d columbic efficiency was 72%. The following several cycles showed a reversible capacity was stabilized around 800 mAh/g. The graphite oxide electrode was cycled between 5mV and 3V for 10 times and no significantly loss was observed.

The typical lithium storage capacities achieved in the NGP010 powders were in the range of 340 - 360mAh/g. No significant capacity increase was observed at either the constant current or the intermittent testing mode. However, it was discovered that NGP010 specimen delivered a high lithium removal rate capability in comparison with the graphite of large dimensionalities, such as Micro450 and HMP850 (provided by Ashbury) having thicknesses of several micrometers. The lateral in-plane dimensional sizes for Micro450 and HMP850 were 1-2 μm and 10 μm, respectively. The lithium removal rate capability was assessed by increasing the charge current from C/20 to 3C while maintaining the discharge current at the same rate of C/10 during each consecutive cycle. Figure 7 presented the reversible charge capacities of the three graphitic structured specimens and overvoltages at the specific capacity of 100 mAh/g changing with the charge rates. The charge capacities of NGP010 decreased slightly from 340 mAh/g to 320 mAh/g upon increasing the rate from 0.05C to 3C. Contrastingly, the charge capacities of Micro450 reduced dramatically from 320 mAh/g to 270 mAh/g and those of HMP850 from 320 mAh/g to 230 mAh/g in the C-rate range of 0.05C to 3C. As a consequence, only 72% of the full lithium capacity recovered out of the graphite structure at 3C charging rate in HMP850 specimen. In addition, the potential polarization, as can be seen from figure 7, increased significantly for HMP850 upon increasing the current, suggesting a slow lithium removal kinetic limitation. At 3C rate, the potential polarization was 0.52 V for HMP850 but less than 0.1 V for NGP010. It is, hence, concluded that lithium removal kinetics were much better in the nano-graphene platelets with a thickness of around 10 nm than that in the micro-sized graphite flakes. The decrease of the in-plane diffusion length reduced charge overvoltage, and hence, increased rate capability. The low potential polarization will ensure a high power density in the Li-ion battery with NGP010 anodes.

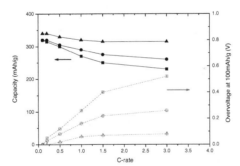

Figure 7. Reversible charge capacity (solid symbol) and overvoltage at the capacity of 100 mAh/g (hollow symbol) as a function of the charging rate, achieved from three different graphitic structured powders. Note: triangle symbol for NGP010; square for HMP850; and circle for Micro450.

4. CONCLUSION AND REMARKS

Lithium storage capacities in nano graphene paltelets (NGPs) reached values over twice of those in graphitic carbons and were higher than those in majority pyrolyzed hard carbons. The decrease of the in-plane diffusion length and thickness of graphite to nanometer scales reduced charge overvoltage, and consequently increased charge rate capability. NGPs are advantageous in lithium storage over hard carbons, nanotubes, and graphites due to the NGP's unique nano-scale structures, and hence, are promising alternative anodes for lithium-ion batteries. It is noteworthy that NGPs are not limited to serve as the active anode alternatives. NGPs may be adopted as the highly conductive supporting substrates for other anode or cathode active materials.

At this infancy stage, systematic experimental and fundamental mechanistic understandings on Li storage in NGPs have yet been addressed. Many fundamental questions are unsolved, which include 1) the maximum lithium storage capacities in mono- bi- and/or tri-layered NGPs; 2) the rates of lithium absorption, diffusion, and desorption on NGPs; 3) the relationships between NGPs' dimensionalities (thickness and basal area) and lithium diffusion coefficients; 4) the formation mechanism and composition of Solid Electrolyte Interphase (SEI) layer on NGPs; 5) lithium storage characteristics and mechanism in nano-structured (e.g NGP-supported or graphene-coated) NGP-composite anodes; 6) mass production of the optimal nanostructured NGP-composite anodes at low cost, and so on so forth. Answers to these questions are indispensible to gain insights in high-capacity lithium storage mechanisms towards development of novel anode materials.

The current progresses on graphene technologies facilitate our direct experimental investigations on lithium storage monolayer graphene, the building block of all kinds of carbons. Moreover, the bandgap of graphene can be tuned via adjusting the width of graphene ribbon, the number of layers, electrical field, dopants, as well as sublayer materials [59-62]. The well-controlled structure and unique bandgap tunability grant graphene an ideal platform to accurately monitor lithium storage processes and to investigate lithium storage mechanisms, kinetics, and SEI formation as a function of dimensionality, bandgap, and foreign element etc. The information may aid the rapid advancement in anode materials chemistry.

ACKNOWLEDGEMENT

The authors would like to acknowledge Drs. B. Jang and A. Zhamu, Angstron Materials, for providing NGP001 and NGP010 samples. We would also like to acknowledge Ohio Space Grant Consortium and Wright State University for the financial supports in this project. We thank Dr. Peter Wolf from Veeco for providing AFM imaging service on NGP001 specimen.

REFERENCES

[1] A. S. Arico, P. Bruce, B. Acrosati, J-M. Tarascon, and W. van. Schalkwijk, Nanostructured Materials for Advanced Energy Conversion and Storage Devices, *Nature Materials*, **4**, 366-377 (2005).
[2] J.-M. Tarascon, and M. Armand, Issues and Challenges Facing Rechargeable Lithium Batteries, *Nature*, **414**, 359-367 (2001).
[3] G.-A. Nazri, and G. Pistoia, *Lithium Batteries: Science and Technology*, Springer; 1 ed., (2009).
[4] M. S. Wittingham, Lithium Batteries and Cathode Materials, *Chem. Rev.*, **104**, 4271–4302 (2004).
[5] T. Ohzuku and R. J. Brodd, An Overview of Positive - Electrode Materials for Advanced Lithium-ion Batteries, *J. Power Sources*, **174**, 449–456, (2007).
[6] K. S. Novoselov, A. K. Geim, S. V. Morozov, D. Jiang, S. V. Dubonos, I. V. Grigorieva, and A. A. Firsov, Electric Field Effect in Atomically Thin Carbon Films, *Science*, **306**, 666-669 (2004).
[7] Y. Zhang, Y.-W. Tan, H. L. Stormer, and P. Kim, Experimental Observation of the Quantum Hall Effect and Berry's Phase in Graphene, *Nature*, **438**, 201-204 (2005).

[8] S. Stankovich, D. A. Dikin, G. Dommett, K. Kohlhaas, E. J. Zimney, E. Stach, R. Piner, S. T. Nguyen, and R. S. Ruoff, Graphene-based Composite Materials, *Nature*, **442**, 282-286 (2006).

[9] B. Z. Jang, and A. Zhamu, Processing of Nanographene Plateltes (NGPs) and NGP Nanocomposites: a Review, *J. Mat. Sci.*, **43**, 5092-5101 (2008).

[10] Z. Chen, Y. M. Lin, M. J. Rooks, and P. Avouris, Graphene Nano-Ribbon Electronics, *Physica E*, **40**, 228-232 (2007).

[11] A. Sakhaee-Pour, M. T. Ahmadian, and A. Vafai, Applications of Single-Layered Graphene Sheets as Mass Sensors and Atomistic Dust Detectors, *Solid State Communications*, **145**, 168-172 (2008).

[12] A. A. Balandin, S. Ghosh, W. Bao, I. Calizo, D. Teweldebrhan, F. Miao, and C. N. Lau, Superior Thermal Conductivity of Single-Layer Graphene, *Nano Lett.*, **8**, 902-907 (2008).

[13] X. Sun, Z. Liu, K. Welsher, J. T. Robinson, A. Goodwin, S. Zaric, and H. Dai, Nano-Graphene Oxide for Cellular Imaging and Drug Delivery, *Nano Res.*, **1**, 203-212 (2008).

[14] C. Stampfer, F. Schurtenberger, E. Molitor, J. Guttinger, T. Ihn, and K. Ensslin, Tunable Graphene Single Electron Transistor, *Nano Lett.*, **8**, 2378-2383 (2008).

[15] M. D. Stoller, S. Park, Y. Zhu, J. An, and R. S. Ruoff, Graphene-based Ultracapacitors, *Nano Lett.*, **8**, 3498-3502 (2008).

[16] S. R. C. Vivekchang, C. S. Rout, K. S. Subrahmanyam, A. Govindraraj, and C. N. Rao, Graphene-based Electrochemical Supercapacitor, *J. Chem. Sci.*, **120**, 9-13 (2008).

[17] X. Wang, L. Zhi, and K. Mullen, Transparent, Conductive Graphene Electrodes for Dye-Sensitized Solar Cells, *Nano Let.* **8**, 323-327 (2008).

[18] Y. Si, and E. T. Samulski, Exfoliated Graphene Separated by Pt Nanoparticles, *Chem Mater*, **20**, 6792-6797 (2008).

[19] R. Kou, Y. Shao, D. Wang, M. H. Engelhard, J. H. Kwak, J. Wang, V. V. Viswanathan, C. Wang, Y. Lin, I. A. Aksay, and J. Liu, Enhanced Activity and Stability of Pt Catalysts on Functionalized Graphene Sheets for Electrocatalytic Oxygen Reduction, *Electrochem.Comm.*, **11**, 954-957 (2009).

[20] B. Z. Jang, A. Zhamu, and J. Guo, Process for Producing Carbon-Cladded Composite Bipolar Plates for Fuel Cells, *USPTO*, Appl.# 20080149900 (2008).

[21] L. Song, J. Guo, A. Zhamu, and B. Z. Jang, Highly Conductive Nanoscale Graphene Plate Nanocomposites and Products, *USPTO*, Appl.# 20070158618 (2007).

[22] E. J. Yoo, J. Kim, E. Hosono, H. Zhou, T. Kudo, and I. Honma, Large Reversible Li Storage of Graphene Nanosheet Families for Use in Lithium Ion Batteries, *Nano Lett.*, **8**, 2277-2283 (2008).

[23] G. Wang, X. Shen, J. Yao, and J. Park, Graphene Nanosheets for Enhanced Lithium Storage in Lithium Ion Batteries, *Carbon*, **47**, 2049-2053 (2009).

[24] S. Chou, J. Wang, M. Choucair, H. Liu, J. A. Stride, and S. Dou, Enhanced Reversible Lithium Storage in a Nanosized Silicon/Graphene Composite, *Electrochem. Comm.*, **12**, 303-306 (2010).

[25] J. Yao, X. Shen, B. Wang, H. Liu, and G. Wang, In Situ Chemical Synthesis of SnO_2 - Graphene Nanocomposite as Anode Materials for Lithium-ion Batteries, *Electrochem. Comm.*, **11**, 1849-1852 (2009).

[26] G. Wang, B. Wang, X. Wang, J. Park, S. Dou, H. Ahn, and K. Kim, Sn/Graphene Nanocomposite with 3D Architecture for Enhanced Reversible Lithium Storage in Lithium Ion Batteries, *J. Mater. Sci.*, **19**, 8378-8384 (2009).

[27] P. Guo, H. Song, and X. Chen, Electrochemical Performance for Graphene Nanosheets as Anode Materials for Lithium-ion Batteries, *Electrochem. Comm.*, **11**, 1320-1324 (2009).

[28] J. R. Dahn, T. Zheng, Y. Liu, and J. Xue, Mechanism of Lithium Insertion in Carbonaceous Materials, *Science*, **270**, 590-594 (1995).

[29] S. Faldrois, and B. Simon, Carbon Materials for Li-ion Batteries, *Carbon*, 37, 165-180 (1999).

[30] Z. Wang, X. Huang, and L. Chen, Lithium Insertion/Extraction in Pyrolyzed Phenolic Resin, *J. Power Sources*, **81–82**, 328–334 (1999).

[31] J. Lee, K. An, J. Ju, B. Cho, W. Cho, D. Park, and K. S. Yun, Electrochemical Properties of PAN-based Carbon Fibers as Anodes for Rechargeable Lithium Ion Batteries, *Carbon*, **39**, 1299–1305 (2001).

[32] Y. Liu, J. S. Xue, T. Zheng, and J. R. Dahn, Mechanism of Lithium Insertion in Hard Carbons Prepared by Pyrolysis of Epoxy Resins, *Carbon*, **34**, 193-200 (1996).

[33] M. Noel, and A. Suryanarayanan, Role of Carbon Host Lattices in Li-ion Intercalation/De-intercalation Processes, *J. Power Sources*, **11**, 193-209 (2002).

[34] T. Zheng, Y. Liu, E. W. Fuller, S. Tseng, U. von Sacken, and J. R. Dahn, Lithium Insertion in High Capacity Carbonaceous Materials, *J. Electrochem. Soc.*, **142**, 2581-2590 (1995).

[35] T. Zheng, J. N. Reimers, and J. R. Dahn, The Effect of Turbostratic Disorder in Graphitic Carbons on the Intercalation of Lithium, *Phys. Rev. B*, **51**, 734-741 (1995).

[36] K. Tokumitsu, H. Fujimoto, A. Mabuchi, and T. Kasuh, High Capacity Carbon Anode for Li-ion Battery: a Theoretical Explanation, *Carbon*, **37**, 1599-1605 (1999).

[37] M. Endo, Y. Nishimura, T. Takahashi, K. Takeuchi, and M. S. Dresselhaus, Lithium Storage Behavior for Various Kinds of Carbon Anodes in Li Ion Secondary Battery, *J. Phys. Chem. Solids*, **57**, 725-728 (1996).

[38] K. Sato, M. Noguchi, A. Demachi, N. Oki, and M. Endo, A Mechanism of Lithium Storage in Disordered Carbons, *Science*, **264**, 556-558 (1994).

[39] R. Yazami, Surface Chemistry and Lithium Storage Capability of the Graphite – Lithium Electrode, *Electrochimica Acta*, **45**, 87-97 (1999).

[40] E. Frackowiak, and F. Beguin, Electrochemical Storage of Energy in Carbon Nanotubes and Nanostructured Carbons, *Carbon*, 40, 1775–1787 (2002).

[41] H. Zhang, G. Cao, Z. Wang, Y. Yang, Z. Shi, and Z. Gu, Carbon Nanotube Array Anodes for High-Rate Li-ion Batteries, *Electrochimica Acta*, in press

[42] H. Shimoda, B. Gao, X. P. Tang, A. Kleinhammes, L. Fleming, Y. Wu, and O. Zhou, Lithium Intercalation into Opened Single-Wall Carbon Nanotubes: Storage Capacity and Electronic Properties, *Phys. Rev. B*, **88**, 015502 (2002).

[43] J. Zhao, A. Buldum, J. Han, and J. Lu, First-Principles Study of Li-Intercalated Carbon Nanotube Ropes, *Phys. Rev. Lett.*, **86**, 1706-1709 (2002).

[44] V. Meinier, J. Kephart, C. Roland, and J. Bernholc, Ab Initio Investigations of Lithium Diffusion in Carbon Nanotube Systems, *Phys. Rev. B*, **88**, 075506 (2002).

[45] C. K. Yang, A Metallic Graphene Layer Adsorbed with Lithium, *Appl. Phys. Lett.* **94**, 162115 (2009)

[46] M. Khantha, N. A. Cordero, L. M. Molina, J. A. Aloso, and L. A. Girifalco, Interaction of Lithium with Graphene: an Ab Initio Study, *Phys. Rev. B*, **70**, 125422 (2004).

[47] F. Valencia, A. H. Romero, F. Ancilotto, and P. L. Silverstrelli, Lithium Adsorption on Graphite from Density Functional Theory Calculation, *J. Phys. Chem. B*, **110**, 14832-14841 (2006).

[48] K. Rytkonen, J. Akola, and M. Manninen, Density Functional Study of Alkali-Metal Atoms and Monolayers on Graphite (0001), *Phys. Rev. B.*, **75**, 075401 (2007).

[49] K. T. Chan, J. B. Neaton, and M. L. Cohen, First Principle Study of Metal Adatom Adsorption on Graphene, *Phys. Rev. B,* **77**, 235430 (2008).

[50] H. Tachikawa, and A. Simizu, Diffusion Dynamics of the Li Ion on a Model Surface of Amorphous Carbon: a Direct Molecular Orbital Dynamics Study, *J. Phys. Chem.*, **109**, 13255-13262 (2005)

[51] L. M. Viculis, J. J. Mack, O. M. Mayer, H. T. Hahn, and R. B. Kaner, Intercalation and Exfoliation Routes to Graphite Nanoplatelets, *J. Mat. Chem.*, **15**, 974-978 (2005).

[52] H. Shiyoyama, Cleavage of Graphite to Graphene, *J. Mat. Sci. Lett.,* **20**, 499- 500 (2001).

[53] G. Wang, J. Yang, J. Park, X. Gou, B. Wang, H. Liu, and J. Yao, Facile Synthesis and Characterization of Graphene Nanosheets, *J. Phys. Chem. C*, **112**, 8192-8195 (2008).

[54] D. Li, M. B. Muller, S. Gilje, R. B. Kaner, and G. G. Wallace, Processable Aqueous Dispersions of Graphene Nanosheets, *Nature Nanotech*. **3**, 101–105 (2007).

[55] S. Gijie, S. Han, M. Wang, K. L. Wang, and R. B. Kaner, A Chemical Route to Graphene for Device Applications, *Nano Lett.*, **7**, 3394–3398 (2007).

[56] Y. Q. Wu, P. D. Ye, M. A. Capano, Y. Xuan, Y. Sui, M. Qi, J. A. Cooper, T. Shen, D. Pandey, G. Prakash, and R. Reifenberger, Top-Gated Graphene Field-Effect Transistors Formed by Decomposition of SiC, *Appl. Phys. Lett,.* **92**, 092102 (2008).

[57] K. S. Kim, Y. Zhao, H. Jang, S. Y. Lee, J. M. Kim, K. S. Lim, J. H. Ahn, P. Kim, J. Y. Choi, and B. H. Hong, Large-Scale Pattern Growth of Graphene Films for Stretchable Transparent Electrodes, *Nature*, **457**, 706-710 (2009).

[58] A. Reina, X. Jia, J. Ho, D. Nezich, H. Son, V. Bulovic, M. S. Dresselhaus, and J. Kong, Large Area, Few-Layer Graphene Films on Arbitrary Substrates by Chemical Vapor Deposition, *Nano Lett.*, **9**, 30-35 *(2009)*.

[59] T. Ohta, A. Bostwick, T. Seyller, K. Horn, and E. Rotenberg, Controlling the Electronic Structure of Bilayer Graphene, *Science*, **313**, 951-954 (2006).

[60] S. Y. Zhou, G. H. Gweon, A. V. Fedorov, P. N. First, W. A. de Heer, D. H. Lee, F. Guinea, A. H. Castro Neto, and A. Lanzara, Substrate - Induced Bandgap Opening in Epitaxial Graphene, *Nature Materials*, **6**, 770 – 775 (2007).

[61] N. Gorjizadeh, A. A. Farajian, K. Esfarjani, and Y. Kawazoe, Spin and Band-gap Engineering in Doped Graphene Nanoribbons, *Phys. Rev. B*, **78**, 155427 (2008).

[62] M. Y. Han, B. Oezyilmaz, Y. Zhang, and P. Kim, Energy Band-gap Engineering of Graphene Nanoribbons, *Phys. Rev. Lett.*, **98**, 206801-206804 (2007).

IN-SITU IMPEDANCE SPECTROSCOPY OF LIMN$_{1.5}$NI$_{0.4}$CR$_{0.1}$O$_4$ CATHODE MATERIAL

Karina Asmar[1], Rahul Singhal[1], Rajesh K. Katiyar[1], Ram S. Katiyar[1]
Andrea Sakla[2], and A. Manivannan[2]

[1]Physics Department, University of Puerto Rico, Rio Piedras, San Juan, PR, USA
[2]US DOE, National Energy Technology Laboratory, Morgantown, WV, USA

ABSTRACT

LiMn$_{1.5}$Cr$_{0.4}$Cr$_{0.1}$O$_4$ cathode material has been synthesized using sol-gel method for possible application in high energy density Li ion rechargeable batteries. The half cells were tested using cyclic voltammetry and charge discharge cycling. Doping with 5% Cr seems to improve operating voltage (4.88V) and cell stability compared to un-doped material (4.7V). In-situ electrochemical impedance spectroscopic measurements were performed before and after each charge and discharge cycle, respectively. After several cycles the diffusional impedance for lithium at the cathode electrolyte interface was found to be high compared to the electron transfer impedance that followed a slight decrease in discharge capacity.

INTRODUCTION

In recent years, various researchers have focused their efforts to identify new cathode materials for high energy density Li ion rechargeable batteries because of increasing demand for the development of advanced energy storage devices for various electronics, automobile, and statioanry power applications. Several lithium insertion compounds have been synthesized for their use in Li ion rechargeable batteries [1-8]. Among all cathode materials, spinel LiMn$_2$O$_4$ is considered promising for Li ion rechargeable batteries, because of its low cost and being environmental friendly material. However, LiMn$_2$O$_4$ cathode material showed severe capacity fading upon cycling due to Mn dissolution in electrolyte and it undergoes structural disorder upon cycling. It has already been reported that the electrochemical performance and operating voltage of LiMn$_2$O$_4$ can be improved by substituting a small amount of Mn by other transitional metal ions e.g. Ni, Cr, Co, Cu, Al, Fe etc. [9-12]. The cycleability of LiMn$_2$O$_4$ can be improved [12] by substitution of Mn with Co and Cr (LiMn$_{2-y}$M$_y$O$_4$;M =Cr and Co, where y = 0.0, 0.1, and 0.2). Sun and coworkers [13] have synthesized Li$_{1.15}$Mn$_{1.96}$Co$_{0.03}$Gd$_{0.01}$O$_{4+\delta}$ spinel cathode material by solid-state reaction. It was reported that 1% Gd and 3% Co substitution stabilized the spinel structure upon cycling. The initial discharge capacity at 25°C was found as 126.5 mAhg^{-1} and after 100 charge-discharge cycles, the discharge capacity retention was 98.33%. Oh and coworkers [14] have synthesized Li[Ni$_{0.5}$Co$_x$Mn$_{1.5-x}$]O$_4$ cathode material by co-precipitation method for 5V battery applications. They found that Co doped and undoped materials showed discharge plateau at 4.7 V (Ni$^{2+/3+/4+}$ redox couple) and 4 V (Mn^{3+}/Mn^{4+} redox couple. The initial discharge capacity was found as 140 mAh/g and 130 mAh/g for Co free and Co doped Li[Ni$_{0.5}$Co$_x$Mn$_{1.5-x}$]O$_4$ cathode material, respectively. The discharge capacity retention for Co free and Co doped cathode material was 90 % and 97%, respectively. Recently, we have reported [15] electrochemical behavior of LiMn$_{2-x}$Ni$_x$O$_4$ ($0 \leq x \leq 0.5$) cathode material. It was reported that the best electrochemical performance was observed in case of x = 0.5, with the two distinct discharge plateau at around 4.7 V and 4.0 V. The maximum discharge capacity was found as 140.18 mAh/g with 98% discharge capacity retention after 50 charge-discharge cycles. In order to obtain stable high energy density cathode material, we have synthesized Cr doped LiMn$_{1.5}$Ni$_{0.5}$O$_4$ cathode material [16].

The $LiMn_{1.5}Ni_{0.4}Cr_{0.1}O_4$ cathode material showed an extra discharge plateau at 4.87 V as compare to $LiMn_{1.5}Ni_{0.5}O_4$ cathode material, which corresponds to Cr^{3+}/Cr^{4+} redox couple. It was reported that $LiMn_{1.5}Ni_{0.4}Cr_{0.1}O_4$ showed maximum discharge capacity of 126 mAh/g with 98.5% capacity retention after 50 charge-discharge cycles, at a current density of 0.2C. The $LiMn_{1.5}Ni_{0.4}Cr_{0.1}O_4$ cathode material showed a stable charge discharge behavior at a higher current density of 1C. In order to better understand the electrochemical performance of $LiMn_{1.5}Ni_{0.4}Cr_{0.1}O_4$ cathode material, we performed in-situ impedance studies after each charge and discharge cycles. In-situ- impedance analysis was made for a set of 100 charge-discharge cycles.

EXPERIMENTAL:

The precursor materials lithium acetate dihydrate [$Li(CH_3COO)$. $2H_2O$, 99%], manganese(II) acetate tetrahydrate [$Mn(CH_3COO)_2 \cdot 4H_2O$, Mn 22%], nickel (II) acetate tetra hydrate [$Ni(CH_3COO)_2.4H_2O$], and chromium acetate hydroxide (Cr 24%) [$Cr_3(OH)_2(OOCCH_3)_7$] (all procured from Alfa Aesar, USA) were dissolved separately in 2-ethy hexanoic acid at 80°C. After complete dissolution of the precursor materials, all of the solutions were mixed together, followed by heating at 80°C and stirring for about 1 hour. The powder was obtained by drying the solution in a patry dish, kept onto a hot plate at about 200°C. The obtained powder was then grind and heated at 450°C for 4 hours for complete organic removal. After the complete organic removal the material was calcined at 875°C for 24 h in an oxygen atmosphere, to obtained phase pure $LiMn_{1.5}Ni_{0.4}Cr_{0.1}O_4$ cathode material. The phase purity and the crystallinity of the $LiMn_{1.5}Ni_{0.4}Cr_{0.1}O_4$ cathode material were obtained using a Siemens D5000 X-ray powder diffractometer [Cu-K_α radiation, 1.5405 Å]. The cathode was prepared by mixing calcined powder ($LiMn_{1.5}Ni_{0.4}Cr_{0.1}O_4$), carbon black, and polyvinylidene fluoride (weight ratio 80:10:10), and subsequently a slurry was made with n-methyl pyrolidone. The resulting paste was spread onto aluminum foil followed by drying at about 60°C in an oven overnight. The coin cells were fabricated in an argon atmosphere, inside a Glove Box (MBraun Inc, USA), using $LiMn_{1.5}Ni_{0.4}Cr_{0.1}O_4$ electrode as cathode, Li foil as anode, and 1M lithium hexafluoride ($LiPF_6$), dissolved in ethyl carbonate (EC) and dimethyl carbonate (DMC) [1:2 v/v ratio] as electrolyte. The electrochemical behavior of the cells was studied at room temperature by cyclic voltammetry and charge-discharge characteristics, using Solartron battery tester, Model 1470E. The impedance measurements of the cells were carried out using Gamry Instruments potentiostat and EIS 300 electrochemical software.

RESULTS AND DISCUSSIONS:

Our previous study on the basic characterization of this electrode material has been reported elsewhere [19]. Some of those results along with the newly developed in-situ measurements have been presented in this paper. XRD pattern of $LiMn_{1.5}Ni_{0.4}Cr_{0.1}O_4$ cathode material showing all peaks corresponds to Fd3m spinel structure is shown in Figure 1. The sharp peaks indicate the crystalline behavior of the synthesized material, where Li ions occupies tetrahedral (8a) sites. The manganese, chromium and nickel ions occupy octahedral (16d) sites and O^{2-} reside at the general positions (32e) [15]. The lattice parameter of $LiMn_{1.5}Ni_{0.4}Cr_{0.1}O_4$ cathode material were obtained using interactive powder diffraction data interpretation and indexing program POWDMULT [17] and was found as 8.1835 Å. The unit cell volume and the standard deviation were found as 548.04 $Å^3$ and 0.00206 $Å^3$, respectively.

Cyclic voltammetric studies of $LiMn_{1.5}Ni_{0.4}Cr_{0.1}O_4$ cathode material were carried out in $LiMn_{1.5}Ni_{0.4}Cr_{0.1}O_4/LiPF_6+(EC+DMC)/Li$ coin cells at room temperature and in the voltage range of

3.0 V to 5.0 V at a scan rate of 0.1 mV/s. The cyclic voltammogram showed [Figure 2] three well defined oxidation peaks at 4.07 V, 4.76 V, and 4.87 V. The peak at 4.07 V is due to Mn^{3+}/Mn^{4+} redox couple while the peaks at 4.76 V and 4.87 V are due to Ni^{2+}/Ni^{4+} and Cr^{3+}/Cr^{4+} redox couple, respectively [18]. It has been reported earlier in $LiMn_{1.5}Ni_{0.4}Cr_{0.1}O_4$ by XPS data that Mn exists in +3 and +4 states and Cr exist in +3 state [16]. Figure 3 showed charge-discharge behavior of $LiMn_{1.5}Ni_{0.4}Cr_{0.1}O_4/LiPF_6$ (1:1 EC:DMC)/Li coin cells. All measurements were performed at room temperature, at constant charge-discharge current of C/10 (1C = 146mA/g) in 3.0 V to 5.0V range. $LiMn_{1.5}Ni_{0.4}Cr_{0.1}O_4$ cathode material showed initial charge and discharge capacity of 148 mAh/g and 125.4 mAh/g, respectively. The discharge capacity reaches at a maximum value of 129.5 mAh/g after 5 charge-discharge cycles. The initial discharge capacity increases in the first few cycles that may be due to electrochemical activation of the cathode and improved Li ion diffusion channels in the material, resulting in higher lithium utilization during initial cycling stage. The discharge capacity retention with respect to maximum discharge capacity was 98% after 100 charge-discharge cycles. It can be seen from 5th cycle charge-discharge that during discharge there are three plateau regions. The first plateau (4.88 V to 4.7 V) is due to Cr^{4+} to Cr^{3+} reduction, the second plateau (4.72 V to 4.5 V) is due to Ni^{4+} to Ni^{2+} reduction, and third plateau is due to Mn^{4+} to Mn^{3+} reduction. It is clear from the charge discharge curves that the plateau in higher voltage range (4.88 V to 4.7 V) has the maximum length (51 mAh/g), indicating that $LiMn_{1.5}Ni_{0.4}Cr_{0.1}O_4$ cathode material can be used in 5V range.

Figure 4 shows the capacity vs. the cycle numbers of up to 100 for $LiMn_{1.5}Ni_{0.4}Cr_{0.1}O_4$ cathode material indicating an average capacity of 125 mAh/g. Earlier we have reported [15] that Cr undoped sample i.e. $LiMn_{1.5}Ni_{0.5}O_4$ cathode material showed a plateau in 4.7 V range and Cr doped sample showed an extra plateau around 4.88 V, indicating that Cr doped cathode materials could be used for higher voltage applications. In-situ impedance spectroscopic measurements were performed at the end of each charge discharge cycles within the frequency rage of 10mHz to 1MHz. Figures 5 & 6 are the in-situ impedance spectra for cycles 1,10,30,50,75.100 etc. As we can see the impedance spectra are different for the charged cycle and the discharged cycle. It is important to note that the impedance seems to decrease for the first 20 cycles and starts to increase after that and continues to increase up to 100 cycles for both charge and discharge cycles. In the case of charging, two semicircles were noticed whereas only one semicircle with an extended tail is observed for the discharge cycle. The radius of the first semicircle is also reduced for the first 20 cycles for the charging and discharging cycles. After 100 cycles, the overall impedance is higher at the end of the discharge compared to end of charge. An impedance value of 60 ohms for the discharge cycle and 10 ohms for the charging cycle was observed. The high frequency semicircle is governed by the anode processes and the low frequency loop is governed by the cathode processes. At the end of charge, the electron transfer processes at the cathode become rate limiting, and hence a semicircle is observed for the cathode. The anode loop grows slightly with cycling and this is to be expected because of the changes at the anode/electrolyte interface; lithium metal at the anode does undergo a surface area increase and subsequent reaction with the electrolyte resulting in a larger polarization resistance. The low frequency region associated with the cathode now appears to be governed by diffusion processes as indicated by the slanting line instead of the semicircle in Figure 5. As the discharge proceeds to completion, the cathode becomes more and more lithiated and the diffusion coefficient of lithium ions becomes slow compared to the electron-transfer process at the cathode electrolyte interface. This indicates the diffusional impedance at the cathode electrolyte interface is high compared to the electron transfer impedance. The structural changes at the cathode could also result in decrease of diffusion coefficient for lithium during discharge.

CONCLUSION:

Sol-gel synthesized $LiMn_{1.5}Ni_{0.4}Cr_{0.1}O_4$ cathode materials have been characterized by XRD, XPS, cyclic voltammetry, charge-discharge cycles, and in-situ impedance spectroscopy. The XRD measurements confirmed the phase pure materials and the cyclic voltammetry supported the reversible reactions of the Li intercalation and also the redox potentials due to Cr, Mn, Ni. The in-situ impedance measurements allowed us to understand the possible differences in the impedance at the end of 100 charged and discharged cycles. The cathode electrolyte interface showed higher impedance for the lithium ions transfer during discharge and showed less impedance for the lithium ions that are transferring from the cathode to the anode. Further, measurements and post analysis are being performed in order to understand the nature of cathode electrolyte interface.

ACKNOWLEDGMENTS:

The authors gratefully acknowledge the research grants (NNX08AB12A) and NASA-URC (NNX08BA48A) received from NASA. The XRD measurements were carried out utilizing UPR Materials Characterization Center (MCC) facilities.

REFERENCES:

1. T. Ohzuku, A. Ueda, M. Nagayama, Y. Iwakoshi, and H. Komori: Comparative study of $LiCoO_2$, $LiNi_{1/2}Co_{1/2}O_2$ and $LiNiO_2$ for 4 volt secondary lithium cells. Electrochim. Acta 38, 1159 (1993).
2. J. M. Tarascon and D. Guyomard: New electrolyte compositions stable over the 0 to 5 V voltage range and compatible with the $Li_{1+x}Mn_2O_4$/carbon Li-ion cells. Solid State Ionics 69, 293 (1994).
3. M. S. Whittingham: Lithium batteries and cathode materials. Chem. Rev. 104, 4271 (2004).
4. J. Reed and G. Ceder: Role of electronic structure in the susceptibility of metastable transition-metal oxide structures to transformation. Chem. Rev. 104, 4513 (2004).
5. S. Choi and A. Manthiram: Synthesis and electrode properties of metastable $Li_2Mn_4O_{9-\delta}$ spinel oxides. J. Electrochem. Soc. 147, 1623 (2000).
6. J. Kim and A. Manthiram: Synthesis and lithium intercalation properties of nanocrystalline lithium iron oxides. J. Electrochem. Soc. 146, 4371 (1999).
7. J. Jiang and J.R. Dahn: Electrochemical and thermal studies of $Li[Ni_xLi_{(1/3-2x/3)}Mn_{(2/3-x/3)}]O_2$ ($x = 1/12$, $1/4$, $5/12$, and $1/2$). Electrochimica Acta 50, 4778 (2005).
8. S. H. Kang, J. B. Goodenough, and L. K. Rabenberg: Effect of ball-milling on 3-V capacity of lithium–manganese oxospinel cathodes. Chem. Mater. 13, 1758 (2001)
9. P. Arora, B.N. Popov, and R.E. White: Electrochemical Investigations of Cobalt-Doped $LiMn_2O_4$ as Cathode Material for Lithium-Ion Batteries. J. Electrochem. Soc. 145, 807 (1998).
10. J. M. Amarilla, J. L. Martın de Vidales, and R. M. Rojas: Electrochemical characteristics of cobalt-doped $LiCo_yMn_{2-y}O_4$ ($0 \leq y \leq 0.66$) spinels synthesized at low temperature from $Co_xMn_{3-x}O_4$ precursors. Solid State Ionics 127, 73 (2000).
11. M. Okada, Y.S. Lee, and M. Yoshio: Cycle characterizations of $LiM_xMn_{2-x}O_4$ (M=Co, Ni) materials for lithium secondary battery at wide voltage region. J. Power Sources 90, 196 (2000).

12. Y. P. Fu, Y. H. Su, S. H. Wu, and C. H. Lin: LiMn$_{2-y}$M$_y$O$_4$ (M = Cr, Co) cathode materials synthesized by the microwave-induced combustion for lithium ion batteries. J. Alloys and Compounds 426, 228 (2006).
13. X. Sun, X. Hu, Y. Shi, S. Li, and Y. Zhou: The study of novel multi-doped spinel Li$_{1.15}$Mn$_{1.96}$Co$_{0.03}$Gd$_{0.01}$O$_{4+\delta}$ as cathode material for Li-ion rechargeable batteries. Solid State Ionics 180, 377 (2009).
14. S. W. Oh, S. T. Myung, H. B. Kang, and Y. K. Sun: Effects of Co doping on Li[Ni$_{0.5}$Co$_x$Mn$_{1.5-x}$]O$_4$ spinel materials for 5 V lithium secondary batteries via Co-precipitation. J. Power Sources 189, 752 (2009).
15. R. Singhal, J. J. Saavedra-Aries, R. Katiyar, Y. Ishikawa, M. J. Vilkas, S. R. Das, M. S. Tomar, and R. S. Katiyar: Spinel LiMn$_{2-x}$Ni$_x$O$_4$ cathode materials for high energy density lithium ion rechargeable batteries. J. Renewable and Sustainable Energy 1, 023102 (2009).
16. R. K. Katiyar, R. Singhal, K. Asmar, R. Valentin, and R. S. Katiyar: High voltage spinel cathode materials for high energy density and high rate capability Li ion rechargeable batteries. J. Power Sources 194, 526 (2009).
17. E. Wu, POWDMULT, version 2.1, an interactive powder diffraction data interpretation and indexing program, School of Physical Science, Flinders University of South Australia, Bedford Park, South Australia 5042.
18. M. Aklaloucha, J. M. Amarillaa, R. M. Rojas, I. Saadoune, and J. M. Rojo: Chromium doping as a new approach to improve the cycling performance at high temperature of 5 V LiNi$_{0.5}$Mn$_{1.5}$O$_4$-based positive electrode. J. Power Sources 185, 501 (2008).
19. 19. R. K. Katiyar, R. Singhal, K. Asmar, R. Valentin, R. S. Katiyar, J. Power Sources **194** 526 (2009).

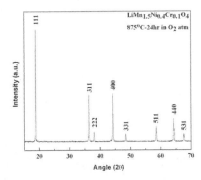

Figure 1. Powder diffraction patterns of LiMn$_{1.5}$Ni$_{0.4}$Cr$_{0.1}$O$_4$ cathode material, annealed at 875°C for 24 h in O$_2$ atmosphere.

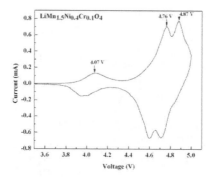

Figure 2. Cyclic Voltammogram of LiMn$_{1.5}$Ni$_{0.4}$Cr$_{0.1}$O$_4$ /LiPF$_6$+(EC+DMC)/Li coin cell in 3.0 V– 5.0 V range, at a voltage scan rate of 0.1mV/sec.

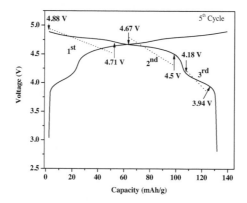

Figure 3. Room temperature charge discharge behavior of LiMn$_{1.5}$Ni$_{0.4}$Cr$_{0.1}$O$_4$ /LiPF$_6$+(EC+DMC)/Li coin cell in 3.0 V – 4.9 V range at a constant current rate of C/10.

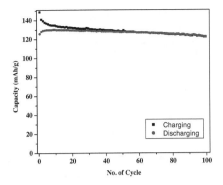

Figure 4. Cyclability of LiMn$_{1.5}$Ni$_{0.4}$Cr$_{0.1}$O$_4$/LiPF$_6$+(EC+DMC)/Li coin cell.

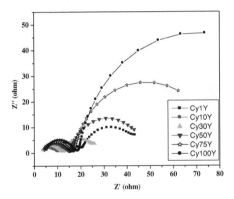

Figure 5. Impedance data at the end of charging.

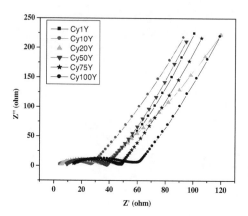

Figure 6. Impedance data at the end of dis-charging

$Cu_2(Zn_xSn_{2-x})(S_ySe_{1-y})_4$ MONOGRAIN MATERIALS FOR PHOTOVOLTAICS

E. Mellikov[1], M. Altosaar[1], J. Raudoja[1], K. Timmo[1], O. Volobujeva[1], M. Kauk[1], J. Krustok[1],
T. Varema[1], M. Grossberg[1], M. Danilson[1], K. Muska[1,2], K. Ernits[2], F. Lehner[2], D. Meissner[1,2]
[1] Dpt. Mat. Science, Tallinn University of Technology, Ehitajate tee 5
Tallinn 19086, Estonia, ennm@staff.ttu.ee, ph: +372620-2798, fax. -3367
[2] crystalsol OÜ, Akadeemia tee 15a, 12618 Tallinn, Estonia

ABSTRACT
In monograin solar cells powders replace wafers or thin films. This allows for cheaper and much
more efficient materials production and minimize materials loss. The separation of materials
formation from the module fabrication - allowing for all temperatures and purity precautions– is the
very important advantage of the monograin layer based technology. Large area module fabrication
proceeds without any high temperature steps in a continuous roll-to-roll process. No up-scaling
problems arise as is typical of thin-film technologies since a homogenous powder leads to
homogenous modules. The influence of different technological parameters to the materials
properties was studied and from $CZTS=Cu_2(Zn_xSn_{2-x})(S_ySe_{1-y})_4$) monograin powders solar cells
have produced with parameters V_{oc} up to 690 mV, fill factors up to 65 % and I_{sc} more than 20
mA/cm². It was founded that changing of the sulfur/selenium ratio allows a change of 1.04 and 1.72
eV the band gap of $Cu_2(Zn_xSn_{2-x})(S_ySe_{1-y})_4$.

INTRODUCTION
Climate change and insecurity of power supplies are among the greatest problems for
humankind at the moment and solutions are to be found within only a few decades. Large-scale
introduction of renewable power systems in combination with a strong increase of power efficiency
and saving will be the only way out. For example, in order to stabilize the world's climate a
continuous growth of PV by at least about 20 % per year will be needed during the coming decades
that will lead to PV installation of at least ten TW_p before 2050. To realize these plans new
abundant materials and cheap technologies are needed [1].
The technologies used today for manufacturing solar cells are the wafer and the thin film
technologies [2]. The wafer technology is based on the growth and use of very large monocrystals
or the casting of ultrapure silicon or A3B5 materials. This is a very expensive process not only in
terms of money but also in terms of energy input. Later these large single or polycrystals with a
high degree of chemical purity and physical perfection have to be sawn into wafers which are
subjected to sophisticated methods of oxidation, diffusion and chemical treatment in order to form
localized regions of different types of conductivity needed for highly efficient solar cell structures.
The idea of making crystalline solar cells by growing very large crystals and then cutting them into
thin wafers is in fact not the most intelligent way to obtain perfect materials and structures. Thin
film technologies of solar cells production can be applied, but the technical parameters of thin film
solar cells are as a rule much worse than those of crystalline solar cells. On the other hand, thin film
technologies are, in general, much cheaper than monocrystalline materials based technologies, both
with respect to financial expenses and energy costs.
Although silicon as semiconductor material is still dominating even in the thin film solar cell
production, photovoltaic cells based on more complex compound semiconductor materials are
becoming increasingly important. Unconventional solar cell materials that are abundant but much
cheaper to produce than silicon could substantially reduce the overall cost of solar photovoltaics.
Currently the most widely used compound solar cell materials, cadmium telluride (CdTe) and
copper indium diselenide, copper indium gallium diselenide or the respective sulfides (CIGS),
contain resource-limited elements (Te and In) that are already today about ten times more expensive
than other metals and do not allow for the provision photovoltaic energy in amounts needed for the
very large scale applications in future [3]. One of the currently available most promising new
unconventional materials is CZTS, here meaning all copper zinc tin sulfo-selenides including e.g.

Cu$_2$ZnSnS$_4$, Cu$_2$ZnSnSe$_4$ and Cu$_2$ZnSn(S$_x$Se$_{1-x}$)$_4$ with their abundant and nontoxic constituents. With a direct band gap that depends on the ratio of sulphur to selenium it can be tuned to the optimum for solar energy conversion even for tandem structures. These materials have also a high optical absorption coefficient (> 10^4 cm^{-1}) [4]. This makes these materials ideal as adsorber layers for low cost high efficiency terrestrial photovoltaic devices [5].

A number of reports have been published on the properties of CZTS materials prepared as thin films by various vacuum as well as low-cost chemical preparation techniques. The most important are sputtering and vacuum evaporation of metals or constituent metal binaries followed by chalcogenization [6, 7, 8], electrodeposition [9, 10], chemical spray pyrolysis [11], and soft chemical technologies [12, 13]. The efficiency of developed polycrystalline films solar cells is lower than theoretically possible. However, in a recent paper Todorov, Reuter and Mitzi achieved a 9.6 % solar efficiency confirmed by NREL [13]. As a result the optimization of growth conditions are regularly discussed by different authors [14, 15]. In addition to the above, the manufacturing of thin films with a large surface area, suitable stoichiometric composition and good reproducibility is a technically complicated task and is not up to end solved even on the laboratory level.

In addition to the monocrystalline and the thin film approach, the third alternative to prepare solar cell structures is the use of powder materials [16]. Powder technologies are the cheapest technologies for materials production. At the same time, although several companies and research institutions have made considerable efforts, powder methods for solar cell applications have not found widespread use yet. It has been shown that isothermal recrystallization of initial powders in different molten fluxes appears to be a relatively simple, inexpensive and a convenient method to produce CIS and CZTS powders with an improved crystal structure, that are perquisite for solar cell use [16-19]. Powders consist of small single crystalline grains. In monograin solar cells powders replace monocrystalline wafers or thin films. This allows for cheaper and much more efficient materials production that minimizes materials loss. The separation of materials formation from the module fabrication - allowing for all temperatures and purity precautions– is an additional very important advantage. Large area module fabrication could be done without any high temperature steps in a continuous roll-to-roll process. Homogenous powders lead to homogenous modules and do not result in up-scaling problems.

EXPERIMENTAL

Monograin powder materials were synthesized from metal binaries (CuSe(S), ZnSe(S), SnSe(S) and elemental selenium or sulphur, respectively) in a molten flux using an isothermal recrystallization process. The precursors were thermally annealed in evacuated quartz ampoules. The evolution of crystal shape and morphology of the monograin powders was analyzed by electron imaging using a high-resolution scanning electron microscope (SEM) Zeiss ULTRA55 with the compositional contrast detector EbS. The chemical composition and the distribution of components in powder crystals were determined using an energy dispersive X-ray analysis (EDX) system). XRD patterns were recorded by using a Bruker AXS D5005 diffractometer using monochromatic Cu Ka-radiation.

Monograin layer (MGL) solar cells (graphite/CZTSSe/CdS/ZnO) were made from grains with diameters of 56–63 µm [19-21]. Powder crystals were covered with chemically deposited CdS buffer layers. For the MGL formation a monolayer of CZTS powder crystals was glued together by a thin layer of epoxy. After polymerization of this epoxy, i-ZnO and ZnO:Al were deposited by RF-sputtering onto the open surface of the layer. Solar cell structures were completed by vacuum evaporation of 1–2 µm thick In grid-contacts onto the ZnO window layer. Subsequently, the layer was glued onto glass substrates. The opening of the back contact areas of the crystals that were originally inside the epoxy was done by etching the epoxy in H$_2$SO$_4$ and by abrasive treatment. Graphite paste was used for the back contacts. The solar cell efficiency was measured in an Oriel class A solar simulator. I–V curves were measured using a Keithley 2400 source meter.

RESULTS AND DISCUSSION

The structural, morphological, electrical and optical properties of $Cu_2ZnSn(S,Se)_4$ materials depend strongly on the composition and on the additional thermal and chemical treatments. Fig. 1 shows the morphologies of CZTS powder crystals grown in the KI flux. The crystals have tetragonal shape with rounded grain edges. A change in the chemical nature of the flux leads to variations in crystals shape.

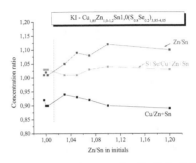

Fig 1. SEM micrographs of $Cu_2ZnSn(S,Se)_4$ monograin powder crystals grown in KI flux.

Fig. 2. Dependence of $Cu_2ZnSn(S,Se)_4$ monograin powder composition on precursor composition.

Fig. 2 shows the results of the compositional analyses of as-grown $Cu_2ZnSn(S,Se)_4$ materials as a function of the initial precursor composition. With the Zn/Sn compositions ratio increasing in the precursors the obtained powders contain an increased ratio of Zn to Sn but a decreased amount of Cu. At the same time the content of chalcogens (S+Se) in the materials does not depend on the Zn/Sn ratio. At stoichiometric ratios of 25%, 12.5%, 12.5% and 50% for Cu, Zn, Sn, and S+Se, in $Cu_2ZnSn(S,Se)_4$ respectively, all studied materials were Zn-rich and Cu-poor and have an stoichiometric content of (S+Se). An overstoichiometric chalcogen amount in the starting materials leads to an increase of the relative contents of Zn and Se in the as-grown powders (Fig. 3).

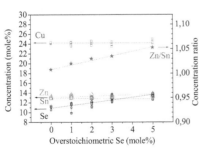

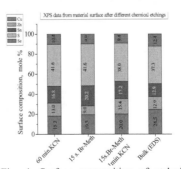

Fig. 3. The influence of the chalcogen (Se) content on the composition of $Cu_2ZnSn(S,Se)_4$ powders

Fig. 4. Surface composition of etched in different etchants $Cu_2ZnSn(S,Se)_4$ powders

By comparing as-grown powders crystal surface compositions with the bulk compositions it was found that the surface of as-grown $Cu_2ZnSn(S,Se)_4$ crystals was Sn-rich while the bulk of crystals was Zn-rich. This was the result of the contamination of the powder crystals by components of Cu$_2$ZnSn(S,Se)$_4$ dissolved in the flux. In order to remove these contaminations (other phases) from the surfaces and to improve the developed Cu$_2$ZnSn(S,Se)$_4$ solar cells performance, the chemical treatments in different etchants (HCl, KCN, Br-MeOH and NH$_4$OH) were performed. It was found that etching with KCN or HCl increases the ratio of Zn to Sn on the surface. At the same time HCl etching leads to a slightly chalcogen-poor surface. Treatment with KCN dissolves mainly Cu, Sn and chalcogen, and ammonia solution remove selectively Cu and chalcogen in an approximate ratio of 1:2. Fig. 4 shows the results of this influence of different etchants on the surface composition of powder crystals.

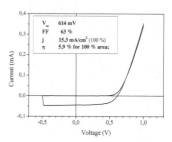

Fig.5. EBIC picture of MGL solar cell from material composed of Cu$_{1.85}$Zn$_{1.0}$Sn$_{0.95}$ (S$_{0.8}$Se$_{0.2}$)$_{3.95}$

Fig.6. I-V dependence of the Cu$_2$ZnSn(S,Se)$_4$ monograin layer cell (certified in the Calibration Lab of Fraunhofer ISE, Freiburg, 2009)

Electron beam induced current (EBIC) investigations of MGL solar cells indicate that all crystals in MGL operate with nearly the same efficiency. Fig. 5 represents EBIC picture for the MGL solar cell from the material composed of Cu$_{1.85}$Zn$_{1.0}$Sn$_{0.95}$ (S$_{0.8}$Se$_{0.2}$)$_{3.95}$. Current–voltage dependences of Cu$_2$ZnSn(S,Se)$_4$ MGL solar cells show V$_{oc}$ values of over 690 mV, fill factors of up to 65 %, and short circuit currents of up to 20 mA/cm^2. Efficiencies of the active area of these solar cells of up to 6.4 % were determined, total area efficiencies of 5.9 % were certified by calibration lab measurements (fig. 6). Varying the S/Se ratio in the material allows one to control the band gap of Cu$_2$ZnSn(SSe)$_4$ materials and with this the quantum efficiency of the solar cell made on their base.

CONCLUSIONS

It is shown that the Cu$_2$ZnSn(S,Se)$_4$ monograin technology allows solar cell production with conversion efficiencies of currently up to 6 %. The solar cells developed here have V$_{oc}$ up to 690 mV, fill factors of up to 65 % and I$_{sc}$ of about 20 mA/cm^2. Changing the sulfur/selenium ratio in materials allows a change the change of the band gap of Cu$_2$ZnSn(S,Se)$_4$ between 1.04 and 1.72 eV.

ACKNOWLEDGEMENTS

This work was supported by the target financing by HTM (Estonia) project SF0140099s08 and by the Estonian Science Foundation grants G-8147, G-8282 and G-7678.

REFERENCES
1. A vision for photovoltaic technology for 2030 and beyond, European Commission, 2004
2. A. Luque, S. Hegedus, Handbook of Photovoltaic Science and Engineering (Wiley), 2006
3. A. Feltrin, A. Freindlich, Material considerations for terawatt level deployment of photovoltaics, *Renewable Energy*, **33**, 180 – 185 (2008).
4. K. Ito, J. Nakazawa, Electrical and optical properties of stannite-type quaternary semiconductor thin films, *Jpn. Appl. Phys*, **27**, 2094 – 2097 (1988).

5. K. Jimbo, R. Kimura, T. Kamimura, S. Yamada, W. S. Maw, H. Araki, K. Oishi, H. Katagiri, Cu$_2$ZnSnS$_4$-type thin film solar cells using abundant materials. *TSF*, **515**, 5997 – 5999 (2007).

6. O. Volobujeva, J. Raudoja, E. Mellikov, M. Grossberg, S. Bereznev, R. Traksmaa, Cu$_2$ZnSnSe$_4$ films by selenization of Sn–Zn–Cu sequential films, *J. Phys. Chem. Solids* **70**, 567 – 570 (2009).

7. O. Volobujeva, E. Mellikov, J. Raudoja, M. Grossberg, S. Bereznev, M. Altosaar, R. Traksmaa, SEM analysis and selenization of Cu–Zn-Sn sequential films produced by evaporation of metals, in: *Proceedings: Conference on Optoelectronic and Microelectronic Materials and Devices, IEEE Publishing* , 257 – 260 (2009).

8. P. M. P. Salome, F. A. Fernandes, A. F. de Cunha, Morphological and Structural characterization of Cu$_2$ZnSnSe$_4$ thin films grown by the selenization of elemental precursor layers, *TSF*, **517**, 2531 – 2534 (2008).

9. J. J. Scragg, P. J. Dale, L. M. Peter, G. Zoppi, I. Forbes, New routes to sustainable photovoltaics: evaluation of Cu$_2$ZnSnS$_4$ as alternative absorber material, *Phys. Stat. Sol. (b)* **245**, 1772 – 1778 (2008).

10. A. Ennaoui, M. Lux-Steiner, A. Weber, D. Abou-Ras, I. Kötschau, H.-W. Schock, R. Schurr, A. Hölzing, S. Jost, R. Hock, T. Voß, J. Schulze, A. Kirbs, Cu$_2$ZnSnS$_4$ thin film solar cells from electroplated precursors: novel low-cost perspective, *TSF*, **517**, 2511 – 2514 (2009).

11. N. Kamoun, H. Bouzouita, B. Rezig, Fabrication and characterization of Cu$_2$ZnSnS$_4$ thin films deposited by spray pyrolysis technique, *TSF*, **515**, 5949 – 5952 (2007).

12. T. Todorov, M. Kita, J. Carda, P. Escribano, Cu$_2$ZnSnS$_4$ films deposited by soft-chemistry method, TSF, **517**, 2541-2544 (2009).

13. T. K. Todorov, K. B. Reuter, D. B. Mitzi, High-Efficiency Solar Cell with Earth-Abundant Liquid-Processed Absorber, *Adv. Mater.* **22**, 1 – 4 (2010).

14. H. Katagiri, K. Jimbo, W. S. Maw, K. Oishi, M. Yamazaki, H. Araki, A. Takeuchi, Development of CZTS-based thin film solar cells, *TSF*, **517**, 2455 – 2460 (2009).

15. A. Goetzberger, C. Hebling, H. W. Schock, Photovoltaic materials, history, status and outlook, *Mat. Sci. Engin.* Rev. 40 (2003), 1 - 46.

16. M. Altosaar, A. Jagomägi, M. Kauk, M. Krunks, J. Krustok, E. Mellikov, J. Raudoja, T. Varema, Monograin layer solar cells, *TSF*, **431–432**, 466 – 469 (2003).

17. M. Altosaar, E. Mellikov, CuInSe$_2$, Monograin Growth in CuSe-Se liquid phase, *Jpn. J. Appl. Phys.*, **39**, 65 – 66 (2000).

18. Mellikov, E., Altosaar, M., Krunks, M., Krustok, J., Varema, T., Volobujeva, O., a.o., Research in solar cell technologies at Tallinn University of Technology. *TSF* **516**,7125–713 (2008)

19. D. Meissner, E. Mellikov, M. Altosaar, Monograin Powder and Monograin Membrane Production, *European Patent* 1097262, Aug. 21, 2002.

20. M. Altosaar, J. Raudoja, K. Timmo, M. Danilson, M. Grossberg, J. Krustok, E. Mellikov, Cu$_2$Zn$_{1-x}$Cd$_x$Sn(Se$_{1-y}$S$_y$)$_4$ solid solutions as absorber materials for solar cells, *Phys. Status Solidi A* **205**, 167 – 170 (2008).

21. E. Mellikov, D. Meissner, T. Varema, M. Altosaar, M. Kauk, O. Volobujeva, J. Raudoja, K. Timmo, M. Danilson, Monograin materials for solar cells, *Solar Energy Mat. Solar Cells* **93**, 65 – 68 (2008).

DETERMINATION OF THE DIFFUSION COEFFICIENT OF LITHIUM IONS IN GRAPHITE COATED WITH POLYMER-DERIVED SICN CERAMIC

Andrzej P. Nowak[a,b], Magdalena Graczyk-Zając[a], Ralf Riedel[a]

[a] Technische Universität Darmstadt, Institute of Materials Science, Petersenstr. 23, 64287 Darmstadt, Germany

[b] Gdansk University of Technology, Chemical Faculty, Department of Chemical Technology, Narutowicza 11/12, 80-233 Gdansk, Poland

*e-mail: riedel@materials.tu-darmstadt.de (R. Riedel), Tel.: +49 6151 166347; fax: +49 6151 166346.

ABSTRACT

The diffusion coefficients of lithium ions (D_{Li+}) in graphite/SiCN ceramic composites, pyrolyzed at temperatures between 850°C and 1300°C, were determined by cyclic voltammetry (CV), galvanostatic intermittent titration technique (GITT) and electrochemical impedance spectroscopy (EIS). The diffusion coefficient values calculated by these techniques are of the same order of magnitude. Depending on the potential the values vary from 10^{-15} to 10^{-08} cm^2/s. Electrochemical measurements show that D_{Li+} depends on the material composition, pyrolysis temperature and charge/discharge state. For composites of graphite to ceramic ratio 50:50 the diffusion coefficient is constant and not temperature dependent. The composite materials of 75:25 ratio pyrolyzed at different temperatures exhibit an increase in the D_{Li+} with the highest diffusion coefficient of lithium ions of the order
$\sim 10^{-10}$ cm^2/s for material pyrolyzed at 1300 °C. Cyclic voltammetry curves demonstrate that lithium ions intercalate into graphite and free carbon segregated from the SiCN phase. The presence of SiCN improves diffusion transport of lithium ions into the graphite and the derived free carbon. These findings strongly indicate that the SiCN ceramic network creates a pathway for lithium ion transfer.

INTRODUCTION

Lithium ion batteries are widely used in small portable electronic devices like laptops, cameras and mobile phones. Due to the fact that lithium is the most electronegative element and one of the lightest elements, it can be used as anode material in batteries. However, metallic lithium was replaced by graphite owing to high corrosivity, forming of undesirable dendrites and safety problems. The problem with graphite is that its theoretical capacity amounts to only 372 mAh/g and graphite based anodes suffer from structural degradation during the charging and discharging process [1]. To overcome these problems many various materials have been proposed such as tin-based oxides, lithium alloys, transition metal oxides [2 and ref therein]. Thus, there is still a need for active anode materials with active sites for lithium ions.

Recently, there has been a growing interest in SiCN polymer-derived ceramics (PDCs) as anode materials for lithium ion batteries [3,4]. We showed that polyvinylsilazane (PVS)-derived SiCN ceramics exhibit low discharge capacity (about 44 mAh/g) while the reversible capacity of graphite/polymer-derived SiCN composite is 10 times higher [3]. Li et al. studied SiCN materials derived from polysilylethylenediamine (PSEDA). The authors showed that the capacity of such anode material is between 170 and 230 mAh/g after 30 cycles [4].

Polymer-derived SiCN materials are known to have an amorphous network consisting of Si, C, and N atoms and finely dispersed nanocarbon clusters in the SiCN matrix [5,6]. The free carbon in the SiCN materials is an active site for lithium intercalation while the amorphous structure of SiCN provides a pathway for lithium ion transfer [4,7].

However, the transport kinetics of lithium in SiCN polymer-derived ceramics is still unclear. The diffusion coefficient of lithium ions in the electrodes is an important parameter for determining the charge and discharge rate of lithium ion batteries. The common electrochemical methods applied for determination of D_{Li+} are cyclic voltammetry (CV) [8-10], electrochemical impedance spectroscopy (EIS) [11-13] and galvanostatic intermittent titration technique (GITT) [14-16]. The CV can provide only apparent values of D, related to the potential range around the voltammetric peak. Highly resolved data on diffusion coefficients are obtained from GITT and EIS. It is known that the diffusion coefficient of lithium in carbonaceous materials depends on the intercalation degree of lithium in the lithium–carbon compounds. The higher the intercalation degree, the lower the diffusion coefficient. Moreover, D_{Li+} values in the carbon based anodes obtained by different authors are in the range of $10^{-07} - 10^{-14}$ cm^2/s [17].

In the present work the electrochemical properties of graphite/SiCN composite materials were studied. The samples were derived from a powder mixture of graphite and poly(organosilazane) forming a graphite/SiCN composite after pyrolysis. The diffusion coefficients of lithium in graphite and graphite/SiCN materials were measured and compared by the use of cyclic voltammetry, electrochemical impedance spectroscopy and galvanostatic intermittent titration technique methods.

EXPERIMENTAL

The graphite/SiCN composites were produced by pyrolysis of a commercially available polysilazane (HTT 1800 KiON®, Clariant, Sulzbach, Germany) with commercially available graphite powder (SLP50 – Timrex®). The materials were mixed at a weight ratio graphite to SiCN 50:50 and 75:25, respectively. After mixing the material was partially crosslinked at 200 °C for 2 h in inert argon atmosphere using the Schlenk technique and finally pyrolized at different temperatures (850, 950, 1050, 1200 and 1300 °C) for 1 h at a heating rate of 100 °C/min.

The compositions of the graphite/SiCN composites were determined by means of quantitative elemental analysis using a carbon analyzer (Leco C-200, Leco Instrumente GmbH, Mönchengladbach, Germany) for the determination of the carbon content and an N/O analyzer (Leco TC-436, Leco Instrumente GmbH, Mönchengladbach, Germany) for the nitrogen and oxygen traces. The amount of Si was calculated as the difference between the total amount of elements minus the sum of C/N/O amount. Prior to electrode preparation each powder was grinded for 15 minutes. Subsequently, the material was mixed with a 10 % polyvinylidene fluoride (PVdF, SOLEF) solution in N-Methyl-2-pyrrolidone (NMP, BASF). The ratio of active material/PVdF was 9:1 and was constant for all the samples. NMP was added in order to form homogenous slurry (about 0.8 g of solvent for 1g of solution). The slurry was spread on the rough side of copper foil (10μm, Copper SE-Cu58 (C103), Schlenk Metallfolien GmbH & Co KG) using the hand blade coating technique and dried at 80 °C for 24 h under vacuum. The active material loading was always between 2 and 4 mg/cm^2. After drying the circles (electrodes) of 10 mm in diameter were cut. The weight of the electrodes was measured after drying them under vacuum at 80 °C for 24 h in a Buchi oven and then transferring them directly to the glove box (MBraun Glove Box Systems, H$_2$O, O$_2$ < 1 ppm) without contact with air.

All the electrochemical measurements were performed in two-electrode Swagelok® type cells, with active material as working electrodes. The lithium foil (99.9% purity, 0.75 mm thick, Alfa Aesar) was used as a counter and reference electrode. A high purity solution of 1 M LiPF6 in ethylene carbonate (EC) and dimethyl carbonate 1:1 (LP30, Merck KGaA) was used as an electrolyte. Porous polypropylene membrane (Celgard® 2400. Celgard) was used as a separator.

The cyclic voltammetry (CV) and galvanostatic intermittent titration technique (GITT) were measured on the VMP®3 multichannel instrument by Bio-Logic. The GITT was employed at a pulse of 37 mA/g for 10 min and with a 30 min interruption between each pulse. Electrochemical impedance measurements at different charge-discharge states were performed with a Solartron 1287 frequency response analyzer combined with a Solartron 1286 electrochemical interface in the frequency range from 100 kHz to 10 mHz and the amplitude 5 mV. To obtain the equilibrium state the constant potential was applied for 30 min before the measurements.

RESULTS AND DISCUSSION

Cyclic voltammetry was performed to examine the behavior for the lithium ion intercalation/extraction process as well as for determination of the diffusion coefficient of lithium ions.

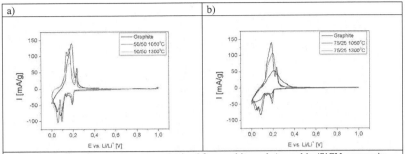

Fig. 1. Cyclic voltammetry curves obtained for graphite and a) graphite/SiCN composite (50:50), b) graphite/SiCN composite (75:25) electrodes pyrolyzed at different temperatures.

Figure 1 presents cyclic voltammetry curves obtained for graphite and composite materials of different ratios, prepared at different temperatures. It is observed that the shape of the curves is similar for carbonaceous materials. Thus, one may conclude that lithium ions intercalate between grapheme sheets in graphite and in free carbon derived from the SiCN phase. The redox couple activities are seen at the same potential while the peak current maxima are related to change in the stoichiometry. According to results obtained by Ohzuku et. al these maxima in the case of graphite reveal redox transfer reactions [18]:

$$LiC_{72} + Li = 2\ LiC_{36}\ \text{(high redox potential)} \tag{1}$$
$$3\ LiC_{36} + Li = 4\ LiC_{27}\ \text{(middle redox potential)} \tag{2a}$$
$$2\ LiC_{27} + Li = 3\ LiC_{18}\ \text{(middle redox potential)} \tag{2b}$$
$$2\ LiC_{18} + Li = 3\ LiC_{12}\ \text{(middle redox potential)} \tag{2c}$$
$$LiC_{12} + Li = 2\ LiC_{6}\ \text{(low redox potential)} \tag{3}$$

It is known that the electrochemical intercalation process is controlled by changing the potential scan rate of the electrode. Depending on the scan rate the intercalation/extraction mechanism may be controlled by electron transfer (slow scan rate) or diffusion transport (high scan rate). If the redox process is controlled by diffusion, then the diffusion coefficient can be calculated from the modified Randles-Sevcik equation:

$$i_p = 2.69x10^5 \cdot n^{3/2} \cdot S \cdot D^{1/2} \cdot c^* \cdot v^{1/2} \tag{4}$$

where i_p is the peak current, n is the number of electrons per species reaction, S is the apparent surface area of the electrode (geometric area), D is the diffusion coefficient of Li^+ in the solid state and c^* is the change in Li concentration in the material due to the specific step to which the peak is related. The Li concentration in graphite/SiCN composites was calculated using elemental analysis data of the C, N, O contents combined with electrochemical measurements according to equation (5):

$$x_{Li} = \frac{3,6xQ_{rev}xM_{mol}}{F} \tag{5}$$

where x_{Li} = molar ratio of Li atoms in the host, Q_{rev} = reversible capacity in mAh/g (galvanostatic charge/discharge with i = 18 mA/g), M_{mol} = molecular mass of the host (100 g/mol), F = Faraday constant (96485 C/mol).

Fig. 2. shows cyclic voltammetry curves of graphite/SiCN composite (50:50 ratio; pyrolyzed at 1050 °C) electrode measured at potential scan rates from 10 to 100 μV/s.

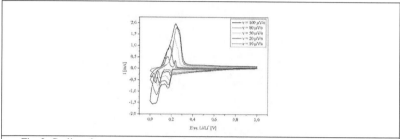

Fig. 2. Cyclic voltammetry curves obtained for graphite/SiCN composite pyrolyzed at 1050 °C (50:50).

Relationships between the current peak values and $v_{1/2}$ or v are depicted in Fig. 3.

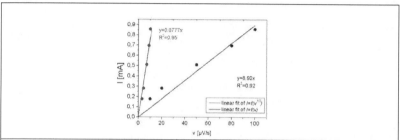

Fig. 3. The relation between peak current height and sweep rate (—) or square root of sweep rate (—) for the intercalation process.

Comparison between the correlation coefficient (R^2) for $I = f(v)$ and $I = f(v^{1/2})$ demonstrates that more linear dependence is observed for latter relation. It proves that semi-infinite diffusion of lithium ions inside the electrode material is a limiting step for the intercalation mechanism of Li ions. On the basis of eq. (4) and the slope of $I = f(v^{1/2})$ diffusion coefficients were calculated for the intercalation and extraction process. The results obtained for graphite and graphite/SiCN composites are summarized in Table I.

Table I. Chemical diffusion coefficient values for Li$^+$ D_{Li+} determined for graphite and graphite/SiCN composites.

Graphite/SiCN composite	D_{int} [cm^2/s]	D_{extr} [cm^2/s]
Graphite	7.89x10^{-12}	3.69x10^{-11}
50:50 at 850 °C	1.39x10^{-12}	1.48x10^{-12}
50:50 at 950 °C	7.58x10^{-12}	9.70x10^{-12}
50:50 at 1050 °C	6.02x10^{-12}	1.60x10^{-11}
50:50 at 1200 °C	7.55x10^{-12}	9.12x10^{-12}
50:50 at 1300 °C	5.76x10^{-12}	6.51x10^{-12}
75:25 at 850 °C	6.73x10^{-12}	1.54x10^{-11}
75:25 at 950 °C	6.55x10^{-12}	8.36x10^{-12}
75:25 at 1050 °C	5.43x10^{-11}	5.40x10^{-11}
75:25 at 1200 °C	4.17x10^{-10}	1.99x10^{-09}
75:25 at 1300 °C	1.95x10^{-10}	6.62x10^{-10}

The apparent diffusion coefficient of lithium ions in graphite is calculated to be $\sim 10^{-12}$ and $\sim 10^{-11}$ cm^2/s for the intercalation and extraction process, respectively. It is in agreement with data obtained by NuLi et al. for highly oriented pyrolytic graphite [17]. The graphite/SiCN composite (50:50) exhibits D_{Li+} values in the same order of magnitude as for graphite. Table 1 shows that the diffusion coefficient of lithium ions in graphite/SiCN composite (50:50) is constant and not temperature dependent. It seems that for the graphite/SiCN ratio 50:50 the annealing process does not influence the diffusion process. In the case of graphite/SiCN composites (75:25) it is shown that constant values and not temperature dependent D_{Li+} values were calculated while the material was pyrolyzed at 850 and 950 °C. These values are in the order of $\sim 10^{-12}$ cm^2/s. However, a rise in the pyrolysis temperature causes the diffusion coefficient to increase. The highest D_{Li+} is obtained for material pyrolyzed at 1200 and 1300 °C and reached the value of 10^{-10} cm^2/s for lithium intercalation. The intercalation process is noticeably slower than that of the extraction process for all materials. In comparison with removing from the sites during the deintercalation in anode materials this finding is due to fact that lithium ions consume more time to find active sites during intercalation process [19,20]. The electrochemical impedance spectra obtained for graphite/SiCN composites (50:50) at different charge states are shown in Fig. 4.

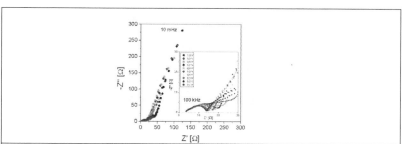

Fig. 4. Complex plane impedance plot for graphite/SiCN composite (50:50 pyrolyzed at 1050 °C) at different charge states.

As illustrated in Fig. 4, the plots consist of two depressed semicircles in the high and intermediate frequency region, respectively. These semicircles correspond to charge transfer reactions at the electrolyte/electrode material interface. The semicircle at a high frequency range ascribes a solid passivating layer formation and the semicircle in the medium frequency range is ascribed to lithium intercalation reaction [21]. There are parallel straight lines in the low frequency region with a slope ~ 45°. These lines suggest that the lithium intercalation process is controlled by diffusion. The diffusion coefficient can be calculated from the equation given by Ho et al. [22]:

$$D = \frac{1}{2}\left[\frac{1}{\sigma_w}\frac{V_M}{SFn}\left(\frac{dE}{d\delta}\right)\right]^2 \qquad (6)$$

where V_M is the mole volume of material (5.33 cm^3/mol], σW – Warburg coefficient and $dE/d\delta$ – the slope of galvanostatic charge–discharge curves, the notation of other symbols is given in Eq. (4). The Warburg coefficient σ_w can be estimated from the slope of $Z''=f(\omega^{-1/2})$ line (see Fig. 5.) in the frequency range responsible for the diffusion process using the equation

$$Z'' = -\sigma_w \times \omega^{-1/2} \qquad (7)$$

where Z'' and ω are the imaginary part of impedance and the angle frequency, respectively.

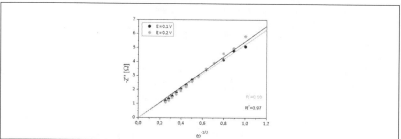

Fig. 5. Warburg impedance, $-Z''$ as a function of $\omega^{1/2}$ for graphite/SiCN composites at the applied potential extracted from Fig. 4.

The diffusion coefficients of lithium ions in graphite and graphite/SiCN composite as a function of the applied potential are shown in Fig. 6.

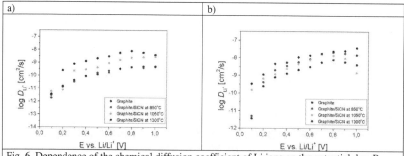

Fig. 6. Dependence of the chemical diffusion coefficient of Li ions on the potential, log D versus E for a) graphite/SiCN 50:50 ratio, b) graphite/SiCN 75:25 ratio obtained by EIS

Fig. 6 shows that the diffusion coefficient of lithium in electrode materials decreased with a lower applied potential. This behavior is due to the lithium intercalation process and narrow passageway for lithium diffusion in carbon based electrode material [17]. As seen in Fig. 1. lithium intercalation occurs at the potential below 0.25 V. This is the reason why D_{Li+} values are about 3–4 orders of magnitude higher ($\sim 10^{-08}$ cm^2/s) for the potential range from 0.2 V to 1.0 V. The diffusion coefficient measured in this range is related to the stage phases and stoichiometry of the material according to reactions (1) – (3). The region at the potential above 0.2 V is ascribed as dilute stage phase and does not have staged structure [23]. It influences on order and disorder of the host material with the potential. The higher potential, the smaller the order and the diffusivity of lithium ions increases with the increase of disorder [24].

The diffusion of lithium ions in the material takes place below 0.2 V. The D_{Li+} values obtained for graphite/SiCN (50:50) are constant, lower than the values for graphite and are not temperature dependent. The diffusion coefficient of lithium ions at E = 0.1 V is estimated to be $\sim 10^{-12}$ cm^2/s. This confirms the results obtained by CV (see Table I). In the case of graphite/SiCN (75:25) the D_{Li+} at 0.1 V changes with the annealing temperature. It is observed that an increase in pyrolysis temperature produces a higher value of the diffusion coefficient of lithium ($\sim 10^{-10}$ cm^2/s). This is again in good agreement with the results from the CV method. However, the most reliable technique to determine the chemical diffusion coefficient of lithium with varying intercalation levels of the ions is the galvanostatic intermittent titration technique (GITT) [14, 25, 26].

Fig. 7. presents the galvanostatic intermittent charge and discharge curves of graphite/SiCN composites (50:50 pyrolyzed at 1050 °C).

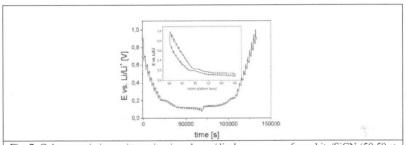

Fig. 7. Galvanostatic intermittent titration charge/discharge curve of graphite/SiCN (50:50 at 1050 °C) as a function of time and as a function of the intercalation level (inset).

Assuming that the lithium transport in the electrode is controlled by the diffusion process, in accordance with Fick's second law, the diffusion coefficient of lithium ions can be calculated from the equation given by Weppner and Huggins [14]:

$$D = \frac{4}{\pi}\left(\frac{V_M}{SFn}\right)^2\left[I_0\left(\frac{dE}{d\delta}\right)\Big/\left(\frac{dE}{d\sqrt{\tau}}\right)\right]^2 \qquad (8)$$

where I_0 – current pulse, $dE/d\sqrt{\tau}$ – slope of the voltage change vs. square root of the time during the current pulse. The notation of the other symbols is given in Eqs. (4) and (6).

Fig. 8a-d. show the diffusion coefficients of lithium ions in graphite/SiCN calculated from GITT studies.

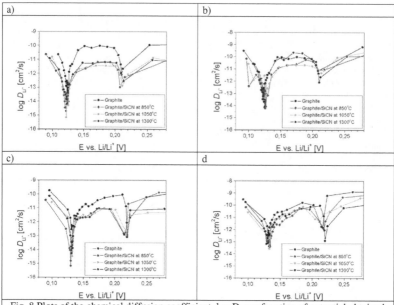

Fig. 8 Plots of the chemical diffusion coefficient, log D as a function of potential obtained for a) graphite/SiCN composite 50:50 (charge curve) b) graphite/SiCN composite 75:25 (charge curve), c) graphite/SiCN composite 50:50 (discharge curve) d) graphite/SiCN composite 75:25 (discharge curve) analyzed by GITT.

Fig. 8 shows that log D vs. E function has its minima at E values in which the CV curve (see Fig. 1.) has its maxima. These peaks relate to the charge transfer between different Li-C intercalation compounds. This phenomenon is in agreement with results obtained for graphite electrodes by Levi and Aurbach [27].

The chemical diffusion coefficients given in Fig. 8a-d show some difference between charging and discharging. The D_{Li+} values for the intercalation process are lower than those for extraction. This tendency is in line with the results found from the EIS technique. Moreover, the graphite/SiCN composite (50:50) exhibits D_{Li+} values which are not temperature dependent and lower than the values found for graphite. In the case of graphite/SiCN composite (75:25) D_{Li+} values vary with the annealing temperature. The diffusion coefficient increases with rising pyrolysis temperature. The material obtained at 1300 °C exhibits a higher D_{Li+} of ~ 10^{-10} cm^2/s. It is noteworthy that the results given by GITT are in agreement with the results obtained by the others techniques, CV and EIS.

CONCLUSIONS

We investigated the influence of the material composition and pyrolysis temperature on the diffusion coefficients of lithium ions in graphite/SiCN composites by CV, EIS and GITT methods. The D_{Li+} values obtained by these methods are in the same order. The diffusion coefficients of Li calculated by CV is ~ 10^{-12} cm^2/s and 10^{-10} cm^2/s for graphite/SiCN (50:50) and graphite/SiCN (75:25) pyrolyzed at 1300 °C, respectively. This suggests that the transport of lithium ions through the composite material is dependent on the graphite/SiCN ratio and on the annealing temperature. The same order of magnitude of D_{Li+} was obtained by EIS and GITT. The GITT curves exhibited "W" type shape which confirmed changes in the Li-C stoichiometry during the charging and discharging process. The presence of the polymer-derived SiCN ceramic improved the diffusion transport into the composite anode. The CV curves showed that lithium intercalates into graphite and free carbon obtained from the SiCN phase. It was confirmed that the free carbon in the SiCN materials acted as an active site for lithium intercalation while the amorphous structures of SiCN provided a pathway for lithium ion transfer.

ACKNOWLEDGMENT
We gratefully acknowledge the financial support by the Deutsche Forschungsgemeinschaft (DFG), Bonn, Germany (SFB 595/A4). We are also grateful to Claudia Fasel for the elemental analysis of the C, N, O content.

REFERENCES
[1] M. Wakihara, O. Yamamoto, Lithium Ion Batteries, Kodansha/Wiley-VCH, Tokyo, 1998.
[2] H. Li, Z. Wang, L. Chen, X. Huang, Research on Advanced Materials for Li-ion Batteries, *Adv. Mater.*, **21** 4593-4607 (2009).
[3] R. Kolb, C. Fasel, V. Liebau-Kunzmann, R. Riedel, SiCN/C-ceramic composite as anode material for lithium ion batteries, *J. Eur. Ceram. Soc.*, **26** 3903-3908 (2006).
[4] D. Su, Y.-L. Li, Y. Feng, J. Jin, Electrochemical Properties of Polymer-Derived SiCN Materials as the Anode in Lithium Ion Batteries, *J. Am. Ceram. Soc.*, **92** [12] 2962–2968 (2009).
[5] A. Saha, R. Raj, D. L. Williamson, and H. J. Kleebe, Characterization of Nanodomains in Polymer-Derived SiCN Ceramics Employing Multiple Techniques, *J. Am. Ceram. Soc.*, **88** [1] 232–4 (2005).
[6] G. Mera, A. Tamayo, H. Nguyen, S. Sen, R. Riedel, Nanodomain Structure of Carbon-Rich Silicon Carbonitride Polymer-Derived Ceramics, *J. Am. Ceram. Soc.*, **1–7** (2010), DOI: 10.1111/j.1551-2916.2009.03558.x.
[7] G. Gregori, H.-J. Kleebe, H. Brequel, S. Enzo, G. Ziegler, Microstructure evolution of precursorsderived SiCN ceramics upon thermal treatment between 1000 and 1400°C, *J. Non-Cryst. Solids*, **351** 1393–1402 (2005).
[8] M.D Levi, D. Aurbach, The mechanism of lithium intercalation in graphite film electrodes in aprotic media. Part 1. High resolution slow scan rate cyclic voltammetric studies and modeling, *J. Electroanal. Chem.*, **421** 79–88 (1997).
[9] M.D Levi, D. Aurbach, Simultaneous Measurements and Modeling of the Electrochemical Impedance and the Cyclic Voltammetric Characteristics of Graphite Electrodes Doped with Lithium, *J. Phys. Chem. B* **101** 4630-4640 (1997).
[10] N. Ding, J. Xu, Y.X. Yao, G. Wegner, X. Fang, C.H. Chen, I. Lieberwirth, Determination of the diffusion coefficient of lithium ions in nano-Si, *Solid State Ionics*, **180** 222–225 (2009).
[11] C. Ho, I.D. Raistrick, R.A. Huggins, Application of A-C Techniques to the Study of Lithium Diffusion in Tungsten Trioxide Thin Films, *J. Electrochem. Soc.*, **127** [2] 343-350 (1980).
[12] P. Yu, B.N. Popov, J.A. Ritter, R.E. White, Determination of the Lithium Ion Diffusion Coefficient in Graphite, J. Electrochem. Soc., **146** [1] 8-14 (1999).
[13] M.D Levi, D. Aurbach, Impedance of a Single Intercalation Particle and of Non-Homogeneous, Multilayered Porous Composite Electrodes for Li-ion Batteries, *J. Phys. Chem. B*, **108** 11693-11703 (2004).
[14] W. Weppner, R.A. Huggins, Determination of the Kinetic Parameters of Mixed-Conducting Electrodes and Application to the System Li3Sb, *J. Electrochem. Soc.*, **124** [10] 1569-1578 (1977).
[15] M.D. Levi, K. Gamolsky, D. Aurbach, U. Heider, R. Oesten, Determination of the Li ion chemical diffusion coefficient for the topotactic solid-state reactions occurring via a two-phase or single-phase solid solution pathway, *J. Electroanal. Chem.*, **477** 32–40 (1999).

[16] T.L. Kulova, A. M. Skundin, Balance between Reversible and Irreversible Processes during Lithium Intercalation in Graphite, *Russ. J. Electrochem.*, **42** [3] 251-258 (2006).

[17] Y. NuLi, J. Yang, Z. Jiang, Intercalation of lithium ions into bulk and powder highly oriented pyrolytic graphite, *J. Phys. Chem. Solids*, **67** 882–886 (2006).

[18] T. Ohzuku, Y. Iwakoshi, K. Sawai, Formation of Lithium-Graphite Intercalation Compounds in Nonaqueous Electrolytes and Their Application as a Negative Electrode for a Lithium Ion (Shuttlecock) Cell *J. Electrolchem. Soc.*, 140 **[9]** 2490-2498 (1993).

[19] Z.-W. Fu, Q.-Z. Qin, Lithium Ion Diffusion Behavior in Laser-Deposited TiO_2 Films, *J. Phys. Chem. B*, **104** 5505-5510 (2000).

[20] T. D. Tran, J. H. Feikert, R. W. Pekala, K. Kinoshita, Rate effect on lithium-ion graphite electrode performance, *J. Appl. Electrochem.* **26** 1161-1167 (1996).

[21] N. Takami, A. Satoh, M. Hara, T. Ohsaki, Structural and Kinetic Characterization of Lithium Intercalation into Carbon Anodes for Secondary Lithium Batteries, J. Electrolchem. Soc., 142 **[2]** 371-379 (1995).

[22] C. Ho, I.D. Raistrick, R. A. Huggins, Application of A-C Techniques to the Study of Lithium Diffusion in Tungsten Trioxide Thin Films, *J. Electrochem. Soc.*, **127** [2] 343-350 (1980).

[23] A. Funabiki, M. Inaba, Z. Ogumi, S.-I. Yuasa, J. Otsuli, A. Tasaka, Impedance Study on the Electrochemical Lithium Intercalation into Natural Graphite Powder, *J. Electrochem. Soc.*, **145** [1] 172-178 (1998).

[24] A. Funabiki, M Inaba, Z. Ogumi, A.c. impedance analysis of electrochemical lithium intercalation into highly oriented pyrolytic graphite, *J. Power Sources*, **68** 227–231 (1997).

[25] G. Maurin, Ch. Bousquet, F. Henn, B. Simon, Determination of the Chemical Diffusion Coefficient of Lithium inMultiwall Carbon Nanotubes, *Ionics*, **5** 156-160 (1999).

[26] D.W. Dees, S. Kawauchi, D.P. Abraham, J. Prakash, Analysis of the Galvanostatic Intermittent Titration Technique (GITT) as applied to a lithium-ion porous electrode, *J. Power Sources*, **189** 263–268 (2009).

[27] M.D Levi, D. Aurbach, The mechanism of lithium intercalation in graphite film electrodes in aprotic media. Part 2. Potentiostatic intermittent titration and in situ XRD studies of the solid-state ionic diffusion, *J. Electroanal. Chem.*, **421** 89–97 (1997).

NANO-AGGREGATE SYNTHESIS BY GAS CONDENSATION IN A MAGNETRON SOURCE FOR EFFICIENT ENERGY CONVERSION DEVICES

E. Pauliac-Vaujour*, E. Quesnel, V. Muffato, O. Sicardy, N. Guillet, R. Bouchmila, P. Fugier, H. Okuno, L. Guetaz
CEA-Grenoble, DRT/LITEN/DTNM, rue des Martyrs, 38054 Grenoble Cedex 9, France

ABSTRACT

We report technological advances in the fabrication of nanocatalysts by means of a dedicated inert-gas-condensation nanocluster source. We operate the source in order to synthesize nanoparticules whose nature and structure comply with very strict specifications in terms of size, size dispersion, crystalline structure, morphology and chemical composition. We aim to reduce the fabrication cost of energy devices by increasing the catalyst's active surface area while decreasing the actual amount of material (e.g. for Pt compounds and alloys in proton exchange membrane (PEM) fuel cells). The commercial magnetron-based reactor that we use offers an alternative to conventional CVD deposition and limits material consumption by reducing the catalyst load by a factor of at least 10 while increasing the texturing of the active surface area. 4nm alloyed PtCo nanoparticles (NPs) were deposited onto a fuel cell gas diffusion layer with an excellent control on size dispersion. Even in the case of the highest loads and material layer thicknesses, no coalescence of the spherical NPs was observed following deposition, so that the fine texturing of the surface was preserved. As was demonstrated for several materials, the amount of deposited material may be accurately varied from a few ng/cm^2 (i.e. far less than a monolayer) up to tens of µg/cm^2. In addition to fuel cells, this technology finds immediate applications in the elaboration of surface plasmon enhanced-absorption photovoltaic cells or carbon nanotubes growth catalysis for heat dissipation in microelectronics, among others.

INTRODUCTION

Nanomaterials and in particular nanoparticles (NPs), exhibit unique properties and an uncommonly high reactivity. Systems of semiconducting NPs, conducting core-insulating shell NPs, metallic NPs, functionalized NPs, etc. are novel illustrations of fundamental physical phenomena such as single electron conduction[1,2], plasmon-enhanced absorption[3-7] or directed self-organization[8,9] to name a few. NPs also commonly play the role of germs, or 'catalysts', in the growth of one-dimensional nano-objects[10]. Research in the field of nanoparticle science has gone one step further with the inclusion of these reactivity-enhanced entities – property related to their high surface to volume ratio – into active layers of performance-enhanced devices[11-13].

Synthesis techniques include purely chemical procedures[14-16], which produce large amounts of polydisperse NPs at a relatively low cost and are favored notably for functionalized particles. Gaseous or liquid precursor-based chemical vapor deposition (CVD) techniques offer a good control on the growth of the particles but are limited to relatively high loads (typically 100µg/cm^2, down to 30µg/cm^2 with liquid-injection MOCVD[17]) and the material yield for costly materials remains unsatisfactory. Dewetting of continuous thin films[18,19] and RF sputtering[20,21] are very well controlled techniques, commonly used for catalyst synthesis, which reproducibly yield well defined – although polydisperse – NPs, at high surface densities (up to 10^{12}/cm^2). However, size control and size-density decoupling remains problematic.

The commercial reactor that we introduce produces particles from a condensed sputtered vapor[22]. NP mean size and size dispersion are accurately controlled by means of the experimental parameters and an in-situ mass filter. NP surface density is function only of the NP flow and the deposition time. It is not correlated to the NP intrinsic characteristics. Material loads can hence be varied accurately from a few ng/cm^2 up to several 100µg/cm^2. This technology offers an unprecedented accuracy in NP synthesis control and addresses the strict specifications required by energy-related applications, such as PEM fuel cells. In such devices, PtCo bi-metallic NPs act as catalysts on the surface of the gas diffusion layer (GDL) of the cell, where the electro-chemical reaction occurs. Pt compounds are excellent catalysts for O_2 and H_2 oxidation-reduction reactions[23-25] but contribute for a large part to the fabrication cost of the device[17].

The main achievement of the present work is to produce highly reactive well-controlled catalysts while reducing considerably and controllably the Pt load in the device. With synthesis yields above a few percent, our NP reactor represents a cost-effective alternative to liquid-injection MOCVD deposition. In addition, a transfer of the process at an industrial production scale is possible.

Assessment of the source performances is achieved by systematic characterization of the deposits, highlighting the versatile character of this technology and the degree of accuracy achieved. In addition to validating the control on the size, size dispersion and density of the NPs, an investigation is carried out in order to determine the variability of the deposit's morphology, structure and chemical composition depending on process conditions. Finally, preliminary electrochemical tests are reported that highlight the interest of the NP vacuum synthesis technology for PEMFC applications.

EXPERIMENTAL PROCEDURE

Nano-aggregate source

Particles are produced in the magnetron-based reactor from a high pressure (10^{-1} mbar) sputtered vapor that is subsequently cooled down and condensates into nano-aggregates[26] (Fig. 1). Cooling results from a combination of external water cooling and thermalization by an inert-carrier gas, typically argon. The aggregates are driven out of the source by the argon flow and accelerated through a small aperture into the deposition chamber. Deposition takes place in vacuum (10^{-5} mbar) and at ambient temperature. The size distribution of the resulting nano-aggregates is measured via a quadrupole mass spectrometer. Mass filtering can be carried out by application of a selective alternative voltage through the quadrupole. Size dispersion as small as a fraction of nm can be achieved through filtering. NPs are collected onto a sample or a quartz microbalance placed perpendicular to the NP flow at a chosen height. The quartz microbalance gives an estimation of the deposited weight (approximated through the acoustic impedance of the material, with a $1ng/cm^2$ sensitivity).

In principle[22], a relatively large proportion of the nano-aggregates is charged once negatively (the others being electrically neutral). As a result, the +18V-polarised collecting grid of the mass spectrometer measures a 'NP current', I_{grid}, which represents for each selective voltage – i.e. NP size – the flow of charged NPs through the filter. A size distribution curve is deduced from these successive measurements (see Fig. 2).

The shape of the NP size distribution measured by the quadrupole can be varied by adjusting several parameters, namely the magnetron power (through the magnetron current I_{mag}), the carrier gas flow (Φ_{Ar}) and the position of the magnetron source inside the column (L). The NP size, size dispersion and emission flow depend on the nature of the material, of the carrier gas and on the cooling efficiency, as demonstrated by a dedicated nucleation and growth model developed purposefully by our group[26].

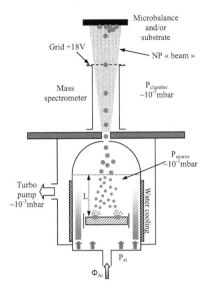

Figure 1. Diagram of the magnetron-based NP source, cooling system, pumping stages and mass spectrometer.

Morphological characterization, structural and composition analyses

Morphological characterization is carried out using a Dimension 3100 AFM from Veeco with super-sharp 2-5nm radius tips from Veeco. Data are analyzed with the WSxM software[27]. Characterization is correlated with transmission electron microscopy (TEM JEOL 2010 on SiOx membranes) and field emission gun scanning electron microscopy (FEG-SEM LEO 1530, Gemeni column) observations. Those techniques allow energy-dispersive X-ray (EDX) analysis, which gives information on the chemical composition of the deposit and/or of isolated NPs.

X-ray diffraction (XRD) is used to investigate the crystalline structure of the NPs and determine the mean size of the diffracting domains. It is carried out on Si(100) and Si(111) substrates, using a D8 Bruker Advance diffractometer with a LynxEye linear detector.

RESULTS : CONTROLLED NANOPARTICLE SYNTHESIS

NP Size control

The range of NP sizes that we can obtain with the source depends on the nature of the material and the set of experimental parameters. Typically, obtaining NPs of diameter 1 to 10nm is relatively straight forward in the default configuration of the source (Fig. 1) for materials with good sputtering yields such as copper or silver in argon gas. NP synthesis of strongly magnetic materials or materials with a very weak sputtering yield is possible, but optimizing the NP flow in order to deposit high densities remains an issue. The degree of feasibility of NP synthesis for every pure material can be estimated via the model[26] prior to experiment. This is particularly useful in case of costly materials, such as Pt. For Pt compounds, as well as for alloys in general, the lack of thermodynamic data does not enable to simulate NP nucleation and growth. However, preliminary experiments showed that deposition from a 50-50% PtCo target yielded massive NP flows which allow a precise calibration and systematic characterization of the samples.

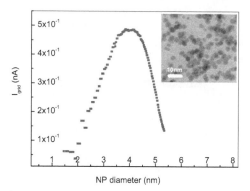

Figure 2. Size distribution, measured by the quadrupole mass spectrometer, of PtCo NPs for standard deposition conditions (I_{mag} = 100mA, d_{Ar} = 40sccm, L = 70mm). (insert) sample TEM image of the deposit.

The size distribution of synthesized NPs is measured through the quadrupole as explained above. An illustration is given in Fig. 2 for a PtCo NP diameter centered on 4nm. For this specific set of parameters (I_{mag} = 100mA, Φ_{Ar} = 40sccm, L = 70mm) and without filtering, NP sizes vary between 2 and 5.5nm, which is confirmed by TEM imaging (insert of Fig. 2). Filtering is achieved by selecting a set of voltage/frequency for the quadrupole, which reduces size dispersion considerably. This is illustrated in Fig. 3 : for a filtered size set to 2nm, TEM characterization yields a mean NP size of 2.4nm with a standard deviation of 0.9nm. However, quadrupole filtering also screens a large part of the NPs, impeding the overall NP flow. As a result, depositing 'monodisperse' NPs at high densities requires a very fine tuning of the experimental conditions and is so far only achievable for materials with a high synthesis yield, such as PtCo.

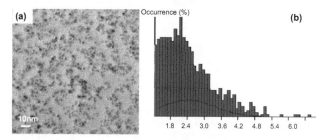

Figure 3. Illustration of mass filtering for 2nm PtCo NPs: (a) TEM micrograph of the deposit (density 3.10^{12}/cm²); (b) histogram of the size distribution (mean NP size 2.4nm).

Surface density control

The NP density on the surface is estimated through three types of measurements. The size distribution curve gives an estimation of the number of NPs contributing to the I_{grid} current (each NP being negatively charged once). However, it does not take into account possible neutral NPs, hence underestimating the overall density by an unknown factor. It needs to be coupled with either morphological analysis (AFM or TEM) or weight measurement. Approximations on the estimation of NP density on the surface are then reliably adjusted by combination of mass spectrometry, weight measurement and morphological characterization.

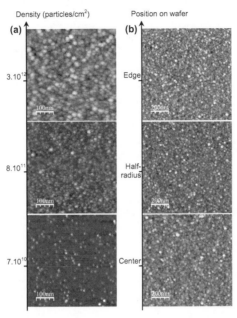

Figure 4. (a) illustration of controlled density deposits on Si(100) of 2nm filtered PtCo NPs imaged by tapping-mode AFM; (b) AFM images taken on a single 200mm Si(100) wafer following non-filtered deposition of 4nm PtCo NPs, at various points along one wafer radius. The NP deposit is homogeneous: it exhibits the same density on the entire surface of the wafer.

Hence, for (sub-)monolayer deposits, NP densities varying from less than 10^{10}/cm² up to several 10^{12}/cm² (Fig. 4a) are obtained on a regular basis, with an extremely good degree of control. Thicker deposits (e.g. 5 to 100µg/cm² for PtCo) are also carried out, for which surface density cannot be reliably estimated from AFM or TEM characterizations. For these, densities are given in terms of mass (µg/cm²) rather than absolute number of NPs (/cm²). This mass is subsequently confirmed by other means of characterization (inductively coupled plasma mass spectrometry). For materials with very high NP flows (e.g. PtCo), deposition can be achieved as far as 70cm above the source aperture, on a 200mm silicon wafer, with a very satisfying homogeneity (Fig. 4b). However, such good homogeneity on such large surfaces becomes difficult when NPs are mass filtered through the quadrupole, unless the deposit thickness is sufficient (equivalent to several monolayers).

Structural characteristics of NPs

We have demonstrated a good control on NP size, size dispersion and density. Furthermore, AFM and TEM images confirm the global spherical aspect of NPs following deposition and the absence of NP coalescence (although some piling up is observed at high surface densities).

In the case of alloys, such as PtCo, it is interesting to carry out chemical composition analysis in order to estimate the proportion of each component within single NPs and possibly to identify a particular organisation of the atoms, like in a core-shell structure for instance. In the case of PtCo NPs synthesized by sputtering of a 50%-50% PtCo target, global EDX analysis shows an average composition of 53% Pt - 47% Co for the deposit. The chemical composition of the target thus seems to be globally preserved.

We used θ-2θ XRD in order to determine the lattice parameters of NPs and the average size of the diffracting (monocrystalline) domains. Fig. 5 shows the XRD diagram for a $360\mu g/cm^2$ load of 4nm PtCo NPs. The broad diffraction peaks are representative of nanometric crystallites. They correspond to a face-centered cubic (fcc) phase with lattice parameter 3.794 ± 0.002 Å, which is smaller than the pure Pt lattice parameter (3.923Å). This implies that Co atoms are substituted to Pt ones in the Pt lattice (insert of Fig. 5) and that the aggregates are indeed alloys, and not core-shell NPs.

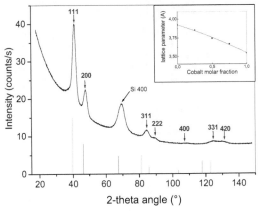

Figure 5. XRD diagram of a $360\mu g/cm^2$ deposit of 4nm PtCo NPs on a Si(100) substrate. The positions of equivalent diffraction peaks for pure fcc Pt are given by the vertical lines (bottom). (insert) Lattice parameter of an fcc $Pt_{1-x}Co_x$ compound as a function of x (molar fraction of Co) – from literature.

The value of the lattice parameter corresponds to a $Pt_{0.6}Co_{0.4}$ compound, as determined via the curve Fig.5, corresponding approximately to the Vegard law. The high diffraction angles lines detected with the $360\mu g/cm^2$ sample (Fig. 5) allow accurate measurement of the parameter and confirm the presence of a single alloy (although a very small dispersion around the 60%-40% composition is possible). Hence, the initial composition of the target (50%-50%) is not exactly preserved. This may be explained by the presence of a cobalt oxide layer on the outside of the NP, which does not contribute to the XRD signal (it is actually possible to visualize this oxide in high-resolution TEM). As a result,

XRD only detects the Co included in the crystallite but does not take into account the Co present in the oxide layer. This explains the larger proportion of Pt within the crystalline domains.

From the X-ray lines broadening and according to the Scherrer law, the size of the diffracting domains is estimated to be 3.6 ± 0.1 nm. This is very close to the mean size (4nm) given by the mass spectrometer and confirmed by AFM and TEM observations. The slight difference may once again arise from the presence of a cobalt oxide layer around the alloyed PtCo. This is a significant result, since it demonstrates that the NPs are single crystals, probably surrounded by a cobalt oxide layer.

DISCUSSION : NANOPARTICLE SPECIFIC PROPERTIES AND APPLICATIONS

The nano-aggregate source enables to reproducibly synthesize NPs with very accurate characteristics, which are particularly useful for energy conversion applications, such as PEM fuel cells. The main challenge regarding NP fabrication for these applications is to increase the number of active conversion sites while reducing the quantity of matter deposited. In other terms, we need to achieve a very fine texturing of the surface, with an average size of the particles near 4nm[28,29]. The chemical composition of the NPs (e.g. PtCo or Pt_3Co) is not our main concern, most Pt compounds generally offering a better conversion yield than Pt alone[30-32]. Loads between 5 and $50\mu g/cm^2$ are required, which is down to ten times lower than the minimum loads of conventional CVD techniques and a hundred times less than commercial cells (typically $500\mu g_{Pt}/cm^2$). However, despite such loads, coalescence of the NPs during deposition must be avoided in order to benefit from the advantages of this novel technology, i.e. 4nm texturing. Fig. 6a shows SEM images for a $50\mu g/cm^2$ NP load on the microporous layer of a commercial GDL (BASF LT1400W). The fine texturing of the surface is visibly preserved, which in addition is also in agreement with AFM images from Fig. 4. PtCo particles are effectively 4nm $\pm$ 1.5nm in size and the spherical NPs do not coalesce, they simply pile-up.

Preliminary electrochemical tests have demonstrated performances that vary coherently as a function of the Pt load. An exhaustive description of these measurements and results is to be given elsewhere. Electrochemically active surface area of four samples loaded with 5, 10, 20 and $50\mu g_{PtCo}/cm^2$ respectively were estimated from hydrogen underpotential region (Hupd - assuming a charge of the electrochemically active Pt surface of $210\mu C$ cm^{-2}_{Pt}) and CO-stripping (assuming $420\mu C$ cm^{-2}_{Pt}). Experiments were performed on a half cell test in H_2SO_4 0,5M at room temperature. The measured values fit with the theoretical values calculated from the developed surface area corresponding to the 4nm NPs (Fig. 6b). Optimization of the repartition of the NPs inside the first few microns of the GDL should help to increase the active surface area far beyond actual PEMFC electrodes and enable to dramatically reduce the noble metals loading.

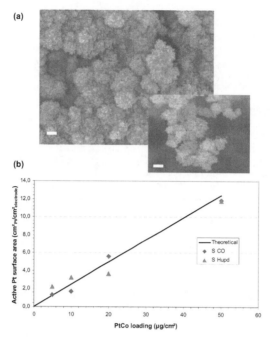

Figure 6. (a) SEM images of a 50 μg/cm² deposit of 4nm PtCo NPs on a flexible GDL substrate. Insert shows one detail of the GDL. Scale bars represent 100nm on both images (courtesy of C. Anglade); (b) Comparison of the Pt active surface area as a function of Pt load, as estimated experimentally (dots) and theoretically (line).

As a result, this technology presents a strong interest for PEM fuel cell development. Despite the use of fundamentally different synthesis techniques, PtCo nano-aggregates and CVD grown nano-catalysts both exhibit the expected catalytic properties once introduced in a GDL.

The nano-aggregate source enables to deposit extremely low Pt loadings, with an electrochemical response that follows predictions. Furthermore, adjustments on the synthesis conditions, size filtering and deposition stage are largely possible, that will lead to a significant improvement in terms of active surface area, conversion rate and, hence, electrochemical measurements.

In other words, the nano-aggregate source eventually provides an ultimate tool to master the deposition of low Pt loaded catalyst layers with pre-required specific properties. This technology is incredibly promising for the development of future PEM fuel cell generations. It also finds immediate applications in various catalytic applications (notably nano-object growth catalysis). Finally, exploiting the nano-aggregate source performances with adapted nano-scale properties materials offers a direct technological solution to the fabrication of plasmon-enhanced absorption and third generation (quantum dot) solar cells[7].

CONCLUSIONS

We have demonstrated well-controlled synthesis of PtCo NPs via the nano-aggregate source for dedicated energy-related applications, and in particular weak-Pt load PEM fuel cells. NP characterization also constitutes a fundamental aspect of NP synthesis, since NP properties are determined by their size, their shape, their spatial arrangement and environment, and their chemical composition.

Electrochemical measurements of devices involving 4nm PtCo NPs deposited on a flexible carbon-coated GDL exhibited very promising performances. Improvement of these performances is expected in the short term thanks to the unprecedented versatility of this technology that yields a very broad range of possible NPs.

This study clearly gives evidence of the possible impact of the nano-aggregate source in the fabrication of new generation devices. It constitutes a very challenging technology for a broad range of applications, notably in catalysis and solar cells.

ACKNOWLEDGMENTS

The authors would like to thank L. Notin and C. Anglade for technical support. We acknowledge funding from National Pan-H project OPTICAT.

FOOTNOTES
*Corresponding author : emmanuelle.pauliac-vaujour@cea.fr

REFERENCES
[1] A.A. Middleton, N.S. Wingreen, *Phys. Rev. Lett.* **71**, 3198 (1993).

[2] M.N. Wybourne *et al.*, *Jpn. J. Appl. Phys.* **36**, 7796 (1997).

[3] S.K. Mandal, R.K. Roy, A.K. Pal, *J. Phys. D: Appl. Phys.* **35**, 2198 (2002).

[4] S.K. Mandal, R.K. Roy, A.K. Pal, *J. Phys. D: Appl. Phys.* **36**, 261 (2003).

[5] A. Pinchuck, G. von Plessen, U. Kreibig, *J. Phys. D: Appl. Phys.* **37**, 3133 (2004).

[6] Y. Yang, S. Matsubara, M. Nogami, J. Shi, W. Huang, *Nanotechnol.* **17**, 2821 (2006).

[7] R. Najjar *et al.*, *45th MIDEM conference proceedings, Postojna , Slovenia* (2009).

[8] C.J. Kiely *et al.*, *Nature* **396**, 444 (1998).

[9] C.P. Martin *et al.*, *Phys. Rev. Lett.* **99**, 116103 (2007).

[10] G.F. Zhong, T. Iwasaki, H. Kawarada, *Carbon* **44**, 2009 (2006).

[11] M. Peuckert, T. Yoneda, R. A. Dalla Betta, M. Boudart, *J. Electrochem. Soc.* **113** (5), 944 (1986).

[12] S. Mukerjee, J. Appl. Electrochem. **20**, 537 (1990).

[13] E. Billy *et al.*, J. Power Sources **195**, 2737 (2010).

[14] J. Turkevitch, P.C. Stevenson, J. Hillier, *J. Discuss. Faraday Soc.* **11**, 55 (1951).

[15] M. Brust *et al.*, *J. Chem. Soc.: Chem. Commun.*, 1655 (1995).

[16] B.J. Hwang *et al.*, *J. Phys. Chem. B* **110**, 6475 (2006).

[17] S. Donet, E. Billy, P. Fugier, A. Morin, *La Revue 3EI* **56**, 5 (2009).

[18] A. Sharma, R. Khanna, *Phys. Rev. Lett.* **81**, 3463 (1998).

[19] Y.Y. Wei, G. Eres, V.I. Merkulov, D.H. Lowndes, *Appl. Phys. Lett.* **78**, 1394 (2001).

[20] G.F. Zhong, T. Iwasaki, H. Kawarada, *Carbon* **44**, 2009 (2006).

[21] K. Kurihara, C. Rockstuhl, T. Nakano, T. Arai, J. Tominaga, *Nanotechnol.* **16**, 1565 (2005).

[22] A.H. Kean, L. Allers, *Proc. NSTI NanoTech* **1**, 749 (2006).

[23] A. Witkowska, *J. Power Sources* **178**, 603 (2008).

[24] V.R. Stamenkovitch *et al.*, *Science* **315**, 493 (2007).

[25] J.G. Chen, C.A. Menning, M.B. Zellner, *Surf. Sci. Rep.* **63**, 201 (2008).

[26] E. Quesnel, E. Pauliac-Vaujour, V. Muffato, *J. Appl. Phys.*, vol. **107**, issue 4 (2010).

[27] I. Horcas *et al.*, *Rev. Sci. Instrum.* **78**, 013705 (2007).

[28] A. Franco *et al.*, *J. Electrochem. Soc.* **156** (3), B410 (2009).

[29] P. Fugier *et al.*, accepted for publication in *J. Electrochem. Soc.* (2010).

[30] H. Yano, M. Kataoka, H. Yamashita, H. Uchida, M. Watanabe, *Langmuir* **23**, 6438 (2007).

[31] V.R. Stamenkovic *et al.*, *Nat. Mater.* **6**, 241 (2007).

[32] V.R. Stamenkovitch, B.S. Mun, K.J.J. Mayrhofer, P.N. Ross, N.M. Markovic, *J. Am. Chem. Soc.* **128**, 8813 (2006).

MODELING NANOPARTICLE SYNTHESIS BY GAS CONDENSATION IN A NANOCLUSTER
SOURCE FOR APPLICATIONS IN PHOTOVOLTAIC AND HYDROGEN FUEL CELLS

E. Pauliac-Vaujour*, E. Quesnel, V. Muffato
CEA-Grenoble, DRT/LITEN/DTNM/LTS, rue des Martyrs, 38054 Grenoble Cedex 9, France

ABSTRACT
    One challenging and relatively cost-effective alternative to MOCVD-deposited nanoparticles
(NPs) is the synthesis of nano-aggregates by inert-gas condensation of a sputtered vapor in a dedicated
magnetron-based reactor. By tuning the collision path length and hence the time of residence of NPs in
the carrier gas phase, spherical aggregates with well-controlled nanometer-scale diameters and
relatively narrow size dispersion are produced from various materials. Such aggregates exhibit
interesting properties related to their high surface-to-volume ratio and surface reactivity (e.g. in
catalysis applications). On the basis of experimental results, a detailed modeling of NP nucleation and
growth based on the classical nucleation theory was developed taking the peculiar geometry and
thermal profile of the NP reactor into account. The simulated curves, calculated by a matlab   program
developed for that purpose, exhibit a good qualitative agreement with experiment and highlight the role
of the process parameters and reactor temperature profile on the NP size distribution. Such calculations
underline the possible optimization of the NP source design in order to improve its efficiency and
reproducibility. The model may also yield a feasibility criterion and facilitate process development for
costly materials by limiting the range of investigated parameters.

INTRODUCTION
    In the field of renewable energy production, a significant enhancement of the generator
performance can be obtained via the nanostructuring of the active surface, i.e. the surface intended to
collect the light in solar cells[1,2] or be the site of chemical reactions in fuel cells[3]. For these applications,
innovation is basically related to the unique catalytic, optical, magnetic or electronic properties of NPs
precisely resulting from their very small size (typically 1 to 50nm).
    At an industrial scale, the relevance of such NPs relies on the accurate control that the synthesis
and deposition processes offer, regarding notably the particle mean size and size dispersion. Among
the different methods developed to produce metal NPs, those based on chemical procedures are the
most commonly used[4-6], mainly because of their relative ease of use and low cost. Vacuum
technologies based on material vaporization and subsequent cooling of the vapor are also commonly
used at the laboratory level for basic research purpose. The various techniques which have been
developed so far differ essentially by the way (i) of producing the vapor – evaporation or sputtering –
and (ii) of cooling down this vapor – adiabatic supersonic expansion of the vapor[7] or simple use of a
cooling gas[8]. To fulfill the industrial market requirements, a commercial NP generator system
(Nanogen 50 from Mantis Deposition Ltd.[9]) based on a magnetron sputtering technique and gas
cooling has been proposed recently. This reactor allows unprecedented and uncorrelated control on
both NP size and density (number of NPs / cm²).
    In the present work, we aimed at studying the physical principles into play during NP synthesis
in such a commercial source. The objective is to estimate the capacities and limitations of the
equipment. In particular, it is to identify the parameters which are the most critical for the control of
NP size distribution and deposition flow. Via the modeling of the NP nucleation and growth processes
occurring in the source, it thus becomes possible to estimate the degree of feasibility for the synthesis
of NPs on the basis of thermodynamic and physical properties of the chosen material. This study is
particularly relevant in the case of costly materials, such as platinum, used as a catalyst in proton
exchange membrane fuel cells (PEMFCs).

In this paper, a model based on the classical nucleation theory (CNT) and describing the nucleation and growth of NPs in the Nanogen source is detailed. A comparison of simulated and experimental curves is established and discussed in terms of sensitivity of the NP size distribution to experimental conditions. Technological improvements for optimized NP synthesis are also discussed.

DESCRIPTION OF THE EXPERIMENTAL ENVIRONMENT

NPs are synthesized by condensation of an atomic vapor produced by sputtering at high pressure. A diagram of the NP source is given in Fig. 1: metallic vapor is generated from a 5cm-in-diameter target, inside a water cooled aggregation chamber. The target, which faces the source aperture, can be moved and its position adjusted from L= 0 to 70mm, thus tuning the length of aggregation (L+R with R=50mm the radius of curvature of the upper part of the aggregation zone, see Fig. 1). Argon gas is introduced for plasma generation and atomic vapor production by magnetron sputtering of the bulk metallic target by ionized argon. As a result, NPs nucleate and grow along the X-axis until reaching the source exit. Because of the gas depression between the plasma source (10^{-1} mbar) and the deposition chamber (10^{-5} mbar), the NPs are accelerated towards the substrate holder.

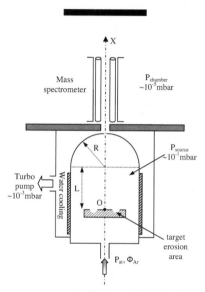

Figure 1. Diagram of the nano-aggregate source. Target position is varied along the X-axis: at x=0, L is maximum (70mm).

Three main process parameters control the NP source: the carrier gas flow rate Φ_{Ar}, the sputtering ion current I and the target position L. Other variables of the model include the pressure within the aggregation column P_{Ar} and that in the deposition chamber P_c. The concentration (N_{metal}) and pressure (P_{metal}) of the metallic vapor can be deduced from the sputtering yield of the metal in argon gas and the calculated emission rate[10].

The sizes of emitted NPs are estimated from their mass measurement through a linear quadrupole mass spectrometer, placed in line with the NP source aperture[10]. A size distribution curve

(NP current vs NP size) is obtained from the charged (1e-) NP flow collected through a polarized grid at the quadrupole exit. Although particular caution should be observed when attempting to give a quantitative interpretation of intensity levels, the signal evolution with size is fully consistent with the actual NP size distributions, as reported in previous works[11]. Note that NP sizes are given in number of atoms n, which is linked to the NP radius r by a simple relationship:

$$n = (r / r_w)^3$$

where r_w is the Wigner Seitz radius of the material. For instance, for Cu in a CFC structure, we calculate r_w to be 1.41Å.

Preliminary experiments with a copper target show that: (i) the NP mean size can be typically varied between 1 and 10 nm by simply adjusting 2 process parameters (L and/or Φ_{Ar}); (ii) the characteristic shape of the NP size dispersion is comparable to an unbalanced Gaussian curve centered around an average size value <n>; (iii) the size distribution broadens as L is increased; (iv) the flow of NPs is generally more important at high values of L and/or Φ_{Ar}.

Additional experiments carried out with other metals, such as silver, tantalum or a platinum-cobalt alloy, lead to comparable results. We observe the same dependency on the source parameters but a noticeable difference in achievable NP flow or NP size range magnitude. That is why, to be more confident in the degree of NP size control that the source actually offers, it is essential to first understand its functioning in the case of Cu where the reproducibility is rather good. The purpose of the following sections is to analyze in greater depth these experimental results on the basis of thermodynamic considerations.

THEORETICAL CONSIDERATIONS AND MODEL

Classical nucleation theory (CNT)

We consider that nucleation constitutes the first step of NP synthesis within our source. Details of the CNT adapted to our NP source have been given elsewhere[10], so that only the main results will be reminded here. The CNT gives the energy variation associated with the aggregation of n atoms into a NP of radius r [12]:

$$\Delta G = 4\pi r^2 \sigma_s - \frac{4}{3}\pi r^3 \rho \Delta \mu \quad (1)$$

with σ_s the surface tension and $\Delta\mu$ the enthalpy variation per mass unit during condensation. According to the Clapeyron relation:

$$\Delta \mu = \frac{H_T}{N_a}\left(1 - \frac{T}{T_{vap}}\right) \quad (2)$$

The above relationship is only valid because the difference H_T between the enthalpies of the metal under gas and solid states is nearly constant over the range 200-1000K (from literature[13]).

The aggregate overcomes the nucleation barrier for a critical value of its radius r*: at this stage, it can further grow to become a NP and we will call it a germ, or nucleus. The critical radius r* is such that:

$$r^* = 2c_s / \rho \Delta \mu$$

which corresponds to a critical number of atoms n*. As a result, inside the metallic vapor, the steady nucleation rate J_0 per unit volume and time is an activated process proportional to the density of nuclei N* and their capture rate $\alpha*$, such that:

$$J_0(T) = \Gamma.\alpha*.N* \quad (3)$$

with

$$N* = N_{metal}\exp\left(-\frac{\Delta G*}{kT}\right)$$

$$\alpha* = \frac{P_{metal}}{\sqrt{2\pi mkT}}.4\pi r*^2$$

$$\Gamma = \sqrt{\frac{\Delta G*}{3n*\pi kT}}$$

The nucleation rate $J_0(T)$ depends on the sputtered material. It is maximum for a well-defined temperature T_{opt} for which the nucleation efficiency is optimum (Fig. 2a). On either side of this maximum, nucleation diminishes, usually steeply (depends on peak width).

As a result of the source geometry, the temperature is not constant in the source column. We thus need to extend the basic CNT – which merely yields the above expression of the nucleation rate – so that it takes into account the temperature profile within the source. The latter can be determined experimentally by placing a thermocouple between the source aperture and the target. In first approximation, this gives the temperature profile T(x) of the source along the x-axis (Fig. 2b). Nucleation will be optimum wherever this temperature profile meets T_{opt} on the x-axis. In the case of copper, as illustrated in Fig. 2b, the optimum configuration is not reached, due to a low vapor cooling efficiency. This point will be discussed below. However, since nucleation can occur over a large range of temperatures but with limited efficiency (corresponding to the area under the Gaussian-like curve), nucleus formation remains possible, notably in the coolest parts of the source.

From this modified CNT calculation, we can deduce the actual concentration of nuclei at each point of the source axis. Indeed, the resulting nucleation rate $J_0(T(x))$ is related to the global concentration profile N(x) of nuclei along the source axis by the instantaneous time of residence of the nuclei $t_i(x)$ at each point of their course. This time of residence is inversely proportional to the displacement speed of the nuclei that we take equal to the carrier gas speed[10]. We obtain:

$$N(x) = t_i(x).J_0(T(x)) \quad (4)$$

N(x) directly reflects the repartition of nuclei along the source axis (Fig. 2c). It is maximum where the vapor temperature approaches T_{opt} and collapses near the source exit, where the time of residence abruptly shortens. The earlier the nucleation occurs (maximum of N(x) near the target), the larger the available aggregation distance for the growth phase (see below) is, resulting in larger NPs.

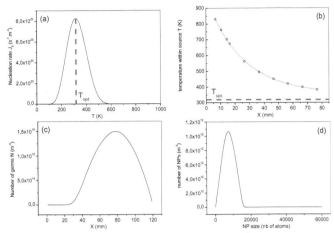

Figure 2. Simulated curves for the case of copper, with experimental parameters I=300mA, L=70mm, Φ_{Ar}=60sccm: (a) Nucleation rate as a function of temperature, $J_0(T)$ showing the optimum nucleation temperature T_{opt}; (b)Experimental temperature profile along the source axis, $T(x)$; (c) Nucleus concentration profile along the source axis, $N(x)$; (d) Global size distribution of synthesized NPs, $D(n)$.

NP growth

The second step of the NP synthesis process proceeds from the above nucleation stage by growing NPs out of the nuclei formed and distributed along the source axis. Growth occurs by successive collisions of nuclei with metallic atoms of the vapor. In first approximation, we consider an atom-by-atom growth and neglect monomer re-evaporation. Assuming that all collisions result in a sticking event, the sticking rate or probability per unit time that an aggregate of size n atoms interacts with a metallic atom is:

$$A_n = \pi r_s^2 \left(1 + n^{\frac{1}{3}}\right)^2 N_{metal} V_n \qquad (5)$$

with V_n the velocity of the aggregate such that :

$$V_n = \sqrt{3kT/m_{at}}$$

m_{at} is the atomic mass of the metal.

For a global population of N particles, $P_n(t)$ is the probability of having $N.P_n(t)$ aggregates of size n at time t. It was demonstrated elsewhere[10,12] that a possible form of $P_n(t)$ was a Poisson distribution of the type:

$$P_n(t) = \left(\frac{1}{n!}\right)\left(At_g\right)^n e^{-At_g} \qquad (6)$$

Note that in the original growth theory, this distribution implies that A is a constant of both n and t_g, which does not reflect the physical phenomena within our NP source. Indeed, in our case the sticking coefficient is a function of the NP size, that is to say A = A(n) according to Eq. (5). We thus make the assumption that Eq. (6) remains valid with A non-constant and update the growth model accordingly. The validity of this assumption will be demonstrated later. It is important to note that t_g in Eq. 6 is the *total* time of residence of the particles inside the source, as opposed to t_i defined above.

According to the assumption of a two-step synthesis process, global NP synthesis can be modeled by correlating the growth probability above with the initial concentration of nuclei formed in the first step process. We obtain the global size distribution D(n) of the particles exiting the NP source (Fig. 2d), which can be normalized to fit the experimental data measured by the quadrupole. This model can be directly applied to all pure materials for which the thermal profile within the source has been measured. It cannot be applied, as is, to alloys and compounds (lack of thermodynamic data). Note that although we have centered our study on the synthesis of metallic NPs so far, the source is also adapted to other types of materials, such as semiconductors for instance (see section 4.2), to which the model is perfectly and directly transposable.

RESULTS AND DISCUSSION

Validation of the model
The simple model described above – based on the CNT adapted to the source geometry and on the updated growth theory – is a powerful tool for predicting the NP synthesis efficiency of the source for all pure materials. We first demonstrate the consistency of the model by comparison with experimental data for two different materials: copper and iron.

Copper is particularly adapted for magnetron sputtering since its sputtering yield in argon is very high and it is a very good electrical and thermal conductor (high cooling efficiency of the target). Fig. 3a-c shows that, for the conditions I=300mA, L=70mm, Φ_{Ar}=60sccm, the predictions of the model are in very good agreement with the experimental data. This set of parameters was determined by preliminary calibration experiments for an optimized flow of NPs, of size centered on 5nm. The model gives an optimum temperature of T_{opt} = 317K with a very high maximum nucleation rate ($\sim 8.10^{20}$ /s/m^3). T_{opt} in this case is only slightly below the lowest temperatures measured in the source ($\sim$350K, Fig. 2b). The maxima of the experimental and calculated size distribution curves are located respectively at 7400 and 7600 atoms, that is to say a NP diameter of approximately 5.5nm. A topographical investigation of the surface by atomic force microscopy (AFM) shows that after only a few minutes of deposition, we obtain more than a monolayer of NPs on the surface, which corresponds to a high ($>10^{12}$/cm^2) surface density. Step measurement even demonstrates that the deposit is several tens of monolayers thick in this precise case.

Iron, on the other hand, is not a good candidate for magnetron sputtering, since it is a magnetic material. As expected, measured NP flows are very weak. Fig. 3d-f shows the comparison with the predictions of the model. The optimum temperature is this time very low: T_{opt} = 192K, which is significantly lower than the lowest achievable temperatures in the column ($\sim$340K for Fe at 200mA, 100sccm). In addition, the maximum nucleation rate is 10^7 times lower than for Cu. Despite the weak signal measured through the quadrupole, experimental and calculated curves are in reasonably good agreement and AFM investigation confirms the weak NP flow (surface density $\leq 10^9$/cm^2).

In both cases, model and experiment yield consistent results. These examples illustrate how the model can qualitatively predict NP synthesis feasibility for various materials. Note that at high pressures, a discrepancy was sometimes observed between simulation and experiment. The model tends to underestimate the NP mean size. This is probably due to the fact that, unlike assumed, the radial distribution of the metallic vapor inside the source is actually not homogeneous and evolves with the argon pressure. At high pressure, the vapor is likely to be mainly confined along the source axis, which would enhance the number of atom collisions and particle growth, and lead to much bigger NPs than expected with a constant radial Cu concentration. Moreover, the increasing particle concentration in the gas phase could promote further NP aggregation or NP growth according to a coarsening mechanism such as Ostwald ripening, for instance[10,14].

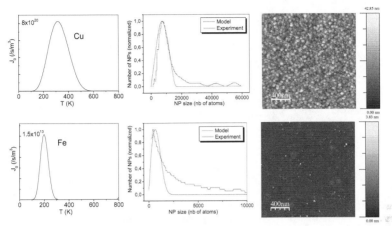

Figure 3. (a), (d) Calculated nucleation rate J_0 as a function of temperature for Cu and Fe respectively. The relative amplitudes reflect the nucleation efficiency of the source. The maximum determines the optimum temperature for nucleation. (b), (e) Calculated (dashed) and experimental (solid line) size distribution curves for Cu and Fe respectively (experimental conditions optimized for 5nm Cu NPs and 2nm Fe NPs). (c), (f) AFM characterization of the surface following NP deposition for Cu and Fe respectively (bright: NPs, dark: substrate).

Usage and set-up optimization

Predictions of the model for NP synthesis of various materials enable to establish the relevance of using our magnetron-based nano-aggregate source for given applications. Energy-related applications – such as plasmon-enhanced-absorption and third generation solar cells – often require an extremely good control on NP size, size dispersion and surface density. It is possible to control all these parameters independently using our NP source, and to estimate the corresponding NP flow achievable with this novel technology. Moreover, these applications require the synthesis of both metallic and semiconducting NPs. For PEMFC applications, our model enables to reduce considerably the range of investigated parameters prior to experiment, hence reducing the consumption of costly materials, such as Pt, used as nanocatalysts of the chemical reaction in the gas diffusion layer. The graph in Fig. 4 establishes a comparative 'feasibility criterion' for various materials for a set of standard (individually calibrated) experimental parameters – typically I=200-300mA, L=70mm, Φ_{Ar}=100-150sccm. It indicates the NP mean size associated with given parameters and the relative amplitudes of the calculated distribution curves, i.e. the number of NPs synthesized for each material in optimized conditions. Approximations on the thermal profiles for Co, Pt and Al had to be made since actual temperature profile measurements remain to be carried out. Fig. 4 shows that, comparatively, the source used in standard conditions is more adapted to the synthesis of Cu and Ag NPs than of Fe NPs, for instance.

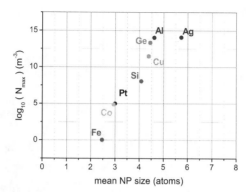

Figure 4. Calculated NP mean size (x-axis) and 'synthesis feasibility' (number of NPs, y-axis) for calibrated experimental parameters for various materials. Exact thermal profiles are known for Cu, Fe, Ag, Si and Ge, have been extrapolated for Co, Pt and Al.

One more advantage of this model is that it highlights the limitations of the nano-aggregate source in its present configuration. Indeed, the results presented here emphasize the predominant role of the cooling efficiency and temperature control within the source. Fig. 2 showed the simulation results for Cu NPs synthesized in conditions I=300mA, L=70mm, Φ_{Ar}=60sccm. In this case, the maximum of nucleation takes place approximately 40mm before the source exit, where the vapor temperature is close to T_{opt} = 317K (Fig. 2b). However, in this curved region of the source column, the gas speed increases dramatically, reducing the time of residence of the NPs. As a result, the nucleus concentration rapidly collapses near the source exit. In order to optimize the nucleation efficiency, it is necessary to increase the length of the 'cool' region in the straight part of the column[10]. Cooling efficiency is thus the main issue as regard to nucleation. Hence, despite the experimentally optimized NP flow, the model highlights the limited nucleation efficiency of the source in this configuration. It emphasizes the essential role of the source temperature monitoring for the emission stability and puts into light the technological potential of this magnetron-based nano-aggregate source in optimized working conditions.

Hypotheses and limitations of the model

Our nucleation and growth model relies on classical and simplistic theories. Its consistency has been demonstrated by the very good qualitative agreement with several experimental results. However, some hypotheses remain to be discussed, such as the introduction of the experimental thermal profile in the simulations or the estimation of the vapor pressure and concentration based on the assumption that the metallic vapor occupies the whole volume of the column, etc. Some of these approximations are discussed below.

Let us remind that our model is based on the CNT, which has been extended to include the source geometry related effects, notably the temperature profile and the speed of the nuclei (instantaneous time of residence), yielding the actual nucleus distribution at each point along the source axis.

Another correction of the CNT lies in the calculation of the nucleation rate and the contribution of the carrier gas. Indeed, the classical CNT only considers isothermal phenomena, and notably excludes metal-carrier gas collisions i.e. local non-isothermal effects. Wedekind et al.[15] have translated these effects through the definition of an 'effective chemical potential' $\Delta\mu_{eff}$ and a correcting coefficient to the nucleation rate $J_0(T(x))$. We have followed their recommendation in our simulations[10]. Calculations both with and without this correction highlight the better agreement between model and experiment in the corrected model.

One more important approximation of our nucleation model lies in the extrapolation to zero of the temperature profile to determine the temperature on the target T_{vap}. Considering the impact of the temperature profile on the results given by the model[10], technical improvements and better accuracy on the measurement of temperature profiles within the source would certainly be profitable.

As stated in the theory section, in the classical growth theory the expression of P_n given in Eq. 6 is supposedly only valid if A_n is a constant[12]. However, we have decided to keep it a function of NP size to offer a more realistic description of the growth phenomenon. To justify this assumption, we have calculated (Fig. 5) size distributions for identical sets of parameters in the cases where the sticking probability A_n is either constant (independent on NP size) or varies (increases with NP size according to Eq. 5). Fig. 5 shows that the main effect of this variation is a slight shift of the mean NP size, by roughly 10% of its value. The general shape of the curve is preserved. For the purpose of a purely qualitative model such as ours, this effect can be considered as negligible and we have knowingly kept A_n a function of NP size.

Finally, we have considered an atom-by-atom growth and a system without evaporation (NP grow only by absorption of monomers). This is not exactly representative of the physical phenomena at play in the source, but does not alter the consistency of the model, as validated above. However, further – much more complex – development of the theory is possible to include NP-NP absorption and atom desorption at the surface of NPs. The nucleation and growth model would certainly benefit even more from these modifications, as it benefits from the ones already discussed so far. Nevertheless, an exhaustive description of the nucleation and growth phenomena within the source is not our main objective. Indeed, the simplified theory constitutes a sufficiently powerful tool for qualitatively describing and predicting NP synthesis in our nano-aggregate source, as was illustrated in the 'Usage and set-up optimization' paragraph.

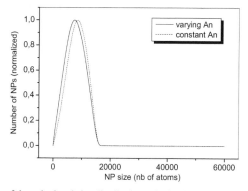

Fig. 5 : Comparison of the calculated size distributions obtained with A_n (the sticking probability) constant (dashed) and A_n a function of the NP size (solid line). The relative error on the NP mean size is about 10%, the general shape of the curve is preserved.

CONCLUSIONS

We have demonstrated that NP synthesis in our nano-aggregate source could be faithfully described, through modified classical theories, by a nucleation and growth process. Although a more complex and accurate description is possible, the consistency of this simplified model was validated through direct comparison of simulated and experimental results.

This model constitutes a powerful predicting tool to evaluate NP synthesis feasibility for all pure materials. By investigating the parameter space, typical NP sizes and flows can be estimated prior to experiment, which is particularly useful in the case of costly materials such as Pt. In conclusion, the model stresses the potential of the nano-aggregate source for applications that require an extremely good and uncorrelated control of NP sizes and flows (i.e. densities), such as energy conversion applications.

ACKNOWLEDGMENTS

The authors would like to thank L. Notin for his technical support in AFM characterization and A. Brenac and R. Morel for useful technical discussions.

FOOTNOTES

*Corresponding author : emmanuelle.pauliac-vaujour@cea.fr

REFERENCES

[1] P. Oelhafen and A. Schüler, Solar Energy **79**, 110 (2005).

[2] S. Pillai, K.R. Catchpole, T. Trupke, and M.A. Green, J. of Appl. Phys. **101**, 093105 (2007).

[3] C.R. Derouin and A. Redondo, J. of Power Sources **22**, 359 (1988).

[4] T. Itakura, K. Torigoe and K. Esumi, Langmuir **11** (10), 4129 (1995).

[5] S. Link, Z. L. Wang and M. A. El-Sayed, J. Phys. Chem. B **103** (18), 3529 (1999).

[6] S. Alayoglu, A.U. Nilekar, M. Mavrikakis, B. Eichhorn, Nature Materials **7**, 333 (2008).

[7] I. Yamada and G.H. Takaoka, Jpn. J. Appl. Phys. **32**, 2121 (1993).

[8] Y.H. Xu, S Hosein, J. Judy and J.P. Wang, J. of Appl. Phys. **97**, 10F915 (2005).

[9] A. H. Kean and L Allers, NSTI Nanotech Conference Proceedings, Boston, May 2006.

[10] E. Quesnel, E. Pauliac Vaujour and V. Muffato, J. Appl. Phys., vol.107, issue 4 (2010).

[11] E. Pérez-Tijerina, M. Gracia Pinilla, S. Mejia-Rosales, U. Ortiz-Mendez, A. Torres Torres and M. José –Yacama, Faraday Discussions **138**, 353 (2008).

[12] P. Feiden, PhD thesis, Univ.Paris XI (2007).

[13] O.Knacke, O.Kubaschewski and K.Hesselmann, Thermo-chemical properties of inorganic substances, 2nd ed., Springer-Verlag (1991).

[14] P.W. Voorhees, J. of Stat. Phys., vol.38, N°1/2 (1985).

[15] J. Wedekind, A.-P. Hyvärinen, D. Brus, D. Reguera, Phys. Rev. Lett. **101**, 125703 (2008).

CARBON ENCAPSULATED-IRON LITHIUM FLUORIDE NANOCOMPOSITE AS HIGH CYCLIC STABILITY CATHODE MATERIAL IN LITHIUM BATTERIES

Raju Prakash,* Christian Kübel and Maximilian Fichtner

Karlsruhe Institute of Technology (KIT), Institute of Nanotechnology (INT), Postbox 3640, 76021 Karlsruhe, Germany. E-mail: raju.prakash@kit.edu

ABSTRACT

Carbon encapsulated-iron LiF nanocomposite has been synthesized from a ferrocene/LiF mixture by pyrolysis under inert condition at 700 °C. The structure and morphology of the material was investigated by scanning electron microscopy (SEM), high-resolution transmission electron microscopy (HRTEM), and X-ray diffraction (XRD) analysis. The composite consists of multi-walled carbon nanotubes and onion-type graphite structures in which Fe and Fe_3C nanoparticles are encapsulated, and LiF is dispersed within the carbon matrix. The electrochemical performance of the nanocomposite has been studied using galvanostatic charge-discharge experiments at different current and potential ranges at room temperature. The nanocomposite exhibited a stable reversible specific capacity of 280 mAh/g. The influence of the carbon framework on the electrochemical performance of the synthesized nanocomposite is revealed.

INTRODUCTION

Lithium ion battery has become the premier technology for portable power applications.[1] The role of lithium batteries in future large-scale energy-storage applications (for instance in electrical vehicles, power backups, and other stationary devices) depends primarily on their ability to provide the required key features: high energy density, high power density, rate capability, safety, and extended cyclic stability. Additional factors, such as costs, natural abundance of the inherent materials, and their toxicity, will play an increasingly important role. In this context, iron-based electrochemical storage systems would be of particular interest. Currently oxides base cathodes are being used for Li-ion rechargeable batteries through Li-ion intercalation process, commonly referred to as the "rocking chair" mechanism. This process has restricted the reversible reaction to at most a single electron transfer per formula unit. Hence, the specific capacities of the state of the art lithium ion batteries are limited only to 150 mAh/g. The performance of a battery depends decisively on the properties of the electrode (cathode and anode) materials, it is therefore obligatory to search for new materials having high reversible storage capacities for lithium.[2] An alternative to the insertion reaction is the conversion reaction in which the active electrode material consumed by Li^+ and reduced to metal according to the Equation (1).[3,4]

$$M^{n+}X_n + nLi^+ + ne^- \Leftrightarrow nLiX + M \tag{1}$$

Where, M: transition metals; X: O^{2-}, N^{3-}, S^{2-}, F^- etc. The reactions can make use of all energetically favorable valence states of the metal cation, enabling a large theoretical energy density. Metal fluorides are the most favorable material for the above conversion reaction. Due to the highly ionic nature of their M–F bond, they should theoretically exhibit a much higher output voltage and lithium storage capacity than any other systems (e.g. TiF_3 767 mAh/g; VF_3 745 mAh/g; MnF_3 719 mAh/g; FeF_3 712 mAh/g; CoF_3 695 mAh/g).[5] However, many of the metal fluorides exhibit a limited electrochemical activity with lithium due to their poor electronic conductivity brought about by their large band gap.[6]

In late 90s, Arai et al. reported the reversible electrochemical activity of TiF_3, VF_3, and FeF_3.[7] These compounds reached a reversible capacity of 80 mAh/g in the potential range between 4.5 and 2.0 V which corresponds to the intercalation portion of the M^{3+}/M^{2+} redox couple. Heterogeneous conversion reactions of TiF_3 and VF_3 at room temperature have also been reported in which the reversible Li storage capacities of TiF_3 and VF_3 observed were as high as 500-600 mAh/g between 4.3 and 0.02 V.[8] Later, Badway et al. demonstrated that an FeF_3/C nanocomposite, prepared by ball milling method, reached an initial discharge capacity of 200 mAh/g between 4.5 and 2.0 V for the Fe^{3+}/Fe^{2+} redox couple.[9] Three-electron reduction (Fe^{3+}/Fe^0) was also reported by using the same nanocomposite at 70°C with capacities >500 mAh/g.[10] In addition, Wu et al. reported that a nanocomposite of FeF_3/MoS_2 or FeF_3/V_2O_5 exhibited a better electrochemical performance than undoped FeF_3 electrode in the potential range of 2.0–4.5 V.[11]

Despite intensive research,[12] application of metal fluoride cathodes has largely hampered by partial reaction irreversibility and poor cycling stability. Hence, an alternative approach to material synthesis is needed.[13-15] Therefore research efforts on encapsulation of metal/metal fluorides in various carbon materials gained much attention. Recently, we have demonstrated that the encapsulation of iron nanoparticles in multiwalled carbon nanotubes/graphitic onion structures led to a significant enhancement of the cycling performance when it applied as cathode in lithium-batteries.[16] Moreover, the three-electron conversion reaction was achieved at room temperature. Herein we report the detailed synthesis and characterization of the novel carbon encapsulated-iron LiF nanocompsite. In addition, its electrochemical performances at room temperature under different potentials as well as current ranges will be reported.

EXPERIMENTAL

Sample preparations and processing steps were carried out in an argon-filled glove box and/or using standard Schlenk techniques under argon atmosphere. Analytical-grade ferrocene and lithium fluoride (Alfa Aesar Co.) were dried at 150 °C under vacuum for 15 h, then ball-milled for 12 h under argon in a planetary ball mill (Fritsch GmbH) using tungsten carbide balls and vials (balls to powder ratio 100:1).

Nanocomposite synthesis

A homogeneous mixture of ferrocene (900 mg, 4.8 mmol) and LiF (375 mg, 14.5 mmol) was filled into a stainless steel reactor (inner diameter, 0.7 cm; length 7 cm) and sealed with VCR fittings at both ends inside a glove box. The reactor was placed horizontally inside a tube furnace (GERO) at room temperature. Then, the furnace was heated at a rate of 5 °C/min to reach the set temperature of 700 °C. After pyrolyzed for 2h, the reactor was cooled to room temperature. The reactor was removed from the oven and transferred into the glove box. Due to various gaseous materials produced during pyrolysis, overpressure had developed inside the reactor. It was opened carefully inside a glove box in order to release the pressure. The product obtained was a dry, very fine, and soft black powder of > 90 wt% of the reactants. The carbon content of the nanocomposite was determined to be 41.5% based on elemental analysis. The ratio between total Fe and LiF contents in the nanocomposite was estimated by EDX analysis as 1: 2.8.

Physicochemical Characterization

The morphology of the sample was observed on a LEO-1530 scanning electron microscopy (SEM). Transmission electron microscopy (TEM) characterization was performed on an FEI Titan 80-300 operated at 300 kV and a Tecnai F20 STwin operated at 200 kV. Both TEM instruments were equipped with a Gatan imaging filter (Tridiem 863, GIF2001), energy dispersive analysis of X-rays-ultra thin window (EDAXs-UTW) energy dispersive X-rays (EDX) detectors, and Fischione high angle angular dark field-scanning transmission microscopy (HAADF-STEM) detectors. TEM samples were prepared by dispersion in dry pentane. A droplet was placed onto a carbon grid and allowed to dry for at least 2 h in the glove box. The grids were sealed and reopened to put them into the TEM, thereby minimizing exposure to atmosphere to about a minute. Powder X-ray diffraction patterns were obtained with a Philips X'PERT diffractometer (Cu Kα radiation, 2 kW, with X'Celerator RTMS detector, automatic divergence slit). PANalytical X'Pert Data Collector and X'Pert HighScore software were used for data acquisition and evaluation, respectively. The samples to be analyzed were spread onto a silicon single crystal in the glovebox and sealed with an airtight hood made of Kapton foil, which is out of the focus of the spectrometer. The patterns were recorded at 25 °C in a 2θ range between 10 and 80°. The crystallite sizes were calculated by using the Scherer formula. Electrochemical experiments were performed with two electrode-Swagelok-type cells. The positive electrode was fabricated by mixing the nanocomposite and poly(vinyldifluoride-co-hexafluoropropylene) (SOLEF 21216/1001) at a weight ratio of 90:10. A pure lithium foil (Goodfellow) was used as negative electrode. A layer of glass fiber (GF/D; Whatman) was used as separator. The electrolyte solution was prepared by dissolving 1 M $LiPF_6$ in a mixture (1:1 by volume) of ethylene carbonate (EC) and dimethyl carbonate (DMC). Galvanostatic charge-discharge cycling tests were performed using an Arbin BT2000 multi-channel battery testing system at various current densities in the voltage range between 4.3 and 0.5 V. After performing one or several charge-discharge cycles, the electrodes were removed from the cells, washed with anhydrous DMC, and dried under vacuum for 2 h for further analyses.

RESULTS AND DISCUSSION

The nanocomposite was synthesized by pyrolysis of an intimate ferrocene-LiF mixture in a closed reactor under argon atmosphere at 700 °C for 2 h. The amount of LiF used (approximately 3 mole ratio to Fe) for the synthesis of nanocompostie was based on the total amount of Fe which can produce from pyrolysis of the given quantity of ferrocene. Energy dispersive X-ray analysis was performed with the nanocomposite and the relative molar ratio of total Fe (Fe/Fe_3C) and F (from LiF) was observed to be 1:2.8. The composite has a carbon content of 41.5% as determined by elemental analysis. To analyze the composition and structure of the nanocomposite, powder X-ray diffraction (XRD) experiments were carried out. Figure 1 displays XRD patterns of the ferrocene-LiF mixture before and after pyrolysis. It is obvious from the Figure that the ferrocene has been converted completely into carbon and iron. The carbon produced by this method exhibits a pronounced graphitic in nature as indicated by the characteristic diffraction peak at 26.4° (002). Furthermore, two distinct iron-rich phases viz. Fe_3C and α-Fe were identified. The lattice parameters of Fe_3C are: Orthorhombic; $a = 5.0910$ Å, $b = 6.7434$ Å, $c = 4.5260$ Å; space group $Pnma$ (62), Joint Committee on Powder Diffraction Standards (JCPDS) powder diffraction file (PDF) 035-0772; The lattice parametars of α-Fe are: Cubic $a = b = c = 2.8664$ Å; space group $Im3m$ (229), PDF 006-0696. The formation of Fe_3C had been expected due to the dissolution of carbon atoms in Fe during nanotube growth at high temperature. The XRD pattern did not show any obvious reflections that correspond to iron oxides or iron fluorides, which indicates that the metallic iron nanoparticles have neither oxidized nor reacted with LiF during pyrolysis.

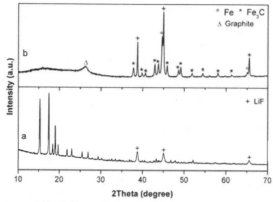

Figure 1. XRD patterns of (a) a ball-milled mixture of ferrocene (unassigned peaks) and LiF, (b) the nanocomposite formed after pyrolysis of the mixture at 700 °C for 2 h under argon atmosphere.

The morphology of the material was examined by using SEM and TEM. Figure 2 shows SEM images of the as-prepared nanocomposite.

Figure 2. SEM image of Carbon-Fe-LiF nanocomposite.

The overview image reveals iron-rich nanoparticles embedded in a graphitic matrix which contains nanotubes. The particle size of the iron nanoparticles are in the range between 2-50 nm, but occasionally up to 100 nm diameters are also observed. High resolution TEM images of the nanocomposite (Figure 3) indicated that the iron-rich particles are typically either located inside graphitic multi-walled nanotubes or embedded inside onion-like graphite spheres. In addition, selected area electron diffraction (SAED) analyses were performed, and it confirms (based on the lattice parameter) that iron (α-Fe) and/or iron carbide (Fe$_3$C) nanocrystals are randomly encapsulated in both the nanotubes as well as onionic graphitie-layers. Closer inspection of the graphitic matrix revealed a significant number of open-ended tube-like structures as well as inhomogeneities in the onion-like graphitic coating indicating also partial encapsulation of the iron-rich particles by the graphitic matrix. HAADF-STEM in combination with local electron energy loss spectroscopy (EELS) analysis reveals small amounts of fluorine (presumably as LiF) present in most of these aggregates. In addition, fairly smooth aggregates exhibiting strong fluorine and noticeable iron signals were present in the material, indicating that these aggregates mainly consist of LiF and iron-rich nanoparticles.

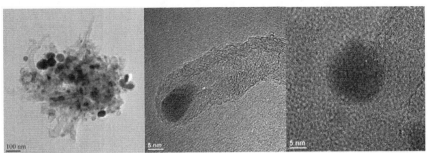

Figure 3. TEM images of Carbon-Fe-LiF nanocomposite. Left: BF-TEM overview image. Center: Iron-rich nanoparticle encapsulated inside a tubular graphitic structure. Right: Iron-rich particles confined inside onion-like graphitic structure.

The charge/discharge experiments were performed at different potential ranges at room temperature to evaluate electrochemical properties of the nanocomposite. Figure 4 exhibits the charge and discharge profiles of the nanocomposite cathode at first and twentieth cycles. The given voltage profiles are consistent with metal fluoride systems those have already been reported in literature.[8-10] In the potential range between 4.3 and 0.5 V, it exhibits three plateaus in the discharging processes. However, the observed short-flat plateaus indicated that the active material is not converted completely. The charge/discharge specific capacities are determined based on the mass of the active material FeF$_3$ (calculated from amount of LiF and Fe present in the composite electrode). At the current density of 20.83 mA/g (0.02C), the first charge and discharge capacities are as high as 372 and 324 mAh/g, respectively. The first plateau during discharge appears between 3.5 and 2.5 V with a capacity of about 80 mAh/g corresponds to the intercalation of Li into FeF$_3$ to form LiFeF$_3$ and the second plateau between 2.5–1.3 V with a capacity of about 120 mAh/g is corresponding to the conversion reaction of LiFeF$_3$ and Li to form Fe and LiF. The third sloped region between 1.3–0.5 V yielded a capacity of about 120 mAh/g can be due to an interfacial interaction of lithium within the M/LiX matrix as proposed in literature.[3,5,6]

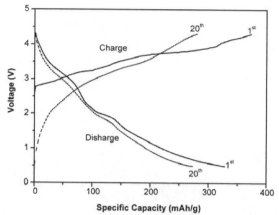

Figure 4. Charge and discharge profiles of Carbon-Fe-LiF nanocomposite which was cycled between 4.3 and 0.5 V at 20.83 mAh/g.

In order to confirm the formation of FeF_3 by a three-electron reaction of LiF and Fe, ex-situ XRD analyses were performed with the nanocomposite electrodes which were at 1, 30, 75 and 125th charged states. The XRD pattern of the composite electrode at the 1st charged state exhibits a distinct Braggs peak at 23.83° which corresponds to FeF_3 (012) phase. Consequently, the Fe signals vanished completely, while the Fe_3C and LiF peak intensities decreased. It suggests that metallic iron (cubic) reacts readily with LiF to form FeF_3. On the other hand, reactivity of the orthorhombic Fe_3C towards lithium fluoride is much less than those of metallic iron. This may be one of the reasons that the specific capacity is much lower than the theoretical value. XRD patterns of the electrode material at 30th, 75th and 125th charged states were reveal that the FeF_3 remained as the only the significant phase developed. However, a small quantity of FeF_2 phase gradually formed during cycling as observed by an increase intensity of the FeF_2 Bragg reflection (110) at 26.8°.

The plot of specific capacity as a function of cycle numbers at different voltage ranges is shown in Fig. 5. It is evident that the capacities decrease over the first few cycles and then stabilizes. In the potential range between 4.3 and 0.5 V, the capacity is maintained at 275 mAh/g after 125 cycles. The loss of capacity between 5 and 125 cycles was less than 3%. The coulombic efficiency (calculated from the ratio of discharge capacity to charge capacity) in the first cycle is 86%. Subsequently, the value always is in the range between 97 and 100%. When the charge/discharge reactions were performed between 4.3 and 1.3 V at the given current density of 20.83 mA/g, it delivered an initial discharge capacity of 190 mAh/g for Fe^{3+}/Fe^0 redox couple. After 5 cycles, the capacity remained stable at about 170 mAh/g, and the capacity retention ratio was 90% after 125 cycles.

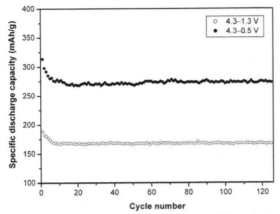

Figure 5. Specific capacity as a function of cycle number at different voltage ranges.

Additionally, the nanocomposite electrode exhibits a good rate capability due to improved diffusion kinetics. At first, the cell was cycled at 20.83 mA/g (0.02C) for 30 cycles, and then the current densities were increased in a stepwise manner to 2.1 A/g (2C) every 30 charge/discharge cycles (Figure 6). It is apparent from the figure that the reversible capacity decreases as the C-rate increases. For instance, at 0.02C rate, the average reversible specific capacity has attained as high as 275 mAh/g, while at 2C rate, it reaches to 92 mAh/g. Moreover, if the current density is lowered again to 0.02C after 180 cycles, a capacity of 260 mAh/g can still be attained. It is interesting to note that the capacity remains stable over the entire 30 cycles at any given C-rate.

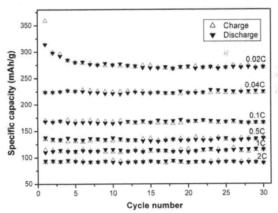

Figure 6. Specific capacity vs. cycle number at various current densities. The cell was cycled between 4.3 and 0.5 V for 30 cycles at every current density.

As a cathode material, the nano nanocomposite exhibits excellent cyclic stability and rate capability than that of the reported carbon-metal fluoride nanocomposites prepared by ball-mill method.[8-10] We believe that the cyclic stability of the system, which comes presumably from the carbon framework which is produced during the pyrolysis. The typical tube- and onion layers-like carbon matrix might have played a crucial role in integrating the active materials during the redox reaction thereby maintaining the cycle stability. In order to check the influence of the carbon framework in maintaining cyclic stability, the original nanocomposite was ball-milled for 6 h under argon atmosphere. The SEM image of the ball milled nanocomposite material (see Figure 7) indicates a complete destruction of the typical carbon framework. The XRD of the ball-milled material shows similar diffraction pattern as that of the original nanocomposite, however, substantial broadening of the Bragg peaks occurred.

Figure 7. SEM image of Carbon-Fe-LiF nanocomposite after ball milling.

The charge/discharge experiments were performed at different potential ranges using the electrode made of the ball-milled nanocomposite at a current density of 20.83 mA/g (Figure 8). It exhibited much higher initial capacities. For example, the first charge and discharge capacities of 488 and 392 mAh/g were achieved between 4.3 and 0.5 V, respectively. Despite the higher initial capacities, the ball milled material exhibits a rapid loss of capacity with cycling. After 50 cycles, the charge and discharge capacities were reduced to 90 and 87 mAh/g, respectively. The cyclic instability in this case can directly be attributed to the destruction of the graphitic nano-encapsulation by ball milling, which in turn leads to the disintegration of the materials during phase transformation resulting in an electrical disconnection of the particles. In contrast, the original nanocomposite electrode under identical conditions shows an excellent cyclic stability.

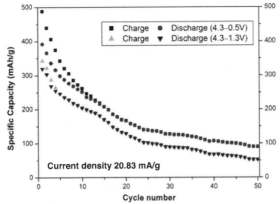

Figure 8. Cyclic performance of the ball milled nanocopmpsite at different voltage ranges.

CONCLUSIONS

The carbon-encapsulated Fe-LiF nanocomposite has been successfully synthesized using ferrocene and LiF by a simple one-step pyrolysis method. The nanocomposite consists of multi-walled carbon nanotubes and onion-type graphite structures in which iron and iron carbide particles are encapsulated, and LiF is dispersed throughout the matrix. The nanocomposite cathode exhibited a reversible capacity of 280 mAh/g in the potential range 0.5–4.3 V and of 170 mAh/g in the range of 1.3–4.3 V at a current density of 20.83 mA/g. The nanocomposite cathode showed excellent cyclic as well as a durable rate capability. The results are clearly indicated that the carbon framework plays a major role in determining the cyclic performance of the battery. But the composite exhibits a moderate capacity, only about 40% of the theoretical value of FeF_3 (712 mAh/g) was achieved. However, efforts are in progress to optimize the Fe and carbon contents of the nanocomposite in order to increase the capacity of the system. In any case, this simple technique can be applied to synthesize a variety of inexpensive carbon-encapsulated nanocomposites.

ACKNOWLEDGEMENTS

Financial support by the BMBF project "*LIB-NANO*", MWK project "*e-drive*" (State of Baden-Württemberg) and "*Verbund Süd*" is gratefully acknowledged.

REFERENCES
[1] X.-P. Gao, H.-X. Yang, Multi-Electron Reaction Materials for High Energy Density Batteries, *Energy Environ. Sci.*, **3**, 174 (2010).
[2] Y.-G. Guo, J.-S. Hu, L.-J. Wan, Nanostructured Materials for Electrochemical Energy Conversion and Storage Devices, *Adv. Mater.*, **20**, 2878 (2008).
[3] P. Poizot, S. Laruelle, S. Grugeon, L. Dupont, J. M. Tarascon, Nano-Sized Transition-Metal Oxides as Negative-Electrode Materials for Lithium-Ion Batteries, *Nature*, **407**, 496 (2000).
[4] S. Laruelle, S. Grugeon, P. Poizot, M. Dolle, L. Dupont, J.-M. Tarascon, *J. Electrochem. Soc.*, **149**, A627 (2002).

THE ORTHO-PHOSPHATE ARROJADITE AS A NEW MATERIAL FOR CATHODES IN LI-ION BATTERIES

C. Kallfaß, C. Hoch and H. Schier
Max-Planck-Institut für Festkörperforschung
D-70569 Stuttgart, Germany

C. Wituchowski, O. Görke and H. Schubert
TU Berlin, Institut für Werkstoffwissenschaften und -technologien
D-10587 Berlin, Germany

ABSTACT

Li-ion batteries as power storage devices for many technical applications have developed tremendously during the last 15 years. Upon analyzing commercially available Li-ion batteries based on different cathode materials, it becomes clear that the durability and the failure of the battery as a whole is closely connected to the most commonly used assembly of the cathode which consists of a redox active material (cathode material), graphite and organic binder. Strain and stress in the cathode material during the cycle process caused by the variable lithium content are the main reasons for these problems.

A novel cathode material based on a transition metal-rich ortho-phosphate is presented utilizing the same chemistry with a different structure. It exhibits an excellent cycle behavior because its lithium content affects its structure negligibly. . In contrast to the commercially used $LiFePO_4$ cathode material in 3.5V batteries this novel ortho-phosphate material offers the opportunity to be used in the so-called "5V-technology". This is possible without the use of expensive and polluting transition metals such as cobalt or nickel.

INTRODUCTION AND LITERATURE

Li-ion Batteries are considered to be the fundamental electrochemical device for mobile communication and transport systems. The arrangement of battery's cathode plays a decisive role in the durability and application possibility of the whole system. A wide range of properties is possible: from high current density to high thermal stability and long durability - in all cases without a significant change in the properties. One of the most desirable properties is a high cycle durability. This is limited to numbers of 500 to 3000 cycles before failure. The reason is hidden in the first order phase transformation of the battery materials (i. e. in the case of $LiFePO_4$).

The following paper highlights some structural aspects of the most common cathode materials. $LiCoO_2$, lithium cobaltate, is a layered structure (α-FeO_2). The Li-ions are located in a layer, followed by cobalt oxygen octahedral layers. The conductivity of lithium cobaltate is due to the mobility of lithium cation in the AB plane, i.e. it is a two dimensional conductor. There is no electronic conductivity. Thus, a battery had to be designed to guide electrons from the surface redox reaction by separate carbon particles. Their role is to increase the cathode`s conductance of electrons towards the metallic connector. The energy density of this material is about 500 Wh/dm^3 [1]. Disadvantages of lithium cobaltate are limited thermal stability, a high price, and the use of toxic cobalt. In 1983 the lithium manganate spinel, $LiMn_2O_4$, was introduced[2]. Increasing the electronic conductivity of these materials for achieving the highest possible capacity was the subject of many research projects over the years. This was attained by mixed occupation of lithium and transition metal postitions in the lattice of the cathode material.

LiFePO$_4$ has been introduced as cathode material in 1997[3,4]. The orthorhombic structure is comparable to olivine. The oxygen anions form a slightly disordered hexagonal sphere packing order (ABAB). The resulting tetrahedral and octahedral coordinated vacancies of this packing order are occupied by Li-, Fe- and P-cations respectively. The voltage of this iron-containing Li-ion battery ranges from 3.2 - 3.5V and a final capacity of c~150 mAh/g[1] can be achieved. LiFePO$_4$ as cathode material is comparably inexpensive and not toxic. The material is thermally stable up to 473 K in air[5]. Pronounce this advantage is the high elastic strain on charging and discharging. If the discharging of LiFePO$_4$ is driven too far, strain of almost 6 vol.% occurs, which is far higher than the fracture strain. Thus, discharging of LiFePO$_4$ to FePO$_4$ with the full extent of lithium is not possible. This process would lead directly to failure and loss of functionality of the cathode material.

Fig. 1 shows the comparison of the LiFePO$_4$ and the FePO$_4$ structure. The redox reaction depending on the transition metal used determines the voltage of the Li-ion battery in the range of 3.2V (for Fe^{2+}/Fe^{3+}) to 4.5V (for Co^{2+}/Co^{3+})[3]. Apart from these types of Li-ion batteries, a technology based on lithium titanate, Li$_4$Ti$_5$O$_{12}$, has been described intensively[6-11]. If lithium titanates are only partially discharged, numbers of charge and discharge cycles as high as 12000 have been achieved and reported, but the usable voltage of this battery type is limited to 2V. An overview on the used materials and their properties is given in table 1.

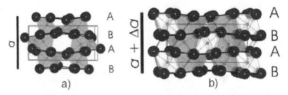

Figure 1. Structures of FePO$_4$ (a) and LiFePO$_4$ (b). The oxygen atoms are drawn in blue. The coordination polyhedra of iron is drawn in green, the phosphate tetrahedra are drawn in grey. The lithium atoms are red colored and shown in colorless polyhedra.

Table 1. Selected characteristics of commercially used cathode materials for Li-ion batteries.

Material	Structure type	Voltage (V)	Redox couple	Number of phases	Toxicity	Price	Reference
LiCoO$_2$	layer	< 4.2	Co^{2+}/Co^{3+}	2	high	expensive	[12] p. 53-55
LiNiO$_2$	layer	< 4.3	Ni^{2+}/Ni^{3+}	2	high	expensive	[13]
LiMn$_2$O$_4$	spinel	~ 2.9	Mn^{2+}/Mn^{3+}	≥ 2	low	inexpensive	[14]
Li$_4$Ti$_5$O$_{12}$	spinel	~ 2.0	Ti^{2+}/Ti^{3+}	1	low	inexpensive	[07-11]
LiFePO$_4$	olivine	~ 3.5	Fe^{2+}/Fe^{3+}	≥ 2	low	inexpensive	[15]

EXPERIMENTAL WORK, RESULTS AND DISCUSSION

Commercial Li-ion batteries using cobaltate, manganate and phosphate cathodes have been used for the cycling test and the following microstructural and X-ray analysis. The commercial available Li-batteries contain often a mixture of different cathode materials. In our case the lithium manganese respectively cobaltate labelled batteries persist of lithium manganate respectively cobaltate (major phase) and Li(Ni,Co)O$_2$ (minor phase). Two types of cycling tests were applied. Both contained charge and discharge cycles from maximal usable voltage to 2V. One test simulated a common usage of Li-ion batteries in technical applications like mobile phones or laptops. The battery was cycled for at least 600 cycles in randomised periods, for instance fully discharged to 2V in one cycle and only half-charged in the next cycle. The batteries of second group were cycled from maximal usable voltage to 2V until the failure of the battery. Microstructural analysis was carried out in the uncycled state, at the end of the cycle test and at the end of their life time. For microstructural analysis of the cathodes and the anodes, SEM-EDX (Zeiss 1540EsB) was used. The powder patterns (XRD) were recorded using a common laboratory powder diffractometer (Philips PW1070) in the range of 10°-90° 2 with LiF-monochromised Cu-Kα1-radiation (step size 0.02° and counting time 20 s per step).

Figure 2a. SEM images of uncycled (left) and cycled (right) LiCoO$_2$ cathodes (top view).

Fig. 2a shows the microstructure of uncycled and uncycled LiCoO$_2$ cathodes. The isometric grains are partly embedded in a polymeric matrix. The powder pattern shows structural composition of the cathode which is a mixture of lithium cobaltate and lithium nickelate (Fig. 2b).

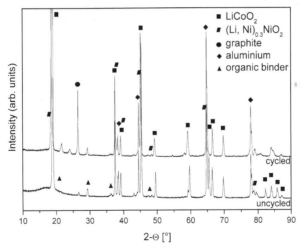

Figure 2b. XRD pattern of uncycled and cycled LiCoO$_2$ cathode.

Fig. 3a shows the microstructure of a lithium manganate cathode, which is composed out of different phases. EDX imaging reveals extended areas of carbon which are in a dense bulk phase (Fig. 3b). In contrast the manganate is agglomerated in cauliflower-like structures. Nickel is located in isolated spots. The powder pattern (Fig. 4) shows a lithium manganate phase and a separated lithium nickelate phase. After cycling the microstructure of the cathode has changed entirely (Fig. 3a). The integrity of the agglomerated cathode structure is more or less lost. The outside of the structure has remained stable because of the lithium concentration. The core however has been altered.

Figure 3a. SEM images of uncycled (left) and cycled (right) LiMn$_2$O$_4$ cathodes (top view).

Figure 3b. SEM-EDX image of uncycled LiMn$_2$O$_4$ cathode (element identification: C-yellow, Mn-red, Co-blue, Ni-cyan, O-green; top view).

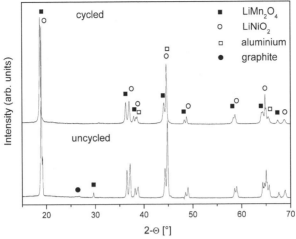

Figure 4. XRD pattern of uncycled and cycled $LiMn_2O_4$ cathodes.

Fig. 5a shows the microstructure of the $LiFePO_4$ cathode in the uncycled and cycled state. The flat morphology is altered towards a more undulated surface. Hence, small particle sizes appeared to be less prevalent in the cycled state. The powder patterns show an increasing amount of $FePO_4$ while cycling the battery. As an additional component a highly conductive iron phosphide Fe_2P phase can be identified[16] (Fig. 5b).

Figure 5a. SEM images of uncycled (left) and cycled (right) $LiFePO_4$ cathodes (top view).

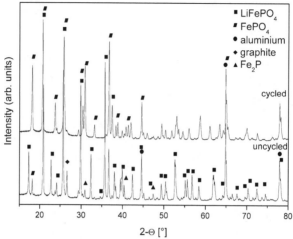

Figure 5b. XRD patterns of uncycled and cycled LiFePO₄ cathodes.

Summarizing the previous investigations, the functionality of a battery is fixed by the electrical properties, but the attainable number of cycles to failure depends on mechanical effects such as strain, strain hindrance, and crack formation.

The key issue with further material research is preparing a material with good electrical properties and low cycle strains. The mineral arrojadite with the simplified formula $(Li,Na,K)_{5.5}(Mg,Ca)_2Al(Mn,Fe,Zn)_{13}(PO_4)_{12}*2H_2O$ is one candidate for this type of new cathode materials. Arrojadite is a transition metal-rich ortho-phosphate mineral with special structural features which were examined by applying modern X-ray and neutron single crystal diffraction and refinement methods. The rigid framework of the structure is made up of phosphate tetrahedra and different coordination polyhedra of iron and manganese. The structure of arrojadite is characterized by an extended thermal stability up to $T \leq 820$ K in oxidizing and inert gas atmospheres[5]. A new and promising feature of the structure, two different types of channels oriented along [010], are described by our group. The occupancy of the atomic positions inside the channels are elucidated and found to be characteristic of the origin of the mineral specimen[17]. The occupancy of these atomic positions inside the channels may be modified by methods of "green chemistry" to achieve a lithium enriched specimen[18]. This process was controlled by using the ⁶Li isotope and following verification of the lithium enrichment by solid state ⁶Li-MAS-NMR.

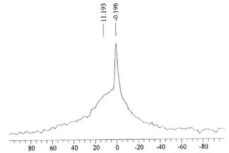

Figure 6. ^{6}Li-MAS-NMR spectra of a lithium enriched arrojadite powder sample.

Fig. 6 shows the ^{6}Li-MAS-NMR spectrum with two separated peaks. This fact answered a key question: lithium cations will occupy atomic positions inside the channels of the arrojadite structure.

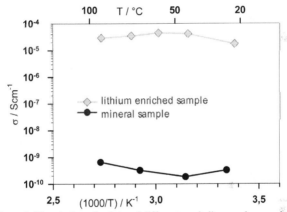

Figure 7. Electrical conductivity of different arrojadite powder samples.

These channels are the migration pathways for one-dimensional lithium cation conductivity. The conductivity of mineral samples is low and enhances with increasing lithium content inside the channels (Fig. 7). The non-Arrhenius behavior of the total conductivity (impedance measurement) versus temperature implies that more than one charge carrier is involved. Further polarization measurements indicate mixed conductance.

These results have been promising enough to build a demonstrator cell. As a first attempt, the battery concept of Chung[19] has been applied. The cell contains four parts in a Swagelok housing: the cathode, which is made of an aluminum foil coated with arrojadite, carbon particles, and an organic binder, the separator (Celagard 2300 film), the lithium metal anode, and the electrolyte (mixture of ethylenecarbonate and dimethylcarbonate). The cathode material arrojadite contains manganese and iron; this cell reaches a final charging voltage of 4.5V for the Mn^{2+}/Mn^{3+} redox couple. This fact offers the possibility of usage as environmentally friendly cathode material in the upcoming "5V-technology" without containing toxic and expensive transition metals like cobalt and nickel.

A cycling test of the arrojadite battery was set up to investigate the maximum number of charge and discharge cycles. While most of the commercially available Li-ion batteries are cycled to a lower cut off voltage of 2V, the arrojadite battery was cycled in the range of 2-4.5V. Fig. 8 shows the results of the cycling tests. A commercially available $LiFePO_4$ battery could be cycled for about 2800 times before failure. In contrast to this, an arrojadite battery works with a high usable capacity (≥ 70 % of the starting value) for more than 19,000 cycles.

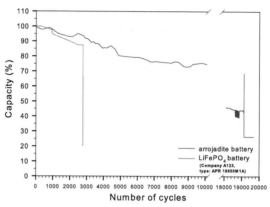

Figure 8. Comparison of a commercially available $LiFePO_4$ battery and our arrojadite battery.

The explanation for this behaviour is given in the rigid framework structure with stable channels. The small changes inside the lattice of arrojadite during the cycling process are the advantage of this material (Fig. 9). The low attainable capacity is a problem in this initial instance.

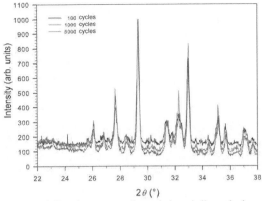

Figure 9. Powder patterns of a cycled arrojadite cathode.

CONCLUSION

The discussion of how the performance of the whole battery will be also influenced by the binder and the carbon particles is induced by analyzing the microstructure of the cathode. There is the possibility of a lower performance due to isolation of the cathode material depending on binder and carbon content of the whole cathode. Further investigations are necessary. However, in our case, arrojadite is a mixed conductor, and therefore a second path for the electron conductance provided by carbon is not necessary. From this it follows that cathode and battery should be designed prospectively for the physical and crystallographic properties of the cathode material used.

REFERENCES

[1]T. Ohzuku, R. J. Brodd, An overview of positive–electrode materials for advanced lithium–ion batteries, *J. Power Sources*, **174**, 449–456 (2007).
[2]M. M. Thackeray, W. I. F. David, P. G. Bruce, J. B. Goodenough, Lithium insertion into manganese spinels, *Mat. Res. Bull.*, **18**, 461–472 (1983).
[3]A. K. Padhi, K. S. Nanjundaswamy, J. B. Goodenough, Phospho–olivines as positive-electrode materials for rechargeable lithium batteries, *J. Electrochem. Soc.*, **144**, A1188–A1194 (1997).
[4]A. K. Padhi, K. S. Nanjundaswamy, C. Masquelier, S. Okada, J. B. Goodenough, Effect of structure on the Fe^{3+}/Fe^{2+} redox couple in iron phosphates, *J. Electrochem. Soc.*, **144**, 1609–1613 (1997).
[5]C. Kallfaß, C. Hoch, R. Dinnebier, H. Schier, H. Schubert, Thermal features of a transition metal–rich *ortho*–phosphate arrojadite: II. Part - high temperature *in-situ* powder diffraction. To be published.
[6]http://www.sequence-omega.net/2009/04/19/toshiba-ups-the-lithium-titanate-battery-ante-with-scib/
[7]http://www.b2i.cc/Document/546/11Ah_Datasheet-012209.pdf
[8]http://www.b2i.cc/Document/546/50Ah_Datasheet-012209.pdf
[9]http://www.altairnano.com/profiles/investor/fullpage.aspf=1&BzID=546&to=cp&Nav=0&LangID=1&s=0&ID=10701
[10]http://www.mobilitaet21.de/innovative-technik/innovative-fahrzeug-technik.html?
    user_umm21_pi1%5Bdetail%5D=71&cHash=124d5fd28f
[11]http://www.golem.de/0502/36525.html
[12]M. Wakihara, O. Yamamoto, Lithium ion batteries, fundamentals and performance. Wiley-VCH, Weinheim (1998).

[13]M. Winter, J. O. Besenhard, M. E. Spahr, P. Novak, Insertion electrode materials for rechargeable lithium batteries, *Adv. Mater.*, **10**, 725-763 (1998).

[14]M. M. Thackeray, P. J. Johnson, L. A. de Picciotto, Electrochemical extraction of lithium from $LiMn_2O_4$, *Mat. Res. Bull.*, **19**, 179-187 (1984).

[15]H. Huang, S.-C. Yin, L. F. Nazar, Approaching theoretical capacity of $LiFePO_4$ at room temperature at high rates, *Electroch. Solid-State Lett.*, **4 (10)**, A170-A172 (2001).

[16]P. Subramanya Herle, B. Ellis, N. Coombs, L. F. Nazar, Nano-network electronic conduction in iron an nickel olivine phosphates, *Nature Mat.*, **3**, 147-152 (2004).

[17]C. Kallfaß, C. Hoch, H. Schier, A. Simon, H. Schubert, The transition metal–rich *ortho*–phosphate arrojadite with special structural features. To be published.

[18]C. Kallfaß, G. Kaiser, C. Hoch, H. Schier, H. Schubert, Chemical constitution and modification of a transition metal-rich *ortho*-phosphate arrojadite. To be published.

[19]S.-Y. Chung, J. T. Bloking, Y.-M. Chiang, Electronically conductive phospho-olivines as lithium storage electrodes, *Nature Mat.*, **1**, 123-128 (2002).

Solar

A NOVEL PURIFICATION METHOD FOR PRODUCTION OF SOLAR GRADE SILICON

Shaghayegh Esfahani[1] and Mansoor Barati[2]

[1]M.A.Sc. Candidate, Department of Materials Science and Engineering
University of Toronto
[2]Assistant Professor, Department of Materials Science and Engineering
University of Toronto
Toronto, Canada, Ontario

ABSTRACT

Purification of metallurgical grade silicon (MG- Si) by a combination of solvent refining and physical separation has been studied. MG-Si was alloyed with iron and solidified under different cooling rates to grow pure Si dendrites from the alloy. The Si dendrites and $FeSi_2$ that were formed after solidification were then separated by a gravity-based method. The separation method relies on significantly different densities of Si and $FeSi_2$, and uses a heavy liquid with specific gravity between the two phases to float the former on the surface of a heavy liquid, while the latter sinks to the bottom. The effect of particle size and cooling rate on the Si yield and separation efficiency of the Si phase was investigated. The floated Si particles were further purified by removing the physically adherent Fe-Si phase, using an acid leaching method. Analysis of the produced silicon indicates that several impurity elements can be efficiently removed using this simple and low-cost technique.

INTRODUCTION

Silicon is the material of choice for manufacture of solar cells, accounting for over 90 percent of today's PV materials (1). Traditionally, Solar Grade Silicon (SoG-Si) was supplied from the off-grade semiconductor grade silicon which is inherently more pure than the requirements for PV applications. Higher cost and insufficient supply of electronic grade silicon rejects together with a growing demand for SoG-Si, has created a substantial thrust for developing a dedicated method of SoG-Si production. One stream of research explores refining of Metallurgical Grade Silicon (MG-Si) to SoG-Si by inexpensive metallurgical refining routes.

MG-Si is a relatively low purity source of Si (~98% Si). However, its availability in large quantities and low price (3.2 \$/kg (2)) favours the use of Mg-Si as the primary feedstock for SoG-Si production. It is believed that a single metallurgical refining process is not sufficient to lower the impurity level of Mg-Si to SoG-Si specifications, due to presence of numerous impurities with different chemical properties. A successful refining process should likely combine several refining stages, each responsible for lowering a certain number of impurities, to meet the purity requirements. However, it is worth bearing in mind that metallurgical purification processes such as oxidation, slagging, and controlled solidification are relatively cheap. Thus a multi-stage process could still provide an economically attractive option for upgrading MG-Si to SoG-Si.

Solvent refining is a purification process in which recrystallization takes place from the supersaturated melt (3). Principally, the method is similar to the well-established crystal growth techniques (such as Zone Refining and Czokralski), that reject impurities of Si to the liquid front, while solidification takes place. In the solvent refining method, however, this effect is promoted by addition of another element that acts as the impurity "getter". When silicon is alloyed with an element such as aluminum, metal impurities will have higher solubility in the alloy melt than in Si if the temperature exceeds the eutectic temperature of that binary system. Therefore, under controlled solidification, pure Si dendrites will grow from the alloy melt, while the impurities are preferentially segregated to the alloy. The binary Si-getter system should be in such a way that the element to be alloyed with silicon has sufficiently low solubility in silicon. The effectiveness of this method for removing different elements depends on the segregation coefficients of the impurities by definition is the concentration of an impurity in solid silicon to the concentration of that impurity in the alloy melt. A recent study by K.Morita and T.Yoshikawa (4) shows that by alloying Si with aluminum, segregation coefficients of the impurities in alloyed silicon (i.e. Al-Si) are reduced substantially compared to those for unalloyed Si.

Although the solvent refining method discussed above has shown success in producing high purity Si dendrites, the separation of the dendrites from the intimately surrounding matrix, i.e. the alloy phase, has been troublesome. Acid leaching is an effective way for the removal of the inter grains composed of Si- Al eutectics between Si grains after the solidification of the alloyed melt. In order to improve the acid leaching one other method is necessary to be applied prior to leaching to collect the Si grains and produce a higher density of the grains. Therefore two methods were thought by Morita's group. One was separation by gravity force and the other Electromagnetic force during solidification.

With regard to the gravity separation it was found that separation of the solidified Si grains is not possible during solidification, since after melting and quenching the Si grains were found to be dispersed uniformly in the sample. The reason of this is the high viscosity of the melt in which particles were dispersed. The other reason could be the low density difference between the eutectic melt (Al-Si) and the Si grains which is reported to be approximately 0.1 gr/cm^3 by the authors. It was thus concluded that the use of gravity force was not effective as a separation method when the density of the matrix and Si grains are very close. The other method was continuous solidification of Si from Si- Al melt under induction heating. In this method, the separation form Si dendrites was investigated by solidification under fixed alternating magnetic field by induction heating (Figure 7). As a result of this solidification the Si dendrites are successfully agglomerated at the bottom of the sample, although the densities of these dendrites are lower than the eutectic phase (5).

In the present study gettering of impurities in MG-Si by Fe was investigated. The novelty of the method is in applying a simple physical method for separating Si dendrites from the alloy matrix. The method takes advantage of difference in specific gravity of silicon and the alloy, and uses a heavy fluid, to float light Si particles while the heavy Si-Fe alloy is sunk. Controlled solidification of Fe-Si alloy, combined with crushing-grinding of the sample, and gravity separation for the basic steps of the process, although additional refining may be carried out to further purify the collected Si particles. This article discuses the efficiency of the Si separation process, as well as the degree of the purification of Si.

EXPERIMENTAL

As illustrated in Figure 1Figure 2, the process consists of four major steps: alloying, controlled solidification, physical separation using heavy media (HMS), and leaching purification. The leaching step is essential to remove the FeSi$_2$ alloy adhered to the Si platelets.

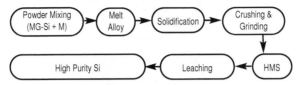

Figure 1.Experimental Process.

The effectiveness of iron as the getter metal was investigated by forming a Fe-Si alloy containing 28 wt pct Fe, that once solidified under equilibrium conditions would yield a 50 wt%Si – 50 wt% FeSi$_2$ sample. MG-Si lumps were crushed and pulverized then mixed with Fe powder and doped with 0.3% P and 0.3% B. The initial MG-Si and Fe powder were analyzed by ICP (Inductive-Coupled-Plasma)-AES (Atomic-Emission- Spectroscopy) with the results shown in Table I.

Table I. Analysis of MG-Si and Fe

MG-Si and Fe analysis		
[X]	Fe	MG-Si
Al	599.9	870.1
B	40.2	0.8
Co	77.6	1.7
Mn	1140.8	158.3
K	11.5	9.3
V	621.8	85.2
Cd	56.0	7.6
Ba	14.1	8.5
Ni	107.7	118.4
Zn	156.3	16.2
Cr	5.2	5.4
P	890.9	65.8

The powder mix was then placed in mullite crucibles sealed with a lid covering the crucible and attached to it by ceramic paste. The crucible was heated in a muffle furnace to 1550°C at the rate of 10°C/min in argon atmosphere, then held for 4 hours for homogenization of the liquid, and then cooled under various cooling rates of 0.5, 1.5, and 3°C/min (S1, S2, and S3 respectively). Once the temperature dropped 100°C below the eutectic temperature (1207°C), the sample was quenched in water. The structure of the sample after solidification was examined by SEM showing Si dendrites (dark phase in Figure 2) within Fe-Si matrix. EDS (Figure 3) analysis of the matrix confirms the stoichiometric composition FeSi$_2$.

Figure 2.SEM image of the solidified Si- Fe alloy

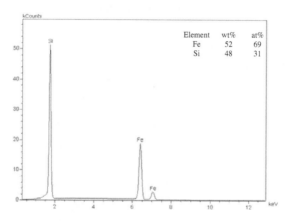

Figure 3.EDS of the bright phase (FeSi$_2$) in Figure 2.

Each alloy sample, weighing about 200 grams was divided into four 50 gram batches for further processing. Each batch was ground to a target particle size range in an agate mortar. The four target size ranges were D (600-800um), C (454-600um), B (212-454um), A (106-212um).

The dendrites size in the three different samples, measured by Image J software, vary with cooling rate as shown in Figure 4.

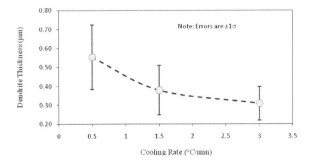

Figure 4.Dendrite thickness vs. cooling rate.

Heavy media used in this study is a neutral liquid also called LST heavy Liquid is manufactured by Central Chemical Consulting in Australia. It is a low toxicity dense liquid containing lithium heteropolytungstates (80-85 wt %) in water commonly used for float-sink particle separations through their buoyancy. The density of HM is 2.85-2.9 gr/cm^3, which is between the density of the silicon dendrites (2.3 gr/cm^3) and $FeSi_2$ matrix (4.74 gr/cm^3) (6). All four samples were suspended in HM in glass tubes then agitated manually and left for 24 hours. The lighter Si particles float on top and the denser $FeSi_2$ particles sink to the bottom. It is expected that under perfect separation, all particles floating to the liquid surface contain 65 wt% or greater silicon.

The floats were scooped out, while the sinks were separated by filtering the remaining liquid. The liquid was cleaned for recycling by passing through several filters. Both the Si and $FeSi_2$ particles were rinsed with DI-water for a few times. Image analysis was performed on float and sink samples, to quantify the degree of separation of the two phases. For removing the adherent $FeSi_2$ from the Si particles the float and sink particles were each leached in HF solution.

RESULTS AND DISCUSSION

Heavy Media Separation

In order to compare the effect of particle size on HM separation efficiency, leaching experiments were carried out by exposing float and sink particles to HF. Unlike Si, $FeSi_2$ dissolves in HF, thus weighting the remainder will yield the amount of Si present or in other word the Si grade in the initial sieved particles. In Figure 5 Si recovery in the floats for four different particle sizes and for the four cooling rates is presented. This value was calculated by the following equation:

$$Recovery\ of\ Si\ in\ float = \frac{float\ wt\% * float\ Si\ grade}{float\ wt\% * float\ Si\ grade + sink\ wt\% * sink\ Si\ grade} \tag{1}$$

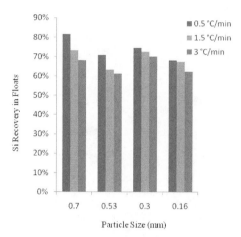

Figure 5.Si recovery in floats for different particle size and cooling rates.

As can be seen, the recovery in most cases is above 60% with the lowest cooling rate and consequently largest dendrites having the best recovery of up to 80%. A combined measure of recovery and grade, commonly known as separation efficiency provides a criteria for selection of the conditions where separation of Fe and $FeSi_2$ is maximized. Separation efficiency (SE) is calculated from equation (2). Figure 7 presents SE against particle size for the three cooling rates studied. As seen, SE is maximum for smaller particles and is generally increased for faster cooling rate rates. This may be attributed to the more liberation of Si grains from the $FeSi_2$ matrix when the sample is crushed to finer particles.

$$Separation\ Efficiency\ \% = Si\ yield - \frac{float\ wt\% * FeSi_2\ Grade}{float\ wt\% * FeSi_2 + sink\ wt\% * Si\ grade} \qquad (2)$$

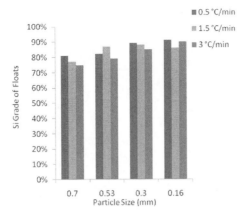

Figure 6.Si recovery in floats for different particle size and cooling rates.

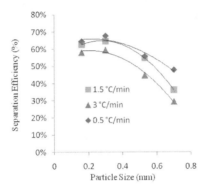

Figure 7.Separation efficiency vs. particle size

Distribution of Impurities

After crushing, sieving and separation of the FeSi$_2$ and Si phase by heavy media, the sinks and floats were analyzed by ICP-AES. In order to analyze the sinks and floats as mentioned before, a leaching step is required to remove the adherent FeSi$_2$ from the floated and sink Si particles.

Si can only dissolve in the presence of an oxidizing agent such as nitric acid (7) but FeSi$_2$ dissolves in HF readily. The assumption that FeSi2 dissolves completely in HF was also confirmed by SEM/EDX and XRF analysis. The dissolution of all the FeSi$_2$ particles by the leaching process was confirmed through mounting different batches from different particle sizes and cooling rates of leached particles. Backscattered images show only one phase which is Si. XRD analysis also only shows Si with no detectable FeSi$_2$. Therefore, respective analysis of FeSi$_2$ and Si may be performed by first dissolving the FeSi₂ in HF solution and later digesting Si particles in a mixture of HF and HNO3 for ICP-OES analysis.

The ratio of the concentration of impurities in FeSi₂ to Si is thus calculated from the ICP analysis and is shown in Figure 8. When compared to the segregation coefficient, it is found that impurities that are removed easier i.e. impurities with higher distribution ratio of FeSi₂ to Si are those with lower segregation coefficient.

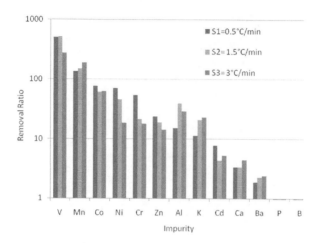

Figure 8.Distribution ratio [i] Alloy/ [i] Si of impurities.

Table II. Amount of impurities in MG-Si and refined Si (ppm).

Element	MG-Si	Refined Si
Al	870	30
B	30	26
Ba	9	9
Cd	7.6	3
Co	1.7	1.0
Cr	5.4	2.0
K	9.3	3.8
Mn	158	9.1
Ni	118	5.0
P	66	52.5
V	85	0.5
Zn	16.2	8.1
Total	1376	160

Table III. Removal sequence of impurities compared to their segregation coefficient

Removal Sequence in current study	Element	Segregation Coefficient
V	V	4×10^{-6}
Mn	Mn	3.2×10^{-6}
Co	Co	2×10^{-5}
Ni	Cr	1.1×10^{-5}
Cr	Zn	1×10^{-5}
Zn	Ni	1×10^{-4}
Al	Al	2×10^{-3}
P	P	0.35
B	B	0.8

The distribution ratio of P and B increases by the increase of particle size in all three samples with different cooling rates as shown in Figure 9. From this it can be presumed that the impurities are mostly concentrated in the dendrite and $FeSi_2$ interface, which could be an effect of impurities being rejected from $FeSi_2$ during eutectic solidification and subsequent cooling. Due to the nature of sample breakage that takes place more pronouncedly at the interface of the two phases, the smaller particles are more generated from the region close to the interface. Therefore, it is expected that the Si particles removed from these regions (i.e. smaller particles contain more impurity).

Apparent from the distribution ratios, it is expected that the distribution ratio for P and B between Si dendrites and $FeSi_2$ favours the segregation of these impurities into the Si. However, thermodynamic evaluation indicates that iron has higher affinity than Si for P and B compared to Si that is inconsistent with the findings. Therefore, it may be argued that the Fe-Si alloy in the liquid state retains P and B while on solidification, these impurities are rejected back to the interface, and subsequently diffuse back to Si. Therefore, it is expected that Si adjacent to the interface contains more of these impurities, which results in higher concentration of the impurities in the smaller particles. Confirmation of this behaviour will be made by measuring concentration profile of the impurities across the interface and the dendrite thickness.

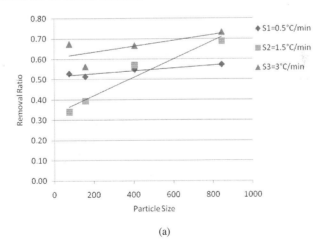

(a)

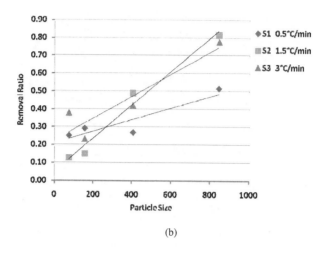

(b)

Figure 9.Distribution ratio of (a) phosphorus (b) boron vs. particle size for three different cooling rates (The lines are just a guide for the eyes).

CONCLUSION

Through solvent refining by Fe, two phases of Si dendrites and $FeSi_2$ were formed which resulted in removal of most impurities in MG-Si which is due to the higher tendency of these impurities to stay in the $FeSi_2$ phase rather than the $FeSi_2$ phase.

Heavy media separation based on gravity was used to separate these two phases by sieving and suspending these particles in HM. To investigate the effect of particle size on the efficiency of separation, leaching experiments were carried out. The results show that the recovery of Si in float particles in most cases is above 60% with the lowest cooling rate and consequently largest dendrites having the best recovery of up to 80%. From the heavy media separation it can also be concluded that the separation efficiency is decreased for larger particles.

The ICP results show that impurities which were studied such as Al, Mn, Cr, Zn, V, Ni, Co, K, Ca and Ba tend to segregate to the $FeSi_2$ phase rather than the Si phase. P and B on the other hand are more retained in the Si. Larger grains of Si contain less of such impurities. This is presumably because during cooling of the sample.

REFERENCES

[1] Johnstone, Bob. Empire of the Sun. Sydney : COSMOS Magazine, **14**, Luna Media Pty Ltd, 2007.

[2] Sollmann, Dominik, Metallurgical silicon could become a rare commodity . Aachen, Germany : PHOTON International, Feb, 2009.

[3] J. Dietl eds. J.A. Amick, V. K. Kapur, On Materials and New processing technologies for Photovoltaic. 1983.

[4] Morita. Takashi, Toshikawa. Kazuki, Thermodynamics on the Solidification refining of Silicon with Si- Al Melts. 2005, EPD Congress , *TMS (The Minerals, Metal and Materials Society)*, p. 549.

[5] Morita, Takeshi Yoshikawa and Kazuki, Refining of Si by the Solidification of Si–Al Melt with. 2005, *ISIJ International*, Vol. **45** , pp. 967–971.

[6] HSC 6.1 Software.

[7] Lide, David R. CRC Handbook of Chemistry and Physics. s.l. : CRC Press, 60th Edition .

METALLURGICAL REFINING OF SILICON FOR SOLAR APPLICATIONS BY SLAGGING OF IMPURITY ELEMENTS

M.D. Johnston and M. Barati
Department of Materials Science and Engineering, University of Toronto
Toronto, Ontario, Canada

ABSTRACT
New measurements have been made on the distribution of impurity elements between molten metallurgical silicon and slags of the type Al_2O_3-CaO-MgO-SiO_2 and Al_2O_3-BaO-SiO_2. Removal efficiencies were determined for B, Ca, K, Mg, Fe and P from metallurgical grade silicon to the magnesia or alumina saturated slags at 1500 °C. The removal efficiencies of the impurity elements were examined and explained in terms of the changing chemistry of the slag. It was found that both the slag basicity and oxygen potential influence the partitioning of impurities between the slag and metal phases. The data were used to estimate overall process efficiencies in relation to the production of photovoltaic grade silicon, with particular focus on the treatment of boron and phosphorus.

INTRODUCTION
A major challenge facing the world today is in generating power to meet ever-increasing demand while minimising the impact on the environment. One aspect of this is in curtailing greenhouse gas emissions by eliminating the use of fossil fuels in favour of renewable alternatives. Solar power is one such alternative and while a growing industry, faces challenges of its own to become competitive in today's marketplace. Presently, the widespread adaptation of solar power is somewhat hampered by the high cost of production of silicon based photovoltaic cells. This is mainly due to the high cost and limited availability of high-purity scrap silicon from the semiconductor industry (99.9999999% Si) that is used as a starting material for producing solar grade silicon (SoG-Si) (99.9999% Si), which is ultimately used to fabricate a photovoltaic cell.

Metallurgical grade silicon (MG-Si) is a cheaper alternative to electronic scrap that is produced on the large scale, but is of considerably lower purity (~98% Si). Furthermore, it contains many impurity elements that must be controlled to precise levels for the photovoltaic cell to operate at optimum efficiency. As such, MG-Si must still be refined by a cost-effective means for it to remain a possible starting material to produce SoG-Si.

Many techniques for refining MG-Si exist and must often be used in conjunction with one another to achieve the requisite purity for solar applications. Of all the impurities present in MG-Si, boron and phosphorus are typically found to be the most difficult to remove, as they are unresponsive to directional solidification, usually the final step in silicon refining. This is a major concern as these elements must be controlled to precise levels for SoG-Si. It is the behaviour of these elements that will be the focus of this study.

Some elements are able to be removed by exploiting their low segregation coefficient between solid and liquid silicon. Studies have shown that some impurities can be collected in a liquid region and discarded by a zone refining plasma-arc method.[1] However, B and P have relatively high segregation coefficients so are not readily removed by this method. One possibility is to first alloy the silicon with aluminium[2] but further refining will still be required. Vacuum refining has been shown to be able to remove P, but long times and high power requirements make this technique unsuitable.[3] Another possibility is to melt the MG-Si in the presence of a flux to produce a slag phase which can take up the impurity elements. Such a technique is attractive compared to other pyrometallurgical processes due to the relatively low cost of fluxes typically used. Further to this, where the main power requirement is in melting of the fluxes, the shorter reaction times necessary help to keep costs comparatively low. Tanahashi et al,[4] proposed the treatment of Si with CaO based fluxes as one part of a more complex overall refining procedure. However, comprehensive data on the behaviour of impurities in terms of temperature and slag composition is quite limited. It is this fluxing step that is to be examined in detail in terms of slag composition in the present study.

For elements such as phosphorus that are considered to have acidic oxides, basic melts such as those based on CaO are expected to be the most appropriate to facilitate their extraction to a slag phase. Generally, impurity elements are stabilised as an oxide in the slag with the inclusion of a suitable flux or gas to create oxidising conditions. Boron, in group IIIA on the periodic table, may be considered less acidic in nature, and so not behave in exactly the same manner as phosphorus. Previous studies have focused on the use of highly basic fluxes containing CaO and CaF_2,[5,6] but also make use of SiO_2 to give the slag suitable oxidising potential. While this will reduce the basicity of the slag, saturating the slag with SiO_2 may also prevent unacceptably high losses of silicon to the slag. Weiss and Schwerdtfeger[7] examined the behaviour of many elements in $CaO-SiO_2$ based ternary systems and showed that the removal of some metallic elements is also feasible by this method. However, it is reported[8] that the addition of excess CaO can increase the concentration of Ca in silicon, which, through ternary interactions, can in turn draw P back from the slag into the metal phase. It is apparent, therefore, that the slag composition must be subject to fine control. Equilibrium studies between silicon and oxide phases have also been carried out to determine thermodynamic quantities for impurities in silicon. A recent review of the thermodynamic data related to refining of SoG-Si is available.[9]

EXPERIMENTAL

Materials

A silicon master alloy was prepared by doping one kilogram of milled MG-Si with 3 g each of 95-97% B (Sigma-Aldrich) and >97% P powders (Riedel-de Haem). This mixture was melted in two portions, approximately 500 g each, in magnesia crucibles under an argon atmosphere inside a muffle furnace at 1485 C for 2.75 hours. Analysis of these master alloys is given in Table I. The final concentration of phosphorus in each is much lower than expected, possibly due to vaporisation of the elemental P on melting of the mixture.

Table I. Chemical analysis of master silicon alloys used in equilibration experiments. All values are in wt pct.

Alloy	Al	Ca	Mg	B	P	K	Fe
1	0.208	0.012	0.238	0.190	0.038	0.012	0.395
2	0.095	0.055	0.114	0.261	0.063	0.009	0.270

Fluxes used to prepare slags were first dried at 105 C for one hour. The powders were weighed and mixed thoroughly in the desired ratio for each particular experiment. Mixtures were designed to give a single molten phase at 1500 C. About 3-4 g of mixture at a time was pressed into a pellet and placed into a crucible with pieces of the alloy. Typically, 5.5 g of doped Si and 7.5 g of fluxes were melted in each experiment.

Methods

In each experiment, the slag and alloy were brought into chemical and thermal equilibrium in a magnesia or alumina crucible in an inert atmosphere inside a vertical tube furnace. The alumina work tube was made gas tight by attaching water-cooled stainless steel caps to the ends of the tube and sealing them by means of rubber "O" rings. A crucible was suspended inside the furnace by means of an alumina pin and lance that was attached to the top end cap. A schematic diagram of the furnace setup is shown in Figure 1. The crucible was suspended inside the cold zone of the furnace which was then flushed with argon for 30 minutes. After flushing, the crucible was slowly lowered to the hot zone and the temperature increased to 1500 °C. The temperature of the hot zone was measured using a B-type (Pt-6% Rh/Pt-30% Rh) thermocouple, held in place by the bottom end-cap at the level of the base of the crucible. Ar was passed through the furnace at a rate of 300 mL/min for the duration of the experiments.

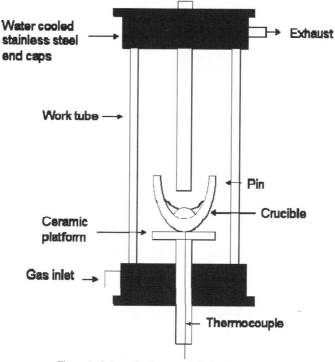

Figure 1. Schematic diagram of the furnace setup.

At the end of each experiment, the crucible containing the melts was removed from the top of the furnace and quenched in a water bath. Care was taken to physically separate the crucible from the slag and metal phases which were then ground to a powder for digestion and analysis.

A series of slag-metal experiments was first carried out to determine the time needed for the system to reach equilibrium. A slag of composition 30% Al_2O_3-40% CaO-10% MgO-20% SiO_2 (by weight) was equilibrated with Si alloy at 1500 C at six reaction times up to 6 hours. Analysis of the Si alloys at the completion of each run showed the concentration of P and B became steady after 2 hours. Similarly, the slag composition (in terms of its major components) did not show any systematic variation after this time. This is shown in Figure 2. On the basis of this, a reaction time of 2 hours was used in subsequent experiments.

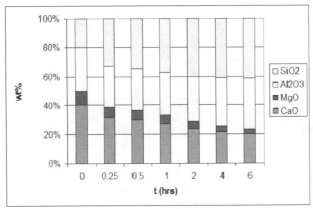

Figure 2. Normalised slag composition. Major oxides comprise ~99.7% of total assay.

Analysis

Silicon alloys were digested in a mixture of concentrated nitric and hydrofluoric acids (trace-element grade, Celadon) in PTFE beakers at about 80 C and the resulting solutions analysed by ICP-OES. Slags were fused with sodium hydroxide (99.99%, Sigma-Aldrich) in an Inconel 601* crucible at around 450 C then dissolved in diluted nitric acid prior to ICP analysis for boron and iron. The slags were analysed by XRF for all other elements reported.

SLAG-SILICON EQUILIBRIA

The effect of slag basicity and oxidising potential was investigated with slags of the type Al_2O_3-CaO-MgO-SiO_2. The effect of slag basicity was further investigated for the Al_2O_3-BaO-SiO_2 system. For the purposes of this study, basicity refers to the ratio, by weight, of $(CaO+MgO)/(SiO_2+Al_2O_3)$, or, $(BaO)/(SiO_2+Al_2O_3)$.

The percent removal from silicon was calculated from the difference between the initial and final mass of the impurity in the metal phase as a ratio to the initial mass in the Si alloy.

$$\% \, R = 100 \frac{m_i^{Si}[M]_i - m_f^{Si}[M]_f}{m_f^{Si}[M]_i} \tag{1}$$

Here, m_i^{Si} is the mass of alloy used in the experiment and m_f^{Si} is the mass of alloy at the end of the experiment as determined by mass balance. Similarly, $[M]_i$ is the concentration of impurity M in the doped alloy and $[M]_f$ is the concentration at equilibrium from the chemical analysis.

Varying $CaO:SiO_2$

The effect of slag basicity on the removal of impurities from silicon was investigated by varying the $CaO:SiO_2$ ratio of the Al_2O_3-CaO-MgO-SiO_2 slag. Slags were prepared in the range $0.31 \leq CaO:SiO_2 \leq 7.86$ with Al_2O_3 and MgO fixed at 35 and 3 wt% respectively. All experiments were conducted in alumina crucibles. The basicity of the slags at equilibrium were in the range $0.18 < B < 0.45$.

The final concentration of B and P in the metal phase was always < 0.1 wt%. The removal efficiencies for phosphorus increase steadily with increasing basicity, although only reach just over 50% in the most basic slag. The efficiencies for boron are higher, reaching 80%; however, they do not follow the same trend as for phosphorus. Instead values appear to increase through a local maximum at $CaO:SiO_2 = 0.66$, before decreasing steadily to just under 70%. This is shown in Figure 3. The removal efficiencies for iron and potassium are generally lower than for boron and phosphorus, but are subject to significant scatter. In the case of potassium, this may be due in part to its low concentration in the MG-Si. Of the slag components, magnesium was the only element that was able to be removed by the slag instead of contaminating the Si. Removal efficiencies for Mg were subject to significant scatter around a value of about 70%. The final concentration of aluminium in the metal phase was generally about 3 times higher than the initial, while that of calcium varied between 2 and 43 times that of the initial, resulting in concentrations in Si of up to 0.5 wt%.

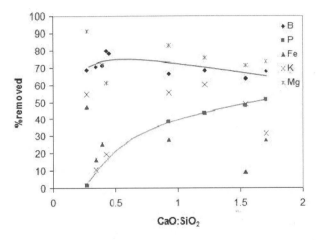

Figure 3. Removal efficiencies for impurities from Si by varying $CaO:SiO_2$ (35%Al_2O_3-CaO-3%MgO-SiO_2) slags.

Varying $SiO_2:Al_2O_3$

To examine the influence of SiO_2 as an oxidising agent on impurity removal, the $SiO_2:Al_2O_3$ ratio of the Al_2O_3-CaO-MgO-SiO_2 slag was varied while keeping the basicity constant. Slags were prepared in the range $0.07 \leq SiO_2:Al_2O_3 \leq 2.2$ with CaO and MgO fixed at 42 and 10 wt% respectively, to give a basicity of 1.08 each time. These experiments were conducted in magnesia crucibles.

The final B and P concentrations of the Si were again always < 0.1 wt% and were generally lower than those recorded in the varying basicity experiments. This is reflected in the higher calculated removal efficiencies for this system, as shown in Figure 4. Values for B again follow a gradual increasing trend with increasing $SiO_2:Al_2O_3$ ratio, reaching 85%. Those for P initially increase sharply with increasing $SiO_2:Al_2O_3$, then level off in the more oxidising slags, eventually reaching 95%. The concentration of iron in the metal phase was again subject to scatter, which is also reflected in the calculated removal efficiencies. Potassium was constant at around 0.01 wt% in the metal phase in most of the experiments, but dropped by an order of magnitude in the two most oxidising slags. The corresponding removal efficiencies are generally low, and increase to only 44%. Interestingly, contamination of the Si by calcium is significantly higher in these slags than in the varying basicity experiments. Concentrations up to 1 wt% in the metal phase were recorded, indicating an increase of over 100 times compared to the starting alloy. There is, however, a strong decreasing trend in contamination with increasing $SiO_2:Al_2O_3$ ratio, indicating that a higher SiO_2 slag would be appropriate to allow for ease of further refining. The degree of contamination of the Si by Al and Mg was consistently relatively low at about 5 times the initial concentration for each.

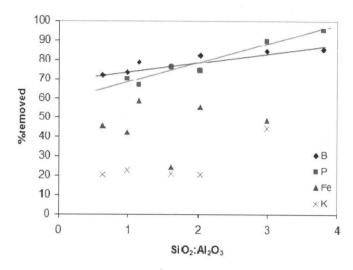

Figure 4. Removal efficiencies for impurities from Si by varying $SiO_2:Al_2O_3$ (Al_2O_3-42%CaO-10%MgO-SiO_2) slags.

Varying $BaO:SiO_2$

The effect of slag basicity on the removal of impurities from silicon was further investigated by varying the $BaO:SiO_2$ ratio of the Al_2O_3-BaO-SiO_2 slag. Barium, below calcium on the periodic table, is expected to have a stronger basic oxide, with perhaps more potential to remove some impurities to a slag phase. Slags were prepared in the range $0.45 \leq BaO:SiO_2 \leq 1.28$ with Al_2O_3 fixed at 20 wt%. All experiments were conducted in alumina crucibles.

Removal efficiencies for impurities from Si with Al_2O_3-BaO-SiO_2 slags are shown in Figure 5. The final concentrations of boron and phosphorus in the silicon metal phase were higher in these experiments than in the BaO-free slags. As a result, the corresponding removal efficiencies for these impurities are significantly lower in this system than those described earlier. The rates for boron increase slightly from 35 to 45% with increasing $BaO:SiO_2$, whereas those for phosphorus are scattered around a value of about 25%. The removal of Ca and Mg from Si was very effective with this CaO and MgO free slag, with both attaining rates above 90% in the most basic slags. Removal efficiencies for potassium were similar in this case to the varying $CaO:SiO_2$ slags, but this time showed a general increase in removal with increasing $BaO:SiO_2$.

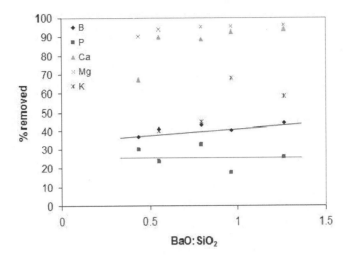

Figure 5. Removal efficiencies for impurities from Si by varying $BaO:SiO_2$ (20%-Al_2O_3-BaO-SiO_2) slags.

DISCUSSION

Contamination of the slag

As the slagging step is to be followed by others in a sequential overall refining method, it will be advantageous to monitor and minimise contamination of the silicon by elements present in the slag and furnace refractories. With the Al_2O_3-CaO-MgO-SiO_2 slags, it is calcium that contaminates the silicon to the highest degree compared to its initial concentration. As mentioned previously, this can possibly be contributing to the retention of P in the metal phase. The importance of the presence of SiO_2 in the slag is further highlighted by the degree of contamination by calcium, which was over 100 times in the lowest $SiO_2:Al_2O_3$ slags. The data for the varying $CaO:SiO_2$ slags appears subject to scatter, but the highest level of contamination was again recorded in the lower SiO_2 slags. Conversely, these slags were able to remove magnesium. The slags from the varying $SiO_2:Al_2O_3$ experiments, however, contaminated the silicon with magnesium by around 4 times its initial concentration. The barium-containing slags on the other hand, were very effective at removing magnesium and calcium, with removal efficiencies for both reaching $>90\%$ where the $BaO:SiO_2$ ratio was ≥ 1. Contamination of the silicon by aluminium was between 2 and 5 times throughout all slags in this study. This could possibly be avoided by replacing the alumina crucibles used with graphite, for example, so that the slag was not saturated in alumina.

Influences on Removal

In the case of boron, the trends seen in the removal efficiencies by Al_2O_3-CaO-MgO-SiO_2 slags suggest that the SiO_2 content is the strongest contributing factor. This is evidenced by the higher values in slags with lower CaO:SiO_2 and higher SiO_2:Al_2O_3 ratio. In the latter case, removal increases with decreasing optical basicity as SiO_2 replaces Al_2O_3. Consequently, it is apparent that a more oxidising slag should be considered, rather than a more basic slag as may be first thought. The reaction for the oxidative take up of boron by the slag is given by

$$[B] + 3/2(O^{2-}) + 3/4O_2 = (BO_3^{3-}) \qquad (2)$$

Here, square brackets indicate a species in the silicon metal phase, while parentheses indicate a species is present in the slag phase. The equilibrium between silicon and silica determining the oxygen partial pressure (p_{O_2})

$$(SiO_2) \rightleftharpoons [Si] + O_2 \qquad (3)$$

Boron, in group IIIA of the periodic table will have a weakly acidic oxide compared to phosphorus in group VA and therefore be expected to be relatively less responsive to changes in slag basicity. In the case of phosphorus, there was a much stronger dependence on slag basicity, but this was not consistent over the range studied. The levelling off of removal efficiencies at higher CaO:SiO_2 suggest again that oxygen potential rather than basicity has a more significant role. The reaction for the oxidative take up of phosphorus by the slag is then given by

$$[P] + 3/2(O^{2-}) + 5/4O_2 = (PO_4^{3-}) \qquad (4)$$

The apparent change in dependence against slag basicity is due to the fact that the basic oxides have two effects on the slag chemistry. While donating free oxygen to the melt, these species also have a high affinity for SiO_2. This has the effect of lowering the activity of silica in the slag (a_{SiO_2}), which, in turn, lowers the oxygen potential of the slag. As lime is continually added to the slag, the effect of decreased p_{O_2} outweighs the increase in basicity, resulting in negative dependence of B and shallow dependence of P on basicity in the more basic region. This is reinforced by the varying SiO_2:Al_2O_3 results, where removal efficiencies for both B and P are higher and always increase with increasing p_{O_2} despite the slag becoming less basic.

The observed low removal of B and P from BaO-containing slags can also be accounted for by a removal mechanism limited by p_{O_2} rather than basicity. Due to the high molecular weight of barium compared to the other alkali earth elements, the number of moles of BaO in the slag is considerably less than the number of moles of CaO+MgO. As a result, the basicity of the BaO slags are less than what may be apparent and so even less efficient at taking up impurities at a given SiO_2 content. This is despite BaO being the strongest single basic component used in the present work based on optical basicities. Furthermore, the deactivation of SiO_2 would be strongest by BaO, leading to slags with low oxygen potential also. Therefore, it is apparent that it is inappropriate to use BaO as a flux as it is expensive compared to the other possibilities and relatively ineffective at removing these impurities.

Impurity Partitioning

Distribution coefficients for boron and phosphorus were calculated as the ratio of concentration of impurity M (in wt%) in the slag phase to that in the metal phase. Assuming no impurity losses by vaporisation, they are directly related to the removal efficiency by

$$L_M = \frac{R}{1-R}\left(\frac{m_f^{Si}}{m_f^{slag}}\right) \qquad (5)$$

where m_f is the mass of impurity M in the denoted phase at equilibrium as determined by mass balance. Thus, a plot of $R/1\text{-}R$ vs. $L(m_{slag}/m_{Si})$ for a particular element should give a straight line with a slope of 1. This is shown for boron and phosphorus in Figure 6. Most points fit the trend fairly well, confirming that, generally, as expected, distribution coefficients and removal efficiencies for these elements follow the same trend. The degree of removal from Si can also be estimated from the distribution coefficient measurement by this method.

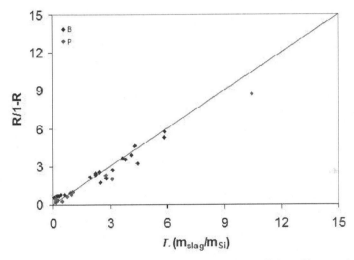

Figure 6. Relation between removal efficiency and distribution coefficient of boron and phosphorus for Al_2O_3-CaO-MgO-SiO$_2$ and Al_2O_3-BaO-SiO$_2$ slags.

By examining the measured distribution coefficients for boron and phosphorus, it is possible to also estimate the amount of slag required to purify MG-Si to SoG-Si specification by this method. For this it is assumed a typical MG-Si containing 15 ppm B and 30 ppm P is to be refined to 1.5 ppm B and 1 ppm P, using distribution coefficients from the slag where the maximum L_P was observed. Thus, for L_B 1.8 and L_P = 8.8, Based on 100 g of silicon, 1.3 10^{-3} g of boron must be taken up by the slag, resulting in a concentration of 2.3 ppm B in the slag. To achieve this, the mass of slag required is roughly 5 times the mass of MG-Si. For phosphorus, 2.9 10^{-3} g must be taken up by the slag to give a concentration of 1.1 ppm P in the slag. To achieve this, the slag requirement is roughly 3 times the mass of MG-Si. A large slag volume is unfavourable from a cost and energy requirement perspective, so further improvement in removal efficiencies will likely be necessary to ensure this is a cost effective method.

SUMMARY AND CONCLUSIONS

In this study, the removal of impurities from metallurgical silicon doped with boron and phosphorus to a slag phase was examined. An equilibrium distribution technique was employed by which silicon was melted in the presence of Al_2O_3-CaO-MgO-SiO_2 and Al_2O_3-BaO-SiO_2 fluxes at 1500 °C. Results for boron and phosphorus showed that the greatest removal from silicon was achieved in relatively highly oxidising (high SiO_2:Al_2O_3) slags at constant basicity. Where basicity was varied, removal efficiencies for boron initially increased, then decreased again through a local maximum with increasing CaO:SiO_2 ratio, which is thought to be due to the presence of excess basic oxide (CaO) reducing the activity of SiO_2 in the melt. This has the effect of lowering the oxygen potential of the slag, which is an important consideration for the oxidative removal of these impurities. Where BaO was present in the slag, the removal of boron and phosphorus was less efficient, although did greatly contribute to the removal of magnesium and calcium. Distribution coefficients were determined for boron and phosphorus and were found to closely follow the same trends as the removal efficiencies.

It is suggested that a suitable slag for purifying metallurgical silicon would be CaO or MgO-SiO_2 based with relatively high SiO_2 and minor Al_2O_3 and BaO. However, due to the low distribution coefficients, the amount of slag necessary to extract boron and phosphorus directly down to levels required for SoG-Si is in excess above the mass of MG-Si and may need to be reduced significantly for such a process to be efficient.

FOOTNOTES

[*] Inconel is a registered trademark of Special Metals Corporation.

ACKNOWLEDGEMENTS

This work was partially funded by NSERC. The assistance of Dan Mathers of the Department of Chemistry, University of Toronto, with the ICP analysis and Prof. Mike Gorton of the Department of Geology, University of Toronto, with the XRF analysis, is gratefully appreciated.

REFERENCES

[1] K. Mimura, M. Kishida, M. Isshiki, S.G. Kim, W.T. Kim and T. Suzuki, Plasma-arc zone refining of silicon, in *Second International Conference on Processing Materials for Properties*, 1059-64 (2000).

[2] T. Yoshikawa and K. Morita, Removal of phosphorus by the solidification refining with Si-Al melts, *Sci. Tech. Adv. Mat.*, **4**, 531-37 (2003).

[3] J.C.S. Pires, J. Otubo, A.F.B. Braga and P.R. Mei, The purification of metallurgical grade silicon by electron beam melting, *J Mat. Proc. Tech.*, **169** (1) 16-20 (2005).

[4] M. Tanahashi, M. Sano, C. Yamauchi and K. Takeda, Oxidation removal behavior of boron and local nonequilibrium reaction field in purification process of molten silicon by the flux injection technique, in *Sohn International Symposium, Volume 1 - Thermo and physicochemical principles*, 173-86 (2006).

[5]M. Tanahashi, H. Nakahigashi, K. Takeda and C. Yamauchi, Removal of boron from metallurgical-grade silicon by applying CaO-based flux treatment, in *Yazawa International Symposium, Volume 1: Materials processing fundamentals and new technologies*, 613-24 (2003).

[6]L. Teixiera, and K. Morita, Thermodynamic properties and structural assessment of boron oxide in $CaO-SiO_2$ and $CaO-SiO_2-CaF_2$ slags for silicon refining, in *Molten 2009 Conference*, 319-26 (2009).

[7]T. Weiss and K. Schwerdtfeger, Chemical equilibria between silicon and slag melts, *Met. Trans. B*, **25B** (4) 497-504 (1994).

[8]D. Lynch, Winning the global race for solar silicon, *JOM*, **61** (11), 41-48 (2009).

[9]K. Morita and T. Miki, Thermodynamics of solar-grade-silicon refining, *Intermetallics*, **11** (11-12), 1111-17.

OCEAN THERMAL ENERGY CONVERSION: HEAT EXCHANGER EVALUATION AND SELECTION

Laboy, Manuel A.J.
Offshore Infrastructure Associates
San Juan, Puerto Rico

Ruiz, Orlando E.
University of Puerto Rico
Mayagüez, Puerto Rico

Martí, José A.
Technical Consulting Group
San Juan, Puerto Rico

ABSTRACT

This study summarizes available data on heat exchanger requirements for closed-cycle OTEC power systems obtained during over thirty years of R&D work and technology demonstration programs, and presents how these requirements can be met using commercially-available heat exchangers used today in other applications by a variety of industries. The study focuses on the following design criteria: configuration (shell-and-tube, compact-type, and others), process performance, surface enhancement, corrosion resistance, biofouling control, manufacturability, ease of operation and maintenance, and over-all cost-effectiveness. Selection of the appropriate working fluid will also be discussed. Data evaluated include previously developed power system designs such as those completed during the 1970's and 1980's by GE, JHU/APL, ANL, and engineering reports from OTEC technology demonstration programs such as the Nauru, Mini-OTEC and OTEC-1 test projects. A critical performance assessment is made between the use of stainless-steel plate heat exchangers and aluminum-brazed plate-fin heat exchangers in the context of present day technology. Alternatives to mitigate and control the adverse effects of biofouling are discussed.

INTRODUCTION

Ocean thermal energy conversion (OTEC) is a base-load renewable energy source that uses the temperature difference between the warm surface ocean water and the cold deep ocean water to generate electricity. OTEC is applicable to most parts of the world's deep oceans between 20° North and 20° South latitude including the Caribbean and Gulf of Mexico, the Pacific, Atlantic and Indian Oceans, and the Arabian Sea, where the temperature difference between the warm surface ocean water and the cold deep ocean water is equal or greater than 20 °C. In essence, OTEC recovers part of the solar energy continuously absorbed by the ocean and converts it into electric power. OTEC does not utilize any fuel. The electricity generated has a fixed cost, thus, it is not susceptible to the volatility resulting from world market fluctuations that affects other energy sources such as petroleum, coal and natural gas. Moreover, environmental impacts are less than those of conventional sources of energy since no products of combustion and no solid or toxic wastes are generated during the power production process. In addition, effluents are essentially similar to receiving waters. All of these aspects have caused a revival of interest in OTEC[1].

An OTEC power system consists of a heat engine cycle that converts thermal energy into mechanical work through the temperature difference between a "heat source" and a "heat sink". Although this temperature difference is relatively small compared to a steam engine, the principle is the same (Rankine thermodynamic cycle). OTEC technology is divided into three major categories: closed, open and hybrid cycles. In the closed-cycle, the temperature difference is used to vaporize (and condense) a working fluid (e.g. ammonia) to drive a turbine-generator to produce electricity (see Figure 1). In the open-cycle, warm surface water is introduced into a vacuum chamber where it is flash-vaporized. This water vapor drives a turbine-generator to generate electricity. Remaining water vapor (essentially distilled water) is condensed using cold sea water. The condensed water can either return back to the ocean or be collected for the production of potable water. The hybrid-cycle combines the characteristics of the closed cycle and the open cycle, and has great potential for applications requiring higher efficiencies for the co-production of energy and potable water[2].

The open and hybrid cycles allows the co-production of potable water through desalination, in addition to electric power. It is possible to produce up to 2 million liters per day for each megawatt of electricity generated[1]. In all of the three cycles, it is required to obtain deep cold water (normally available at depths of 1,000 meters, where the water temperature of is approximately 4 °C) to condense the working fluid. An OTEC plant can be installed on-shore or off-shore depending on the resource characteristics and market conditions of the proposed location. An off-shore plant could be built with a foundation on the ocean bottom (shelf-mounted), or be located on a moored platform or as a grazing plantship, depending on the proposed use for the electrical energy generated by the on-board OTEC power system (transmitted ashore via underwater power cables or stored in the form of chemical energy for periodic transfer to on-land users).

During the 1970's and 1980's R&D projects such as Mini-OTEC and OTEC-1 in Hawaii and the Japanese 100-kW land-based pilot plant at the Republic of Nauru demonstrated the technical viability of OTEC, specifically with a closed-cycle system to generate electric power. Over 40 years of cumulative experience (and more than $500 MM invested in R&D) are available to us today in the form of engineering data, equipment development, environmental studies, conceptual & preliminary designs, and technical information. The information is sufficient to build the first commercial OTEC plants, with a capacity in the range of 50 to 100 MWe[2]. Although the source of OTEC energy is renewable, continuous and fuel cost-free, the OTEC net thermal efficiency of operation is approximately 3%[*].

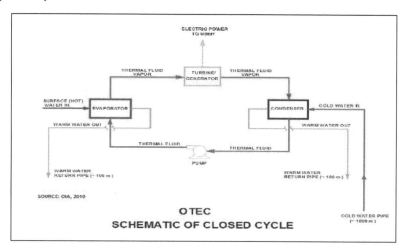

OTEC
SCHEMATIC OF CLOSED CYCLE

Figure 1. Schematic of an OTEC closed-cycle system with ammonia as working fluid.

Extensive research and development has been conducted to determine the optimum configuration, design basis and material of construction for the heat exchangers required for an OTEC process[2][3]. Projects such as Mini-OTEC and OTEC-1 as well as the test facilities at the Argonne National Laboratory (ANL), Keahole Seacoast in Hawaii and the experiments conducted at Punta Tuna in Puerto Rico provided the bases for the development of design methods for OTEC commercial-scale heat exchangers[2]. Lessons learned from these experiences have been applied to the development of heat exchangers which are being used for other applications today. Since large heat exchangers are required for a commercial size plant (due to the large flow rates of water required for the process), design and selection should be based on two major factors: optimum heat transfer rate and low cost. In addition, the material selection should be based on durability (resistance to corrosion and biofouling), compatibility with the working fluid and life-cycle cost.

The objective of this paper is to demonstrate that commercially available heat exchangers used today in other industries can be used for the first generation of OTEC commercial-scale facilities, while meeting cost-effectively the criteria of heat transfer rate and 30-year life expectancy.

The heat exchanger evaluation and selection process presented in this study is based on a closed-cycle system and ammonia as the working fluid. The reason behind this decision is that the majority of the technical studies, research and development, conceptual designs and demonstration projects completed during the past four decades focused on the Rankine closed-cycle system with ammonia as the working fluid[1][2]. In addition, ammonia is the preferred working fluid over other substances such as propane and commonly-known refrigerants (R-12, R-22 and R-114), due to its cost-effectiveness[1], its superior thermal characteristics, proven safety record, and to the extensive operational experience with ammonia refrigeration systems in commercial and industrial applications[1][2][3].

EQUIPMENT DESIGN & CONFIGURATION

The OTEC research and development programs initially focused on shell-and-tube heat exchangers because there was more experience with this design as compared to other configurations[2][3]. Later it was realized that the use of shell-and-tube heat exchangers for OTEC commercial plants would make these equipment a major volume element in the overall plant installation. R&D programs shifted their attention to investigate other designs and configurations with the objective of reducing heat exchanger unit size per kilowatt of power produced. Compact heat exchangers became the central focus of subsequent R&D programs and technical demonstration projects[2]. The volume of the OTEC heat exchangers and associated water and working fluid piping establishes the requirement for minimum construction and materials costs and the implementation of a systems integration strategy during the design of the power modules and its incorporation into the rest of the plant subsystems. Otherwise, the cost of the heat exchangers could become the major factor in the total OTEC plant cost.

Shell-and-tube heat exchangers are the most widely used type of heat exchangers for industrial evaporator and condenser applications (see Figure 2a). They are typically used for high pressure and high temperature applications. Shell-and-tube heat exchangers consist of an array of parallel tubes commonly referred as the core and a cylindrical vessel that encloses the tube bundle. One of the fluids in the process flows through the tubes interior while the other fluid flows over the tubes external surface (shell side) following a tortuous path while heat transfer occurs between both fluids. Configurations evaluated for OTEC include horizontal (flooded bundle and spray) and vertical (falling film and two-phase upflow) installations for both evaporator and condenser.

Plate heat exchangers are widely used in a variety of industries such as chemical plants, oil and gas, pulp and paper, HVAC and power generation (see Figure 2b). These are composed of multiple, thin, slightly-separated plates that are configured in a stack defining flow passages and resulting in very large surface area for heat transfer. Advances in gasket and brazing technology have made the plate-type heat exchanger increasingly practical. Configurations evaluated for OTEC include conventional design, cross-flow and plate-and-shell. Plate designs studied for OTEC include gasket sealing, weld/semi-weld sealing and nickel-brazed sealing.

Brazed aluminum plate-fin heat exchangers have been successfully used a variety of applications (see Figure 2c). The major applications have been in the cryogenic separation and liquefaction of air, natural gas processing and liquefaction, the production of petrochemicals and offgases treatment, and large refrigeration systems. A brazed aluminum plate-fin heat exchanger consists of a core of alternating passages of corrugated fins. The stacked assembly is brazed in a vacuum furnace to produce a rigid core. Configurations evaluated for OTEC include vertical conventional design and horizontal installation (upflow and falling film).

In 1978 a heat exchanger test facility was constructed at Argonne National Laboratory (ANL) with the support from the U.S. Department of Energy. The objective was to test various OTEC heat exchangers design at a large enough scale to provide sufficient and valid design data for demonstration-size OTEC power units[2]. Various designs of evaporators and condensers were tested at this facility: shell-and-tube Linde flooded-bundled evaporator, shell-and-tube Linde sprayed-bundled evaporator, shell-and-tube Linde enhanced-tube condenser, shell-and-tube Carnegie-Mellon (CMU) vertical fluted-tube evaporator, Alfa-Laval and Tranter plate heat exchangers, Saga University plate heat exchangers and Trane brazed-aluminum plate-fin heat exchangers. Another heat exchanger design, folded-tube heat exchanger, was evaluated and tested by John Hopkins University/Applied Physics Laboratory (JHU/APL), and eventually incorporated into their 1980 baseline design for a 40-MW OTEC pilot plant[2]. Nevertheless, neither the CMU fluted-tube or the JHU/APL folded-tube heat exchangers are commercially available today. For this reason the study will concentrate its analysis for equipment evaluation and selection on the shell-and-tube, plate and plate-fin heat exchanger designs.

Process Performance

In general, compact heat exchangers are considered to provide higher heat transfer efficiency than the traditional shell-and-tube heat exchangers in most applications. Tests results from ANL showed that the overall heat transfer coefficients for compact heat exchangers (plate and plate-fin) were 1.5-3 times[2] greater than the values obtained for the shell-and-tube heat exchanger designs for both evaporator and condenser***. This is consistent with industry experience in other applications where overall heat transfer coefficients values for compact designs are typically several times greater than shell-and-tube designs[4]. OTEC-1 (a converted T-2 tanker used for OTEC heat exchanger tests off the coast of Hawaii in 1980) used one shell-and-tube evaporator and one shell-and-tube condenser, each unit with a capacity of 1-MWe. Heat exchangers operation and performance were very similar to the behavior predicted by the ANL test facility for the shell-and-tube designs[2].

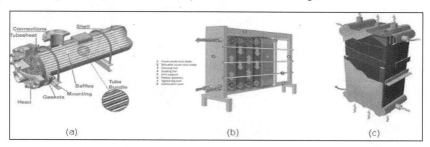

Figure 2. Schematics of commercial heat exchangers: (a) shell-and-tube, (b) plate and (c) plate-fin**.

In 1979, Mini-OTEC (a collaborative research effort involving the State of Hawaii and other private companies) provided the first demonstration of net OTEC power production in the world[2]. It used plate-heat exchangers from Alfa Laval, which performed closely to analytical predictions[2]. In 1983, General Electric (GE) completed a trade-off analysis for compact heat exchanger designs as part of their conceptual design for a 40-MWe OTEC shelf-mounted plant in Hawaii. The survey examined three candidates: brazed-aluminum plate-fin design by Trane, plate design by Alfa Laval and a new compact design by GE. In the category of overall thermal-hydraulic performance (heat transfer/pressure drop), both the plate and the plate-fin heat exchangers were rated "good"[4]. These ratings were based on the predicted overall heat transfer coefficient and low pressure losses, and equipment "compactness" characteristic. However, in the category of OTEC related performance (test experience), the brazed-aluminum plate-fin exchanger was rated "good" and the plate design was rated "fair"[4]. Optimized designs for brazed-aluminum plate-fin heat exchangers have improved their thermal performance when compared to the initial designs tested at ANL[5]. Cross-flow configurations for plate heat-exchangers offer great potential for further improvements in thermal and hydraulic performance[2].

Surface Enhancement

Heat transfer surfaces can be enhanced on both the working fluid and water sides. For the water side, enhancement is attained by the utilization of roughed/porous surfaces, internal fins, corrugations, spirals, flutes and other modifications. For the working fluid side, enhancement is attained with alterations comparable to those on the water side and other techniques such as wire wraps and flame-sprayed aluminum (Hi-Flux®). Most of these surface treatment alternatives apply to shell-and-tube heat exchangers.

The Nauru 100-kW OTEC pilot plant (a 1981 collaborative project between the Republic of Nauru, Tokyo Electric Power Company and Toshiba Corporation) employed shell-and-tube heat exchangers with enhanced surface. Freon was selected as the working fluid, although analysis showed that ammonia would be a better choice for commercial operation[2]. The surface treatment consisted in spraying the evaporator tubes with copper particles and installing spirally grooved tubes in the condenser sealed at intervals to separator plates. The heat exchangers performed as expected, with overall heat transfer coefficient values slightly higher than those recorded at the ANL test facility[2]. The OTEC-1 shell-and-tube evaporator included two independently sections of tube-bundles. The upper section had plain tubes and the lower section had Linde-design Hi-Flux® enhanced tubes. The condenser only contained plain tubes. These enhancements were not found cost-effective[3]. The plate heat exchangers used in the Mini-OTEC project were not provided with surface enhancement.

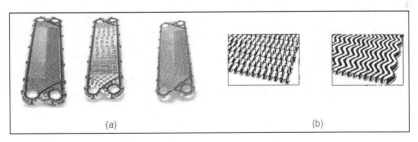

Figure 3. Integrated surface enhancements: (a) chevron pattern, (b) fin corrugations[****].

Compact heat exchangers depend on integral enhancement as part of the unit construction. Plate designs rely on a chevron pattern stamped into the plates (see Figure 3a) for the formation of the water and working fluid passages and for rigidity[3]. The thermal performance of these heat exchangers can be improved by employing high-flux surface on plates as demonstrated by performance tests at ANL[2]. However, applying a coating to the heat exchanger plates may not be cost-effective[3]. It will depend on the material of construction selected for the plate heat exchanger (stainless steel, titanium, etc.). The brazed-aluminum plate-fin design do not provide enhancement on the water extruded passages but aluminum fins such as straight and serrated types provide enhanced passage for the working fluid side (see Figure 3b). Integral enhancement configurations for both plate and plate-fin heat exchangers provide high heat transfer capabilities compared to shell-and-tube designs and are found to be cost-effective[3].

MATERIAL SELECTION

The selection of material for commercial heat exchangers has depended on tradeoffs among durability, fabrication and packaging, thermal conductivity and cost. In the case of OTEC, the selected heat exchanger material must withstand the corrosive action of seawater. In order to maintain the operation, maintenance and repair (OM&R) costs of OTEC plants to a minimum, a service of 30 years has been established for the heat exchangers[3]. For this reason, the selection of materials for the heat exchangers and their biofouling controls is extremely critical to the success of the OTEC operation. To achieve maximum heat transfer, the surfaces of OTEC heat exchangers must be kept free of significant corrosion product films, calcareous deposits, biofouling and any other foreign deposits. Before the 1970's, data on the resistance to corrosion in seawater of metals that required to be clean at all times was very scarce[3]. Since then, the OTEC R&D programs as well as materials research and development from other applications and industries have successfully provided the necessary technical data and commercial experience to design, construct, operate and maintain heat exchangers under OTEC process conditions and requirements[2][3].

Heat exchangers for commercial use have been constructed from alloys of copper, titanium, aluminum and stainless steel. The R&D and test programs at ANL, the Naval Coastal Systems Center in Florida, the LeQue Center for Corrosion Research, the Center for Energy and Environmental Research in Puerto Rico and the Sea Coast Test Facility in Hawaii were established in the 1970's and 1980's to furnish data on the durability of the aforementioned candidate alloys in order to qualify OTEC heat exchangers for 30-year life.

Corrosion-Resistant Alloys

Copper-nickel was included in the OTEC test programs due to its resistance to corrosion in seawater, which is required to provide the desired 30-year life for the heat exchangers. In addition, copper-nickel has antifouling characteristics that make it even more attractive for OTEC application. However, copper-nickel is not compatible with ammonia/water solutions[3]. Nevertheless, if other working fluid is selected, copper-nickel is a strong candidate for an OTEC application.

Titanium has been recognized as a material with exceptional resistance to corrosion under seawater conditions[4]. This metal is available in the form of tubing and in plates (for compact heat exchangers). For many years, titanium has been successfully used in seacoast power plants and has performed "free from corrosion" in all tests conducted under OTEC process parameters[3][4]. Moreover, tests results from the LeQue Center for Corrosion Research showed that biofouling control methods such as mechanical cleaning and intermittent chlorination did not impact the material's performance on corrosion resistance[3]. Titanium was also tested at Punta Tuna, Puerto Rico with warm water. No perceptible corrosion was shown[3]. For these reasons, titanium was qualified very early in the OTEC R&D programs for the required 30-years life.

Stainless steel alloys have been candidates for OTEC heat exchangers based on numerous corrosion tests and vast experience in seacoast power plants. Specifically, alloys AL-6X and AL-29-4C has demonstrated to resist corrosion in seawater applications[3]. These metals are also available in the form of tubing and plates. Crevice corrosion tests were conducted in the LeQue Center for Corrosion Research for both AL-6X and AL-29-4C alloy materials which yielded successful results in terms of validating its corrosion resistance characteristic. In addition, the U.S. Navy sponsored independent crevice corrosion tests on this two alloys showing similar results[3]. The program concluded that the AL-6X and AL-29-4C alloys are strong candidates for the OTEC heat exchangers. Moreover, these two alloys will performed very similar to titanium in that erosion-corrosion will not occur during frequent mechanical cleaning/intermittent chlorination for biofouling control. For these reasons, AL-6X and AL-29-4C alloys have been qualified for 30-year life service.

Aluminum alloys have been extensively evaluated for OTEC heat exchangers due to their potential readily extruded enhancements and shapes, and their lower cost when compared to other metals such as Titanium and stainless steel alloys. Specifically, aluminum alloys 5052, 3003, Alclad 3003 and Alclad 3004 are considered leading candidates. The OTEC R&D and test programs focused on evaluating the tendency of aluminum alloys to pitting corrosion and erosion-corrosion under seawater conditions[3]. Most commercial aluminum applications in seawater are protected with paint or cathodic protection. Since OTEC requires clean heat-transfer surfaces, the erosion-corrosion effect on these aluminum alloys caused by biofouling control methods such as mechanical cleaning and chlorination were viewed as threats for their qualification.

Alclad 3003 and 5052 alloys were tested at the LeQue Center for Corrosion Research with mechanical cleaning and no chlorination. Tests showed that at a maximum cleaning cycle, erosion-corrosion caused early failure of the aluminum alloy[3]. Corrosion tests were performed for several years at the Secoast Test Facility in Hawaii on alloys 5052, 3004, Alclad 3003 and Alclad 3004 using warm water (no biofouling control). No pitting was shown on any of these alloys[3]. Tests conducted using cold water (no biofouling control) indicated that all alloys pitted by corrosion, but at rates lower than those acceptable for OTEC heat exchanger requirements[3].

The DOE conducted extensive tests using warm water at Punta Tuna, Puerto Rico initially on alloy 5052 and later on alloy 3003. The 5052 tests included infrequent mechanical cleaning. Tests showed that no localized attack occurred, although some general corrosion was observed. No erosion had occurred as a result of the mechanical cleaning. For the 3003 alloy, aluminum sections from a plate-fin heat exchanger designed by Trane were tested. A patented cladding method using zinc was implemented on the seawater surfaces of the alloy sections. Various samples with different zinc concentrations were tested including 0.5%, 1% and 45%. Bare alloy 3003 extrusions were tested, too. With brush cleaning, intermittent chlorination and the use of a device to monitor heat transfer, average corrosion rates recorded were below those required for OTEC application[3].

In the GE trade-off analysis for compact heat exchanger designs, the Trane aluminum-brazed plate-fin heat exchanger evaluated had a zinc coating in the water side passage. In the category of corrosion-erosion resistance/protection, the plate-fin heat exchanger was rated "fair"[4]. As anticipated, the titanium plate heat exchanger was rated "excellent". The GE reports concludes that 5052, Alclad 3003 and the zinc-coated 3003 alloys are candidate materials for OTEC heat exchangers. Table 1 shows a summary of test results for aluminum alloys evaluated under OTEC conditions at the aforementioned test facilities and sites.

Long-term research and development programs have proven that these aluminum alloys experience corrosion rates low enough to qualify them for at least 20-year life, and with a great potential for 30-year life for OTEC heat exchangers[2][3].

AL Alloy	Water	Comments	Conclusions
Al-3003 drawn tube	Warm	No pitting	Acceptable
Al-3003 extrusion	Warm	No pitting	Acceptable
Al-3003 extrusion (diffused zinc)	Warm	Some pitting	Acceptable with caution
Al-5052 tube	Warm	No pitting	Acceptable
Alclad 3004 RFW	Warm	Pitting	Not acceptable
Alclad 3004 drawn tube	Warm	No pitting	Acceptable
Al-3003 drawn tube	Cold	Some pitting	Acceptable with caution
Al-5052 drawn tube	Cold	No pitting	Acceptable
Al-3002 extrusion	Cold	Some pitting	Acceptable with caution
Al-3002 extrusion (diffused zinc)	Cold	Some pitting	Acceptable with caution
Alclad 3004 RFW	Cold	Pitting	Not acceptable
Alclad 3004 drawn tube	Cold	Pitting	Not acceptable

Table 1. Summary of test results for aluminum alloys evaluated for OTEC.

BIOFOULING

Biofouling, the undesirable accumulation of microorganisms, plants, algae, and/or animals on wetted structures, has to be prevented or removed in order to achieve maximum heat transfer efficiency in the OTEC heat exchangers. This issue was considered critical at the beginning of the OTEC R&D programs due to the fact that biofouling could cause heat exchangers performance degradation. Nevertheless, as critical as it is, the extent of the potential problems were overestimated since fouling rates in tropical open-ocean waters suitable for OTEC operation are significantly lower than in coastal waters[2]. The result is that biofouling control is more effective for OTEC as compared to typical marine heat exchangers. In terms of biofouling control methods, physical methods can be effective at lower intensities or longer time cycles. Chemical agents can also be used in concentrations that are environmentally safe and in compliance with applicable environmental regulations.

An important conclusion in regards to biofouling is that the fouling effect in the heat exchanger (a percentage reduction in overall heat transfer coefficient) is independent of the change in temperature[2]. In other words, the low thermal efficiency of OTEC does not make its performance particularly sensitive to a heat transfer reduction caused by biofouling. If overall heat transfer coefficients are improved by design, the sensitivity of the heat exchanger performance to biofouling is expected to increase as well[2].

Another important conclusion is that the differences between the organisms that causes biofouling at sites in Puerto Rico, Gulf of Mexico and Hawaii are not significant, which provides enough assurance that the methods adopted to control biofouling for one of these sites will be applicable to the rest[2].

Microfouling

The initial OTEC R&D programs concluded that an acceptable value for the fouling factor for both the evaporator and condenser must be less than 0.000088 m^2 °C/W[2][3]. Multiple experiments and test activities were conducted in Hawaii, the Gulf of Mexico and Puerto Rico to determine rates of biofouling under typical OTEC conditions using a heat transfer monitor (HTM) originally developed by the Carnegie-Mellon University (CMU) and later improved by ANL. Results showed that biofouling would exceed unacceptable levels in the warm water system (fouling factor > 0.000088 m^2 °C/W) after six weeks of operation without fouling control[2][3]. To maintain the fouling factor below the acceptable level, both physical and chemical methods were explored including chlorination (continuous and intermittent), brushing, smooth or abrasive balls, slurries, ultrasonic and ultraviolet. The method found to be the most practical and cost effective is intermittent chlorination[2][3][4]. Since continues chlorination would require more parasitic power, intermittent chlorination was the preferred choice for the majority of the tests. It has been proven that injection of 70 parts per billion of chlorine for one hour per day in the warm water system prevents biofouling formation effectively. This concentration is significantly less than the limits allowed by the Environmental Protection Agency on the discharge of chlorine from coastal power plants and similar industries[2][3]. Intermittent ozonation should be at least as effective as chlorination, and may be more practical for actual plants.

In the case of the cold water system, there was no indication of biofouling formation in all the tests conducted at the above sites. This is consistent with the operational data from the NELHA open-cycle OTEC test facility at Keahole Point in Hawaii, where no biofouling was ever found in the cold water system during the project's five-year operation[2].

The GE trade-off analysis rated the Trane aluminum-brazed plate-fin heat exchanger "good" in the category of biofouling control category due to its ease of chlorination and the smooth, defined water passages[4]. The report states that the Alfa-Laval plate heat exchanger design will require a closed-cycle clean-in-place (CIP) system based on slow circulation of a cleaning solution (e.g. 3% NaOH) through the heat exchanger. Due to this added complexity, the plate heat exchangers were rated "fair"[4] in this category. These recommendations were based on tests results from the Seacoast Test Facility in Hawaii and the tests conducted in Puerto Rico.

Macrofouling

The biofouling control program described above also included macrofouling, the accumulation of coarse matter of either biological or inorganic origin. This can be material suspended in water, and tends to adhere to surfaces and impede flow, as well as the growth of algae and marine organisms that feed on the bacterial film or slime that forms on surfaces exposed to seawater. In addition to the OTEC heat exchangers, other surfaces such as screens, sumps, piping, pumps and valves require macrofouling control or prevention. The degree of macrofouling development was specifically identified during the biofouling and corrosion tests conducted in Puerto Rico in the early 80's. Nevertheless, the power industry has available various macrofouling control and prevention measures for service in seawater such as anti-fouling paint systems and mechanical cleaning. The selection for macrofouling control should be based on reliability and cost-effectiveness. Suitable effective low-cost methods are commercially available today and are applied at coastal power plants and comparable industries to address this issue, which represents a critical factor to achieve the long service life needed for OTEC power systems[1][2].

ECONOMICS & OVERALL COST-EFFECTIVENESS

As mentioned before, the selection of commercially available heat exchangers for the first generation of OTEC power plants should be based on the equipment's capability to meet the following design criteria: maximum thermal and hydraulic performance, 30-year life cycle and low unit cost per net power produced. Establishing the cost-effectiveness of OTEC heat exchangers is not an easy task. Since the beginning of the OTEC R&D and test programs, heat exchangers provided the largest targets for potential cost reductions in order to make OTEC commercially attractive[2][5]. Ample data is available today that can be used for cost comparison between shell-and-tube heat exchangers and compact heat exchangers, and between plate heat exchangers and plate-fin heat exchangers with variations in the material selected for the equipment construction.

Compact heat exchangers, when compared to shell-and-tube design, are more likely to have the lowest cost and meet the scalability requirements to successfully commercialize OTEC[4]. The reduction in labor costs is considerable in going from shell-and-tube heat exchangers to compact design such as plate and plate-fin heat exchangers[3]. ANL predicted potential cost reductions of approximately 40% of total hardware costs if brazed-aluminum plate-fin heat exchangers were used instead of titanium shell-and-tube heat exchangers for their 10-MWe shore-based OTEC plant conceptual design[5]. These design evaluations are based on comparisons between cores only (tubing versus plates versus plate fins) since cost and weight of the other steel components (tube sheets, baffles, shells, waterbox, etc.) are significantly small compared with the cost and weight of the cores[3]. For example, titanium plate heat exchangers reflect approximately 35% in core weight savings when compared with titanium shell-and-tube heat exchangers[3].

In terms of material selection, metal savings of approximately 54% can be expected when selecting plate-fin heat exchangers versus shell-and-tube designs using Alclad 3004; 39% when selecting plate heat exchangers versus shell-and-tube designs using 29-4C stainless steel alloy; 37% when selecting plate heat exchangers versus shell-and-tube designs using Al-6X stainless steel alloy; and 29% savings when selecting plate heat exchangers versus shell-and-tube designs using titanium grade I[3]. Based on these data and due to the volatility and high costs of titanium over the past decades, emphasis has been made on commercially available compact heat exchangers using stainless steel and aluminum alloys, specifically, the alloys extensively evaluated and tested for corrosion resistance and biofouling control as discussed in previous sections. This is supported by the conceptual and preliminary design reports completed by JHU/APL, ANL and GE. Particularly, in their trade-off analysis, GE rated the Trane brazed-aluminum plate-fin heat exchanger design as "excellent" in the category of commercial plant capital cost; the Alfa-Laval titanium plate heat exchanger was rated "fair" in this category[4].

In recent years, ANL has favored commercially available compact heat exchangers such as plate heat exchangers using Al-6X or 29-4C stainless steel alloys and plate-fin heat exchangers using aluminum alloys such as 3003, 3004, 5052 and 6061 to be used in the first generation of OTEC commercial plants[2][5].

Maturity of Design & Manufacturing

The stainless steel plate heat exchangers and the brazed-aluminum plate-fin heat exchangers are both the result of long commercial application. The plate heat exchanger has many years of industrial and commercial application. The present configuration evolved in the 1920's and 1930's with application in the chemical industry, in the power industry in the 1940's and in marine service in the 1950's and 1960's. Continues development in plate design and sealing methods have yielded significant improved performance and cost reductions. Stainless steel plate heat exchangers manufactured by Alfa-Laval (T-50M) and Tranter (Superchanger), with small design modifications, are considered strong candidates for the first generation of commercial OTEC plants. This configuration is widely used today in commercial-scale ammonia refrigeration systems. However, special attention must be given to the potential significant pressure drops that have characterized the plate heat exchangers design in other applications.

Brazed-aluminum plate-fin heat exchangers have over 35 years of accumulated industrial experience, primarily in the cryogenic field for hydrocarbon gas separation, natural gas processing, petrochemical offgases treatment and large refrigeration systems. The Standards of the Brazed Aluminum Plate-Fin Heat Exchanger Manufacturers' Association (ALPEMA) was created in the 1990's to promote the quality and safe use of this type of heat exchanger. This standard contains all relevant information for the specification, procurement, and use of brazed-aluminum plate-fin heat exchangers[6]. Alternate extrusion methods and furnace-brazed fabrication capabilities are being introduced to further improve the product for other commercial applications.

ANL concluded in 1981 that the facilities needed to manufacture the stainless steel plate and the brazed-aluminum heat plate-fin heat exchangers were in place and only modest extension of current technology were required for their use in OTEC power plants[2]. This conclusion was supported by the GE trade-off analysis, where both the Alfa-Laval and the Trane compact heat exchangers were rated "good" in the category of maturity of design[4]. This conclusion remains true today, where manufacturing facilities are available to manufacture the appropriate size compact heat exchangers (plate and plate-fin) that can be used in the first generation of commercial OTEC plants.

Installation & Space Requirements

As mentioned in previous sections, from the beginning of the OTEC R&D and test programs it was realized that the selection of shell-and-tube heat exchangers for OTEC commercial plants would make the heat exchangers a major volume element in the OTEC total plant capital cost. Compact designs reduce the heat exchanger volume requirements by 55-70% in comparison with the shell-and-tube units, with a potential reduction of 45-60% in total module space requirements[2]. Moreover, the active total heat exchanger volume, which is indicative of the complexity of the equipment installation and arrangement, is approximately 10 times greater for the shell-and-tube design than the brazed-aluminum plate-fin heat exchangers. The volume for a shell-and-tube heat exchanger is approximately 4 times greater than the plate heat exchangers[4]. The heat exchangers volume is more critical in floating OTEC plants since the equipment space requirement affects directly the design and final cost of the selected platform configuration (shelf-mounted, moored platform or grazing plantship). For instance, the total equipment footprint for a floating platform with plate-fin heat exchangers could be approximately 20-30% less than the area required using plate heat exchangers[2][5].

The piping requirements for plate and plate-fin heat exchangers diverge in terms of arrangement complexity. For the brazed-aluminum heat exchangers, its design simplifies the piping configuration. However, in the case of the plate heat exchangers, although they are simple in principle, manifolding and ducting requirements may require special attention[2]. In both cases, pipe distribution arrangement need to be optimized in order to minimize friction losses, which leads to an increase in total parasitic power.

Operation, Maintenance & Repair

The cost of heat exchangers must consider its original cost plus the replacement cost if the material selected does not results in an equipment life of 30 years. As indicated previously, titanium and stainless steel alloys Al-6X and 29-4C have been qualified for 30 years without replacement. By comparison, the extruded aluminum alloy Alclad 3003 would need to last at least 20 years to be competitive with titanium and at least 25 years to compete with stainless steel 29-4C[3]. In the case of welded aluminum Alclad 3004, it would need to last at least 15 years to compete with titanium and 20 years to compete with 29-4C[3]. Plate heat exchangers based on Al-6X and 29-4C can perform as designed over a period of 30 years with biofouling control. In the case of the brazed-aluminum plate-fin heat exchangers, long-term R&D has qualified this design for at least 15 years[3]. Some reports even suggest that these heat exchangers can be qualified for 30-years service life[2]. Regardless of this potential discrepancy, results from all the associated R&D programs indicate that the life-cycle cost of aluminum heat exchangers will be lower than the costs of other alternatives, specifically, when compared with titanium heat exchangers[2].

The impact of cleaning systems and their ability to preserve an acceptable level of fouling resistance is vital for OTEC heat exchangers cost-effectiveness. Biofouling formation reduces the heat exchanger's capability to transfer heat efficiently. Some of the consequences of not controlling or preventing biofouling formation in heat exchangers are: reduction of output, additional requirement of chemicals, an increase in parasitic power, unscheduled maintenance and cleaning, additional downtime due to leakages and repairs and overall reduction of system's life. All of these potential scenarios translate into higher operational costs and profit loss. Mechanical cleaning diminishes the life of aluminum heat transfer surfaces but is not a key problem with titanium or the stainless steel alloys. Thus, for brazed-aluminum plate-fin heat exchangers it is important to establish a biofouling control method based on periodic mechanical cleaning and intermittent chlorination in order to reduce the risk of surface erosion-corrosion. In the case of plate heat exchangers, the cost associated to the application of chemicals for biofouling prevention, including such as the management of spent cleaning solutions, need to be closely evaluated, specifically, if a closed-cycle clean-in-place (CIP) system is to be used in an offshore OTEC plant.

CONCLUSION

For a closed-cycle system using ammonia as the working fluid, it is concluded that commercially available compact heat exchangers can be adapted without major difficulty to be used as heat exchangers for the first generation of commercial-scale OTEC power plants. Both the plate heat exchanger and the plate-fin heat exchanger offer high thermal and hydraulic performance and surface enhancement can be integrated into the equipment design to improve the overall heat transfer coefficient. Stainless steel alloys such as Al-6X and 29-4C are qualified for 30-year service without replacement. Plate-fin heat exchangers based on aluminum alloys such as Alclad 3003 and 3004 can be qualified for at least 15 years service, which offer great potential for a competitive life-cycle cost. Biofouling control can be achieved successfully with intermittent chlorination and periodic mechanical cleaning under OTEC process conditions. Final selection between plate heat exchangers and brazed-aluminum heat exchangers should be determined by a trade-off analysis between material costs, design requirements, manufacturability, installation and space requirements, ease of operation and maintenance, and overall equipment cost-effectiveness.

REFERENCES
[1] R. Cohen, Energy from the Ocean, *Phil. Transactions, Royal Society, London*, A-307, 405-437 (1982).
[2] W. Avery, C. Wu, Renewable Energy from the Ocean: A Guide to OTEC, Oxford University Press, New York, NY (1994).
[3] E. Kinelski, Ocean Thermal Energy Conversion Heat Exchangers: A Review of Research and Development, *Marine Technology*, Vol. 22, No.1, 64-73 (1985).
[4] General Electric Company (Advanced Energy Programs Department), Conceptual Design Report for Ocean Thermal Energy Conversion Pilot Plant Program (1983).
[5] H. Stevens, L. Genens, C. Panchal, Conceptual Design of a 10-MW Shore-Based OTEC Plant, Argonne National Laboratory (1983).
[6] The Brazed Aluminum Plate-Fin Heat Exchanger Manufacturers' Association (ALPEMA), Second Edition (2000).

FOOTNOTES
[*] This low net efficiency is due to the available temperature gradient for an OTEC plant (20-24 °C), the plant's parasitic power and the expected thermal and hydraulic losses across the system.
[**] Schematic sources: (a) SEC Heat Exchangers web site, (b) Trelleborg web site, (c) Chart Industries web site.
[***] This comparison considers plain and enhanced surfaces for both shell-and-tube and compact heat exchangers.
[****] Schematic sources: (a) Brandex Directory Co. web site, (b) ALPEMA Standards 2nd Edition.

SYNTHESIS OF SOLAR-GRADE SILICON FROM RICE HUSK ASH – AN INTEGRATED PROCESS

*K. K. Larbi, M. Barati, A. McLean and R. Roy

University of Toronto
Department of Materials Science and Engineering
184 College Street, Suite 140
Toronto, Ontario, M5S 3E4
Canada

ABSTRACT

Impurity optimized silicon is needed for the advancement of terrestrial photovoltaic power generation. In this study an approach to synthesis of solar grade silicon using rice husk ash has been pursued. Metallothermic reduction of the purified rice husk ash (RHA) was investigated within the temperature range of 500-950 °C using magnesium in varied amounts. The reduction product was purified by two stage acid leaching sequence. Analysis of the final silicon product by XRD, SEM, and ICP-OES showed crystalline silicon with boron to be less than 3 ppm, corresponding to a reduction by a factor greater than 10 while the phosphorus level was reduced by a factor of over 20 reaching less than 73 ppm. Transition metal impurities and other elements were generally reduced in the processing steps.

INTRODUCTION

Worldwide concerns over energy related climate change coupled with spiralling cost and resource-scarcity of fossil based fuels in recent years has provoked interests in renewable and alternate energy technologies. Solar photovoltaic power is one of the main renewable energy alternatives being actively pursued worldwide. Although solar photovoltaic power production is a proven sustainable energy technology especially in the aerospace industry, development of this technology to meet terrestrial energy demands has been largely limited.

One of the well known hindrances to widespread use of PV technology is the prohibitively high unit cost of solar PV generated electricity which is partly attributed to cost of substrate material used in fabrication of efficient solar cells. [1]

Presently, silicon in both monocrystalline and polycrystalline form is the dominant semiconductor material used in the production of most commercially available high efficient solar cells, commanding over 90% of the market share of existing PV technologies. [2]

Much of the past and present research efforts to produce low-cost solar grade silicon have focussed on either upgrading metallurgical grade silicon by metallurgical refining processes or modification of the conventional Siemens process to produce silicon with purity in the range of 5-7N which is recognized to be suitable for fabrication of high efficient silicon solar cells [3]

Up until now, the possibility of producing solar grade silicon from biomass resources such as rice husk ash which is known to have high purity silica content has only been explored by relatively few research groups.

Singh and Dhindaw [4] reported producing polycrystalline silicon of 6N purity by magnesium reduction of rice husk ash followed by successive acid leaching refining. Subsequently, Bose et al [5],

---

* Corresponding author
E-mail: kingk.larbi@utoronto.ca ; kklarbi@hotmail.com

Banerjee et al [6], and Ikram and Akhter [7] further investigated the production of silicon by magnesium reduction of RHA, but obtained silicon purity significantly less than that reported by Singh and Dhindaw. Hunt et al [8] who investigated the possibility of reducing both purified and unpurified RHA with carbon concluded that purified RHA is promising silica material for synthesis of solar grade silicon. The reduction of purified RHA with calcium has also been reported by Mishra et al [9]. However, the narrow and isolated experimental conditions of previous works have limited the engineering application of such results. The lack of commercially viable process for synthesis of solar grade silicon using silica from rice husk ash therefore provides the motivation to conduct further research into the feasibility of this approach.

Presently an estimated 120×10^6 t/yr of rice husk is generated globally [10] Depending on the combustion conditions, the mineral ash content of rice husk ranges from 13-29% of which 87-97% is known to be amorphous silica [11].

The objective of this research therefore is to develop an integrated process with optimized conditions for the synthesis of high purity silicon from RHA for possible use as solar grade silicon feedstock.

MATERIALS

The RHA used in this research was obtained from a local company in Toronto, Canada (Process Research Ortech Inc). Magnesium turning (99% purity, Fisher) was selected as reductant reagent for similar reasons as outlined by Banerjee et al [6]. Both magnesium turnings and magnesium granules (98% purity, Sigma Aldrich) were used under the conditions of this research. High purity argon (Grade 6.3, Lindy) was used to maintain inert atmosphere in the furnace. Leaching solutions were prepared from Caledon trace metal grade acid reagents (HCl (37.3wt %), HF (48 wt %) and glacial acetic acid (CH_3COOH, 99.7 wt %)

EXPERIMENTAL METHODS

The scheme of experimental approach followed throughout this work is illustrated in Figure 1.

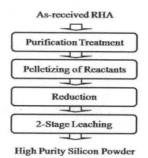

Figure 1: Experimental process steps

The as -received RHA was characterized by XRD for crystal structure, SEM for morphology and ICP-MS/OES for chemical composition. The specific surface area and particle size distribution of the RHA were respectively determined by the BET and laser particle size analysis methods.

The as-received RHA was subjected to purification treatment by acid leaching followed by roasting in air. Leaching of RHA was carried out at 10% solids with 10 wt% HCl at leaching temperatures of 60 and 90 °C. Leaching time at each temperature was varied at 1 and 4 hours.

The leached RHA with the overall least impurity content was selected for roasting in a muffle furnace at temperatures of 700 °C for 2 hours in order to reduce carbon content and thus further increase the purity of the silica content.

The purified white ash and magnesium turnings/granules were pelletized into cylindrical compacts using polyvinyl alcohol solution as a binder. The Mg/SiO_2 (RHA) mole ratio was varied between 2-2.5 which corresponds to 0-25wt% excess Mg. The pellets were reduced in a horizontal tube furnace (Linberg Heavi-Duty) with flowing argon to maintain inert atmosphere.

The schematic of the experimental set-up is shown in Figure 2. Reduction temperature was varied between 500-950 °C.

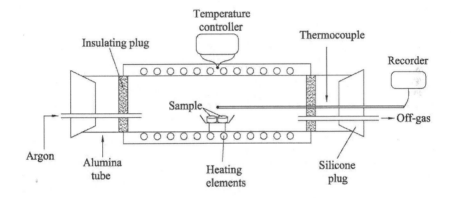

Figure 2: Reduction experimental setup

The reduction product was hand milled and sieved to ≤300μm. The phases in the reduction product were determined by Philips X-ray diffractometer equipped with nickel -filtered CuKα radiation source. The relative amounts of phases were determined by Reitveld Quantitative Powder XRD (QPXRD) method.

Leaching of the previously ground reduction product was performed in two stages. The first stage leaching was carried out using mixture of 1.25M HCl and 25wt% CH_3COOH in a volume percent ratio of 80:20 at temperature of 70 °C for a period of one (1) hour. The equivalent {H^-} to MgO ratio was varied between 2-4 times the stoichiometry requirements. The leached residue obtained after vacuum filtration was thoroughly washed with de-ionized water, dried and then subjected to a second stage leaching.

The reagent for the second stage leaching comprised a mixture of 4.8wt% HF and 25wt% CH_3COOH (acetic acid) in a volume percent ratio of 10:90 respectively. Leaching temperature was kept at 70 °C for a time period of one hour at solid to liquid ratio of 20g/L. The final residue was oven dried. The final residue from the second stage leaching was subjected to XRD, ICP-OES, and SEM/EDX, BET and laser particle size analysis as previously described. The effects of reagent type and temperature on the kinetics of the first stage leaching were also investigated in detail in order to optimize conditions for best silicon purity and recovery.

RESULTS AND DISCUSSIONS

The physicochemical properties of the as-received RHA are summarized in Table I. The SEM micrograph and XRD pattern of the as-Received RHA is shown in Figure 3.

Table 1 Physico-chemical properties of as-received RHA

	RHA Particle Size (μm)			BET Surface area		
Parameter	d_{50}	d_{80}		(m^2/g)		
Value	35	52		39		
	RHA Chemical Assay (wt%)					
Species/Parameter	B	P	SiO_2	Al_2O_3	Fe_2O_3	CaO
Value	5.1×10^{-3}	0.16	91.5	0.62	0.57	0.39
Species/Parameter	MnO	MgO	Na_2O	K_2O	C	LOI
Value	0.04	0.3	0.18	1.23	1.87	3.05

The elemental impurity content of the as-received RHA is relatively higher than the regional average impurities in RHA that were reported by Hunt et al [8]. This is possibly due to impurity pick-up from combustion of the Rice husks in a general purpose Torbed™ reactor. However the relatively higher contents of Phosphorus, Sodium and Potassium are consistent with trend in the regional average impurities. This can be attributed to the wide use of NPK fertilizers on rice farm lands. The SEM micrograph in Figure 3 shows porous and multifaceted particle morphology possibly due to the toroidal motion of the rice husk bed during combustion.

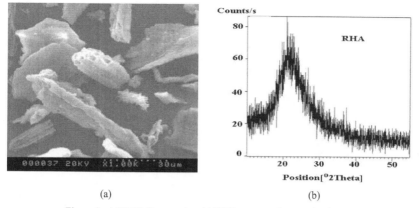

(a)   (b)

Figure 3: a) SEM Micrograph ; b) XRD pattern of as-received RHA

The XRD pattern reveals a characteristic amorphous structure. The diffuse peak in the Bragg-Brentano 2Theta angle of 18-22 is in agreement with that reported by Della et al [12]

PURIFICATION TREATMENT

The effect of purification treatment compared to the as-received RHA is as shown in Figure 4(a,b). It can be seen that the decrease in oxide impurities and carbon content resulted in increase of silica content to slightly over 98wt%. The change in color or appearance of the as-received RHA from a black to a greyish -white color after roasting at 700 C can be associated with the degree of carbon content in the ash and the grey -white color represents low carbon content [13]

PELLETIZING AND REDUCTION

The typical pellet conditions before and after reduction is shown in Figure 5. It can be seen that the cylindrical pellet with initial diameter of 16.3mm and uni-axially compressed at 188.5MPa somewhat maintains their geometrical integrity up to and even beyond the critical reduction temperature. This is potentially good for minimizing material losses due to fragmentation.

The effect of temperature increase on the relative amount of phases formed in the reduction product is compared in Figure 6. Qualitatively, the peak intensity of silicon (Si) in the XRD pattern is observed to increase whilst that of the unwanted by-product magnesium silicide (Mg_2Si) appears to decrease with increasing temperature. This observation was confirmed quantitatively by the Reitveld QXRD technique with the aid of Topas (Version 2.1) software.

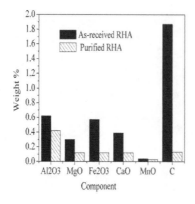 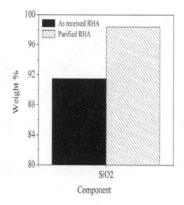

(a) Main oxides and carbon impurities (b) SiO₂ content of ashes

Figure 4: Effect of purification treatment on composition of as-received RHA

Before Reduction   After Reduction

Figure 5: Pellet condition before and after reduction reaction at 800 C

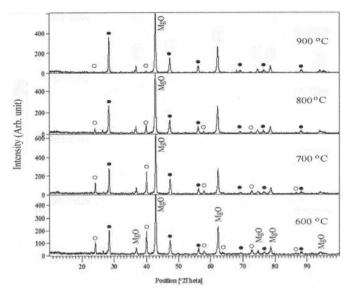

Figure 6: Effect of temperature on phases formed in reduction reaction

Si; O Mg₂Si

LEACHING AND REFINING OF REDUCTION PRODUCT

The XRD pattern of the final silicon powder obtained from 2-stage leaching of the reduction product is compared with the XRD pattern of 99.9985% silicon powder obtained from Alfa Aesar. It can be seen that the silicon powder as produced from Rice husks here after referred as RH-Si shows all the prominent reflections of silicon in the 2Theta angle range considered here and thus in good agreement with the high purity silicon standard. The very small peak immediately before the first reflection of RH-Si pattern was identified as residual carbon which remained inert in process steps subsequent to the roasting step. The range of impurities obtained by chemical assay of the RH-Si is summarized in Table II along with the particle size distribution and BET specific surface area. In a parallel experiment where the residue was washed continuously with warm de-ionized water instead of cold water between the first and second stage leaching, chemical analyses showed high magnesium (Mg) content in the final product as noted in Table 2.

The results feature a product with a very fine particle size. The levels of the impurities most deleterious to PV-silicon (P and B) are relatively low and below the detection limit of the analysis method. Nevertheless, the purity level of the product far exceeds the metallurgical grade silicon that is currently being investigated as the main feedstock for solar grade silicon production. Due to higher quality of RH-Si and the limited supply of metallurgical grade silicon, this material may be considered as an attractive feedstock for production of PV-Si. Since the majority of impurities present in the product consist of transitional and reactive metals, they may be removed by the conventional metallurgical refining methods. This post-refining step is currently being investigated within our research group.

Table 2 Physico-chemical Properties of RH-Si Powder

Parameter	Value		
Particle diameter d_{10} (μm)	4.2		
Particle diameter d_{50} (μm)	18.6		
Particle diameter d_{90} (μm)	45.8		
BET Surface Area (m^2/g)	46.9		
Impurities	ppmw		
B	3	-	18
P	25	-	73
Al	1265	-	1581
Fe	534	-	658
Mn	62	-	313
Mg	1078	-	16266
Ca	276	-	672
Na	422	-	753
K	1447	-	2803
C	1200		
Others (Summed Total)	< 500		
**Si (wt %)	> 99.3		

**Estimated silicon purity excluding Mg and C content.

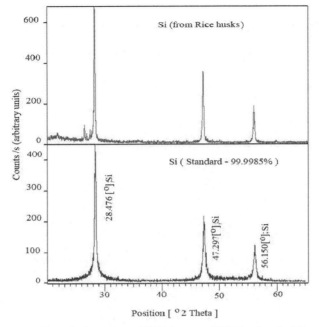

Figure 7: Comparison of XRD Patterns of RH-Si to Standard Si

SUMMARY AND CONCLUSIONS

Feasibility of producing high purity silicon from rice husks has been demonstrated. A multi stage process flowsheet was established for reduction and purification of the silicon. The process involves initial treatment of the ash by leaching and roasting, followed by reduction with magnesium, and post-reduction refining through two stages of leaching.

The results indicate that the process steps in this research reduced boron and phosphorus impurities by a factor greater than 10. The impurities remaining in the silicon are mainly reactive and transition metals that may be removed by a rather simple high temperature refining. The silicon purity may not readily meet solar grade requirement but is of much higher purity than metallurgical grade silicon. The product may be used as a high quality feedstock for PV silicon, by including additional refining steps.

REFERENCES

[1] R. M. Swanson, "A Vision for Crystalline Silicon Photovoltaics," Progress in Photovoltaics: Research and Applications, vol.14, (2006) pp. 443-453,

[2] B. Ceccaroli and O. Lohne, "Solar -grade Silicon Feedstock," in Handbook of Photovoltaic Science and Engineering A. Luque and S. Hegedus, Eds. Hoboken, NJ: Wiley, (2003) pp. 154-160.

[3] Muller A, Gosh M, Sonnenschein R., Woditsch P. "Silicon for Photovoltaic Applications". Materials Science and Engineering B, vol. 134, (2006) pp. 257--262.

[4] R. Singh and B. K. Dhindaw, "Production of High- purity Silicon for use in Solar cells," in Sun, Mankind's Future Source of Energy: Proceedings of the International Solar Energy Congress, (1978) pp. 776-781.

[5] D. N. Bose, P. A. Govindacharyulu and H. D. Banerjee, "Large Grain Polycrystalline Silicon from Rice Husk," Solar Energy Materials, vol. 7(12), (1982) pp. 319-321

[6] H. D. Banerjee, S. Sen and H. N. Acharya, "Investigations on the Production of Silicon from Rice Husks by the Magnesium Method," Materials Science and Engineering, vol. 52(2), (1982) pp.173-179

[7] N. Ikram and M. Akhter, "X-ray Diffraction Analysis of Silicon Prepared from Rice Husk Ash," Journal of Materials Science, vol. 23,(1988) pp. 2379-2381,.

[8] L. P. Hunt, J. P. Dismukes, J. A. Amick, A. Schei and K. Larsen, "Rice Hulls as a Raw Material for Producing Silicon," Journal of the Electrochemical Society, vol. 131,(1984) pp. 1683-1686

[9] P. Mishra, A. Chakraverty and H. D. Banerjee, "Production and Purification of Silicon by Calcium Reduction of Rice-husk White Ash," vol. 20, (1985) pp. 4387-4391

[10] Bronzeoak Ltd, "Rice Husk Ash Market Study," DTI/Pub URN 03/668, (2003).UK, Tech. Rep. ETSU U/00/00061/REP (accessed online)

[11] Shinohara Y., and Kohyama N., "Quantitative Analysis of Tridymite and Cristobalite Crystallized in Rice Husk Ash by Heating," Industrial Health, vol. 42, (2004) pp. 277-285

[12] V. P. Della, I. Kühn and D. Hotza, "Rice Husk Ash as an Alternate Source for Active Silica Production," Mater Lett, vol. 57, (2002) pp. 818-821

[13] D. F. Houston, "Rice: Chemistry and Technology". Edited by D.F. Houston. American Association of Cereal Chemists, (1972) pp. 517.

SUITABILITY OF PYROLYTIC BORON NITRIDE, HOT PRESSED BORON NITRIDE, AND
PYROLYTIC GRAPHITE FOR CIGS PROCESSES

John T. Mariner
Momentive Performance Materials
Strongsville, OH USA

ABSTRACT
Pyrolytic boron nitride (PBN), hot pressed boron nitride (hpBN), and pyrolytic graphite (PG) have been studied for chemical and thermal stability in copper indium gallium selenide (CIGS) processing environments. CIGS is a photovoltaic material well suited to high volume production techniques and can produce thin and flexible films. Evaluations are made from thermodynamic calculations for BN and C relative to Cu, In, Ga, and Se. Vapor pressure data of BN and graphite is presented and compared with the environments common to both thermal evaporation and to sputtering, as used in CIGS thin film production. The conclusion is supported, that boron nitride and pyrolytic graphite are suitable materials for CIGS processing. Proposals are made of suitable product forms for evaporation sources and heaters.

WHAT IS THE BUSINESS EXCITEMENT WITH CIGS?
CIGS is a rapidly growing segment of the photovoltaic (PV) market. A quick review of market studies shows the world PV market grew 110% over the previous year (Solar Buzz [1]); and Gartner reports a 17% compound annual growth rate (CAGR) to 2013 [2]. This enviable growth is for PV as a whole. Thin film and CIGS is even higher growth.

The report Global Solar Photovoltaic Market Analysis and Forecasts to 2020 [3] states that non-silicon, thin film technology will govern future growth. Thin Film Today predicts that the CIGS PV market will touch $2.16bn in 2016, a enviable 39% CAGR[4]. And NREL, "The Potential for PhotoVoltaics" states that CIGS produces laboratory cells with the highest conversion efficiency of any thin film solar cell [5]. Graphs of champion solar cell efficiencies can be found at NREL's website. PV growth is high, thin film growth is higher, and CIGS is a leader in thin film.

The typical CIGS cell consists of top contacts and antireflective coatings, a window/junction layer, the CIGS absorber layer, and back contacts. And, per the NREL report mentioned above [5], "no other technology offers so many ways to grow the absorber layer." Options being practiced today are co-evaporation, metal deposition with post selinization, sputtering, screen printing, and electro deposition. The thermal routes of evaporation, metal deposition, and sputtering, all involve metal vapors at high temperature and reduced pressure. This can be a very corrosive environment and demanding for materials of construction.

So, CIGS is one of the fastest growing segments in a rapidly growing market. And thermal routes to CIGS require demanding properties for the materials of construction.

HOW ARE PBN, PG, AND hpBN MANUFACTURED?
PBN and PG are made via chemical vapor deposition, CVD; while hpBN is made from more conventional dry ceramic powder processing. For PBN and PG, a CVD reactor is employed. A typical CVD system for PBN and PG is a hot wall reactor with a hot zone of about 1m dia. x 1m height, and operates at temperatures up to about 2000°C, at pressures of a few Torr to a fraction of a Torr. Graphite mandrels that mirror the surface of a desired finished article are placed in the reactor. For PBN, the general reaction is $BCl_3 + NH_3$ to $BN + 3HCl$. For PG, methane, CH_4 is used. These gases are introduced through a injector system, react, and coat all surfaces in the reactor, including the mandrels. Since the byproduct of PBN is H and Cl, and the environment is $>\sim 1700$°C, the free hydrogen and chlorine react with likely contaminants to make volatile salts, which are carried out in the exhaust

241

stream. PBN is thus very pure, typically 5-9's (99.999%) pure. The coated mandrels are removed from the reactor. PBN coated graphite can be made; or, with the correct processing, the coating can be removed and provided as a free standing article, a crucible. Or, multilayered structures can be made of PBN and PG, for use in such items as heaters.

Hot pressed BN uses more conventional ceramic processing. A blend is made of boron rich and nitrogen rich powders; the boron from such as boric oxide or boric acid, and the nitrogen from such as melamine. This blend is calcined at an intermediate temperature to initiate the B:N reaction, then heat treated at higher temperatures to further the reaction to BN. Much of the impurity left at this point is boric oxide, B_2O_3, and this can be removed by washing the powder, followed by filtering and drying. Various screening, milling, and additional firing cycles make a range of particles sizes and densities. For hpBN articles, powders can be hot pressed to form billets. BN is a very machineable, high dielectric, but good thermal conductivity ceramic; so crucibles and structures can be machined from hpBN.

WHY ARE PBN, PG, AND hpBN GOOD FOR CIGS PROCESSING?

PBN, hpBN, and PG are compared with other refractory materials, and vapor pressure and thermodynamic analyses are reviewed. Representative test data is shown for a cylindrical heater of PBN/PG laminated structure.

Material Property Comparison

Table 1 shows relative performance of typical refractory materials for consideration in CIGS thermal processing environments. PBN and PG score well on all metrics. Note that for hpBN, there is a balance between thermal stability and purity. Materials are available that can be added to the hot press blend that improve the thermal stability, but these detract from the chemical purity. Quartz softens at ~1100°C and aluminum nitride (AlN) starts to dissociate at ~950°C, so neither of these materials will survive some CIGS thermal processing conditions. Alumina (Al_2O_3) is technically viable, but the purity and machineability are a concern, and thermal shock is poor. PBN, hpBN, and PG are well suited for the application.

Table I. Material Property Comparison.

Need	PBN	PG	BN Hot pressed	Qz	AlN	Al2O3
Chemical stability	++	++	+	0	-	+
Thermal stability	++	++	+ ↓↑	Softens ~1100C	Dissociate ~950C	+
Purity	++	++	0	++	-	0
Machineability	+	+	++	0	--	-
Complex shapes	+	+	++	+	--	0
Thermal shock	++	++	++	+	0	--
PBN, BN and PG are well suited for the application						

Vapor Pressure

Chart I shows vapor pressure of boron nitride and carbon relative to CIGS components. Chemically, both PBN and hpBN will behave the same as boron nitride, but PBN kinetics and dissociation will be much slower than hpBN. PG will behave as carbon, C, but again with slower kinetics due to limited grain boundaries and limited edge plane exposure. The data presented is from various literature sources. BN [6] shows over 4 orders of magnitude lower vapor pressure at typical CIGS conditions than Cu, In, Ga, or Se [7]. Carbon is one of the most stable materials, only subliming at temperatures over ~3500°C [8]. Alumina [8] shows intermediate stability, with Cu, In, and Ga above it, and with BN below it.

Chart I. Vapor Pressure

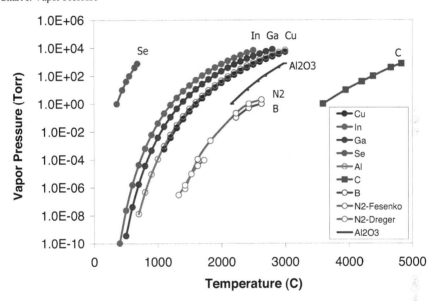

Thermodynamic Analysis

Thermodynamic analysis was performed using a commercially available software, 'HSC chemistry' (http://www.hsc-chemistry.net/) to look at relative stability of pBN with CIGS; these calculations do not include kinetics. While pBN in practice show usable stability to ~1700°C at E-8 bar in CIGS applications, such thermodynamic analysis lends itself to comparing different materials and their relative stability in various processing environments. Two charts follow, only as examples of many charts analyzed: Chart II is of BN alone, for ease of visualization, and chart III is of BN with all components of the CIGS environment. Analyses were also done with BN for reactivity/stability in each of Cu, In, Ga, and Se, individually. The BN decomposition was defined as the temperature at which BN concentration dropped below log[kmol]<-1, and a threshold for usable vapor (for Cu, In, Ge,

Se,) was defined as the temperature at which vapor concentration raised above log[kmol]>-1. These data are presented in Table II.

Table II summarizes the operating margin, the temperature range in which the element is available as a vapor before the decomposition of BN. This is listed by both reactivity of the individual element with BN, and with all components together to show any change in the operating margin due to interaction between materials. For example, the operation margin for Cu alone with BN is 240°C, but increases to 280°C with all the elements present. The operating margin for the other materials show a relative decrease when all components are present. Selenium shows the biggest change, from 600°C when only selenium is present , to 320°C when in a system with all materials present.

Table II, Operation Margin Temperature

@E-8 bar	Operation Margin Temperature* [Thermodynamic decomposition of BN] – [Thermodynamic vapor of element] °C	
	By Element	Whole Process
Cu	240	280
In	520	440
Ga	400	200
Se	600	320

Chart II, Thermodynamic Analysis for BN alone

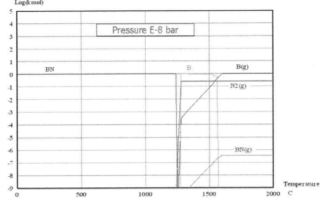

Chart III, Thermodynamic Analysis of BN with Cu, In, Ga, Se

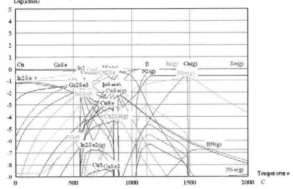

Representative Test Data

A PBN/PG/PBN heater can be produced by making a cylindrical form of PBN, coating with ~50um of PG, removing such PG as is appropriate to hit a designed circuit resistance, then over coating again with PBN to seal the conductor from the process. Figure I shows such a heater using two zones, top and bottom; and the test data for 170 cycles to 1300°C at E-8 bar. The left images show both a schematic of the article and visual picture of the heater at temperature. The upper right image shows one of the heat cycles, with temperature versus time. The lower left hand image shows a composite of all 170 cycles, with resistance versus time. There was no apparent change in the circuit resistance.

Figure I, Representative Test Data

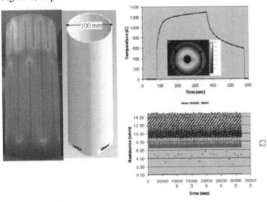

EXAMPLES OF PRODUCTS THAT CAN BE MADE

Examples can be found at http://www.advceramics.com/products/pyrolytic_bn/index.html and http://www.advceramics.com/products/heaters/index.html. PBN crucibles are routinely used for sources for metal vaporization for such industries as MBE, as in Fig. II; and as growth crucibles for GaAs and InP, as in Fig. III. PG can be coated on to PBN, and then sealed with another layer of PBN, as in Fig. IV.

Fig. II, MBE Crucibles   Fig. III, Growth Crucibles   Fig.IV, PG/PBN Crucibles

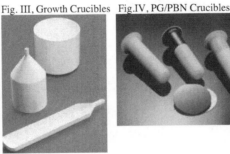

PBN can be coated onto graphite, to seal the graphite from the process. Fig. V shows a 500mm long, PBN coated, graphite boat. This coating technique can also be used to seal a graphite heater from the process. PBN is an excellent dielectric, maintaining good properties at 1500°C and more. Fig. VI shows a heater using graphite as the conductor, sealed with PBN as the dielectric, with the top a continuous plate; but you can see the machined graphite spiral in the mirror under the heater. This is called a "BE Series" heater. Fig. VII shows heaters made using the dielectric properties of PBN, and the high current carrying capability of PG as the conductor. These PBN/PG/PBN heaters can be made in a thin profile of ~2mm, or can also include a machined graphite core for strength and stability. A graphite BE Series heater as in Fig. VI may have a resistance of .2 ohms, due to the large cross section of graphite as the conductor. But the PBN/PG/PBN heaters of Fig. VII, using a thin film of PG as the conductor, may have a resistance of 10 ohms or more.

Fig. V, PBN Coated Graphite Boat   Fig. VI, "BE Series" Heater   Fig. VII, PBN/PG/PBN Heater

  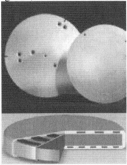

Fig. I shows a complex, 3D heater made by forming a PBN crucible to some geometry, then coating the PBN crucible with PG. The PG is machined to create a resistance heater circuit, and in this case with separate upper and lower connections for a 2-zone circuit; then the whole part is sealed again with PBN dielectric to protect the heater from the process environment. By marrying these technologies, it is feasible to make a heated distribution manifold of sections that fasten together, as shown in Fig. VIII. This would enable control of metal vapor across a web.

Fig. VIII, Heated Distribution Manifold

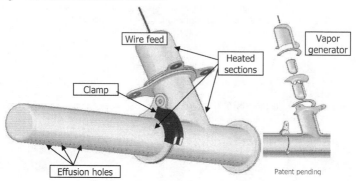

Wire feed

Vapor generator

Heated sections

Clamp

Effusion holes

Patent pending

CONCLUSION

Chemical and thermal stability, machineability, purity, thermal shock resistance, vapor pressure and thermodynamic analysis show that PBN and PG are excellent candidate materials for use in CIGS processes, with safe operating margins of several hundred degrees C. Representative heat cycle data shows reliable performance in CIGS thermal conditions.

REFERENCES

[1] http://www.solarbuzz.com/Marketbuzz2009-intro.htm
[2] http://www.edn.com/article/CA6644419.html
[3] http://www.prlog.org/10198293-global-solar-photovoltaic-market-analysis-and-forecasts-to-2020.html
[4] Thin Film Today, http://social.thinfilmtoday.com/news/cigs-pv-market-touch-216-bn-2016
[5] The Potential for Photovoltaics, B.P. Nelson, NREL/CP-520-44105, B.P. Nelson, Oct. 2008
[6] V.V. Fesenko, Poroshkovaya Metallurgiya, No. 4(4), 80(1961)
[7] C.B. Alcock, V.P. Itkin, and M.K. Horrigan, Canadian Metallurgical Quarterly, 23, 209, 1984
[8] Daniel R. Stull, Industrial and Engineering Chemistry, 39, 517, 1974

Authors:
John Mariner
        440 878 5685, John.Mariner@Momentive.com
Jon Leist
        440 878 5728, Jon.Leist@Momentive.com
Anand Murugaiah, Ph.D.
        440 878 5808, Anand.Murugaiah@Momentive.com
Wei Fan, Ph.D.
        440 878 5703, Wei.Fan@Momentive.com
Xiang Liu, Ph.D.
        440 878 5711, Xiang.Liu@Momentive.com

MATERIALS SELECTION AND PROCESSING FOR LUNAR BASED SPACE SOLAR POWER

Peter J. Schubert
Packer Engineering, Inc.
Naperville, Illinois, USA

ABSTRACT

The ultimate form of solar power is to collect sunlight in space where there is no night, no clouds, and no atmosphere. Low-density radio waves can transmit this power to earth-based receivers, invert the power, and couple it to the grid. This low-pollution power source can be scaled to the entirety of human enterprise. Until recently, economics argued against space solar power (SSP), because launch costs of the megatons of materials are prohibitive. A new approach is to use lunar materials for the bulk of SSP mass. The moon is 21% silicon, which can be formed into solar panels. Several new innovations make possible economical production of electric power from space, provided key materials challenges can be overcome. This paper reviews the ultra-high temperature ceramics and metals required for lunar-based SSP, and some of the laboratory results from experiments to build factories which, when landed on the moon, will produce many times their launch mass in valuable solar panels and array infrastructure.

INTRODUCTION

Humankind uses energy at a prodigious rate, 472 Quads/year as of 2006. Projections indicate a need for 678 Quads/year by 2030, equivalent to an extra 7,500 GW of additional installed baseload capacity[1]. At a linear rate, this means 286 GW of worldwide power plant installations every year for generations. With mega-nuclear facilities at 5-8 GW, taking 8 years to build, at a cost of 25 billion USD each, this amounts to a trillion dollars a year. Clearly, a very large, scalable, and environmentally-sustainable power source is needed.

Renewable energy sources help, however, their availability and accessibility is limited. Table 1 shows the available power in GW for traditional renewable energy sources. Only solar power has the ability to meet worldwide power demand beyond 2030.

RENEWABLE SOURCE	AVAILABLE GWs
Wind	4000
Biomass	7000
Tide/Ocean	2000
Hydroelectric	900
Solar	600,000

Table 1. Available power generating capacity of variable terrestrial renewable energy sources.

Solar power is abundant, yes, but diffuse, less than 1 kW per square meter, and depends on elevation, latitude, season, time of day, weather conditions, and atmospheric attenuation due to airborne particulates. These variable factors conspire to make solar power inconvenient and unreliable, not a good choice for baseload power. Furthermore, the production processes by which solar panels are made produce significant environmental problems, including mining, extraction, discharges of liquid, solid, and airborne pollutants, transportation-related emissions, installation-related emissions, and end-of-life considerations.

Solar power collection on the scale needed to make a significant impact to projected global energy demands must be performed in orbit. Solar power in space is 1.35 kW/m², suffers no attenuation from clouds, dust, or smog; and eclipses last 4 hours (per satellite) twice per year at the equinoxes. First invented in 1968 by Peter E. Glaser[2], space solar power (SSP) is the concept of placing large solar panels in geostationary earth orbit (GEO), and beaming the power back to earth via low-density microwaves. NASA, the US Department of Energy, and other organizations have studied SSP at various times in the last 42 years[3,4,5,6,7]. The National Space Security Office of the US Department of Defense published a study affirming that the underlying technologies are within present engineering capabilities, and that SSP has a significant strategic value for energy security, environmental stewardship, and for forward operating bases[7]. The problem, all seem to agree, is the prohibitive cost of launching such vast structures from earth's surface.

The solution is to build solar panels using materials in space (such as asteroids), a concept advocated strongly by Gerard O'Neill[8,9]. The moon is an attractive site. Lunar soil is 21% silicon. The energy required to launch solar panels into space from the moon is 24 times lower than the energy required to launch from earth. Technologies have been invented for the extraction of the raw materials from lunar soil[10,11,12]. A complete cislunar architecture supporting SSP using lunar-based materials has been published elsewhere[13,14]. The focus of this paper is the materials selections for key enabling technologies for lunar-based SSP.

OXYGEN EXTRACTION

Robotic operations in space are dramatically lower in cost and complexity compared to manual operations. For this reason, much of the future work in space-based manufacturing and assembly will be performed by machines. However, until the suite of needed technologies are sufficiently mature, there is a significant advantage in having humans uncrate, install, operate, and debug lunar factories. The single greatest need for these early pioneers is oxygen, required in large quantities for return trips to earth (3.5 MT per 2 person ascent vehicle), and in modest quantities for life support (air) and chemical processes. A factory for extracting oxygen from lunar soil is shown in Figure 1[15].

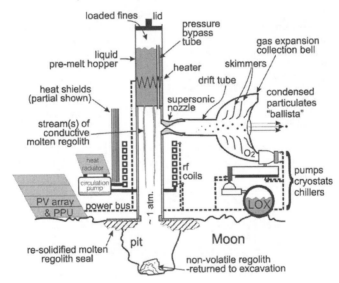

Figure 1. Lunar dust roaster to extract oxygen from regolith.

This is a new method, first proposed in 2007, and adds to the 20 or more known methods for in situ production of oxygen on the moon[16]. What distinguishes this method from others is three key concepts used harmoniously: (1) the use of free-fall heating (in lunar gravity) to hyperheat molten regolith and produce a 100,000 torr atmosphere; (2) creating a supersonic flow to cool the vapor (according to the Mach relations) in a moving beam where particles agglomerate in transit; and (3) introducing an underexpanded flow vessel along the drift tube at a point where agglomerated particulates are massive enough to disengage from the flow stream. Oxygen molecules expanding radially downstream of the underexpanded flow point are skimmed away from the ballista, and impinge on sorption beds, pumps, and chillers to make liquid oxygen (LOX). A demonstration of concept (1) under simulated lunar gravity conditions brought the technology readiness level to 4-6 (see figure 2).

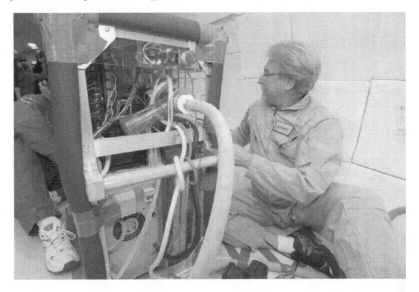

Figure 2. Author conducting free-fall heating experiment under lunar gravity conditions.

Although novel, and scoring well on metric charts (see table 2), this invention faces an incredible challenge in the operating temperature: 3010 °K (2737 °C). Oxidizing environments at these ultra-high temperatures rapidly degrade the performance of mankind's most advanced materials. Missiles, rail gun loads, and hypersonic aircraft face the same challenge, but for briefer periods: seconds or minutes at peak temperature. For the LDR to support a lunar manufactory, this apparatus must operate continuously for 14 to 23 day periods out of every 28 (depending on location), with very little resupply of consumables, and run for years at a time.

Component	Mass (kg)	Power (kw)	Key Assumptions
Hopper	257	49.8	50 stream apertures, 45 minute heat-up time
Free-fall shaft	143	19	Conditions based on temperature, length calculated by matching flow rate to evaporation
SS nozzle	44.8	0	Area relation to determine shape
Drift tube	9.69	0	1.8 m long, shares same area as the exit of the nozzle.
Expansion bell	1.28	0	Half sphere, tube area 5% of total exit area so we capture 95% flow.
Pumps and cryochillers	62	2.16	Mass of pumps is linearly related to the flow rate.
Passive cooling pipes	260	0	Length for radiative cooling from 1200 down to 200 K
Storage	100	0	Mass of large storage tank (buried)
Subtotal	876	**70.9**	
Grand Total	**1302**		Assumes 6 kg/kw, including power processing unit

Table 2. Performance metrics for the Lunar Dust Roaster.

Ultra-high temperature materials (UHTM) are a rare breed. Rarer still are those which withstand oxidizing atmospheres, even for brief periods. A leading candidate for leading edges of nose cones and control surfaces is a diboride-carbide composite. Carbon composites (such as silicon carbide on carbon fibers) coated with diborides of hafnium, yttrium, or zirconium, form protective, self-sealing oxides, which stabilize the matrix, and reduce erosion[17] (note these properties are not generally found with such coatings on refractory metals such as rhenium or iridium). Such thermal protection systems have high thermal conductivity, and can withstand the rapid thermal shock of hypersonic (greater than Mach 5) flight through oxygen-containing air. But this scheme only works to about 2200 °C.

Monolithic structures of hafnia, yttria, and zirconia – the dioxides of their namesake element – are safe, readily available, and are well-characterized. Hot isostatic pressing of their powders, or mixtures of these powders (possibly with some carbides or refractory metals thrown in), can produce near net shapes of vessels, pipes, and de Laval nozzles. However, these materials all have melting points below the LDR maximum temperature. Even if they could be cooled, creep would eventually deform the interior surfaces depending on geometry and load (gravity, pressure, mechanical coupling forces). These materials may be suitable for some components, but may still require replacement rather often.

The ultimate UHTMs is thoria (ThO_2). With a 3300 °C melting point, thoria ranks far above all other oxygen-resistant UHTMs. It is this amazing material, with its cubic crystal structure, that makes LDR operation possible, enabling the first critical step of in situ resource utilization (ISRU) supportive of lunar-based SSP. But Thorium has an Achilles' Heel.

THORIUM

Thorium is radioactive, and everybody in America is afraid of it. With a half life equal to the age of the universe, this material decays so slowly it is almost not radioactive. The type of radiation given off by thorium is alpha particles (equivalent to the nucleus of a helium atom) which can be stopped by a sheet of paper. Thorium may cause cancer if inhaled or ingested, but with suitable precautions in handling and personal protective gear, can be handled safely. The US Nuclear Regulatory

Commission[18] has no regulations for less than 1.5 kg, and allows sites to process up to 15 kg per year without a license. Personal dosimeters are not even required.

On the moon, with the LDR operated out-of-doors, alpha particles are never an issue. Space suits are protection enough. Cosmic rays are far more dangerous, and a constant concern on a lunar base, because they can arrive from any direction. By comparison to other risks of space-based operations, thorium radioactivity is a minor concern. Provided that components can be safely delivered to the moon, thoria is an ideal enabling material for production of lunar oxygen.

Small amounts of thorium metal are available from a few American suppliers. Thoria powder is available from only one or two. Facilities willing to work with thoria are very difficult to find. Since thoria was discontinued as the key ingredient for gas lantern mantles in the US in the 1990s, it has become persona-non-grata of the materials community.

However, thorium is experiencing a resurgence of interest as a nuclear fuel. Although not very radioactive in its native state, when "bred" deep inside a uranium fission reactor, thorium transmutes to U^{233}. This isotope, when used in a fission reactor of its own, produces radioactive byproducts which are far, far less hazardous than those from a uranium reactor. Countries with rapid growth in power demand are developing thorium fuel technology. This makes thorium both easier and harder to obtain – there is more being produced than during the last decade, but there is also greater demand.

For earth-bound experiments, meanwhile, hafnia, or its lesser sisters, can be used to demonstrate the concept at pressures somewhat less than 1 atmosphere, where the required temperatures are within their melting points.

SOLAR CELLS

The LDR effluent is a suboxide mineral particle stream, plus some fugitive oxygen escaping out the aperture in the collection bell of Fig. 1. That particle stream is exactly the input required for silicon extraction as taught in two US patents[10,11]. A mock-up of this arrangement, as conceived by student interns, is shown in Figure 3.

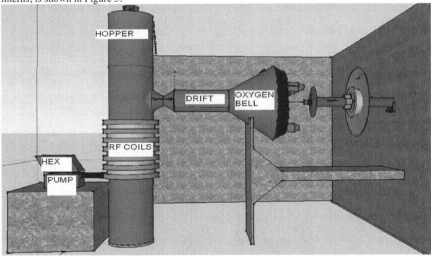

Figure 3. Lunar dust roaster (LDR) of Fig 1. coupled to isotope separator (right).

The isotope separator operates similar to a mass spectrometer, separating ions from a contained plasma beam by their charge-to-mass ratio[19]. Critical surfaces will see high temperatures and aggressive gas or plasma environments. The materials outlined above are available to build these components, and operate this apparatus continuously. Individual isotopes are separated, somewhat similar to the enrichment of uranium, but in this case is used to separate iron, silicon, aluminum, titanium and phosphorus. These are the elements needed to build solar arrays. Phosphorus and aluminum are n- and p-type dopants in silicon respectively, and having both of them makes it relatively easy to build the p-n junctions which are the heart of a solar cell in silicon. Aluminum doubles as a contact material for both n- (of sufficiently-high doping level) and p-type degenerate silicon. Together with oxygen from the LDR, silicon dioxide insulating thin films can be shadow-sputtered selectively on top to serve as an anti-reflective coating and electrical passivation. These solar cells may be held with titanium fixtures, and delivered in tight stacks inside electromagnetically-launched iron payload canisters[20,21]. Upon arrival in geosynchronous orbit, decelerated canisters[22,23] are robotically assembled into sun-tracking solar farms[24] with maximum extent measured in kilometers.

DISCUSSION AND SUMMARY

The high temperatures and aggressive environments involved in extraction of individual elements from regolith require advanced UHTM ceramics. Various combinations and configurations are available to solve the materials needs for a lunar facility capable of producing the most massive components of GEO-located SSP arrays. Slag produced in the LDR can be formed into bricks for shelters. Slag from the isotope separator can be additively-formed on motile workpieces into a plethora of shapes, useful for making furniture, roadways, landing pads, plates, and a host of other objects which would not need to be launched ("upported", a term coined by Dave Dietzler) from earth.

With the technology outlined herein, and the materials identified to make them technically feasible, mankind is capable of meeting future growth in power demands for an indefinite period of time, and sufficient for the entirety of human enterprise. Economic studies of this cislunar architecture show achievable cost parity with mega-nuclear installations[14,19], assuming projected costs by private launch services.

Studies of SSP are diverse and poorly-coordinated. However, there has been a recent focus on SSP in international conferences in 2009 and 2010. Certain Members of the US Congress are aware of SSP, as is the Office of Science and Technology Policy of the Executive Office of the President. An outline for achieving commercial SSP has been developed recently[14]. All that is required is the awareness of the potential for this technology, the political will to fund its development, and the cooperation and coordination of the many talented scientists and engineers who can bring about this ultimate answer to electric power. As an additional bonus, mankind will have a permanent base on another planetary body – a stepping stone for further exploration and exploitation of the many riches available to us in outer space.

ACKNOWLEDGEMENTS

This work is being funded as a Phase II SBIR from NASA, contract NNX09CB52C. Team members who have helped refine the ideas herein are; Rod Burton, David Carroll, Darren King, and Andrew Palla, all of CU Aerospace (Urbana, IL); plus Jason Babcock and Jeff Williams of Packer Engineering. Experiments flown aboard the Zero Gravity Corporation's reduced gravity aircraft were courtesy NASA's Reduced Gravity Office, and the FAST (Facilitated Access to Space Technology) program. The author is forever endeared of his fellow riders in the "Vomit Comet": Jeff Williams and Tom Bundorf. The author is especially grateful for the many student interns who made valuable contributions to this body of work.

REFERENCES

[1] U.S. Energy Information Adminstration, International Energy Outlook 2009, Report DOE/EIA-0484, 27 May 2009.

[2] Glaser, Peter E., "The Future of Power from the Sun," Intersociety Energy Conversion Engineering Conference, Boulder, CO, 13 Aug 1968.

[3] National Research Council, Laying the Foundation for Space Solar Power, (Washington, D.C.: National Academy Press, 2001).

[4] "Satellite Power System Concept Development and Evaluation Program Reference System Report," DOE/ER-0023, October 1978.

[5] Schaefer, John P., et. al., "Solar Power Satellites," Office of Technology Assessment, NTIS #PB82-108846, August 1981.

[6] National Space Security Office, "Space-Based Solar Power as an Opportunity for Strategic Security," October 2007.

[7] "Solar Power from Satellites: Hearings before the Subcommittee on Aerospace Technology and National Needs of the Committee on Aeronautical and Space Sciences, U.S. Senate, 94th Congress," 19 and 21 January 1976.

[8] O'Neill, Gerard K., "The High Frontier: Human Colonies in Space," Space Studies Inst. Press, Princeton, NJ, 1989.

[9] O'Neill, Gerald K., "Space Colonies and Energy Supply to the Earth," Science, Dec 5, 1975.

[10] Schubert, Peter J., "Process and apparatus for continuous-feed all-isotope separation in microgravity using solar power," US 6,614,018, 2 Sept 2004.

[11] Schubert, Peter J., "Process and apparatus for isotope separation in a low-gravity environment," US 6,930,304, 16 Aug 2005.

[12] Schubert, Peter J., "Isotope separation process and apparatus therefor," US 7,462,820, 9 Dec 2008.

[13] Schubert, P.J.,"Synergistic Construction Mechanisms for Habitats in Space Environs," International Space Development Conference 2006, Las Angeles, CA 3-6 May 2006.

[14] Schubert, P.J., "Energy and Mass Balance for a Cislunar Architecture supporting SSP," AIAA Space '09, Pasadena, CA, 14-17 Sept 2009.

[15] Schubert, P.J., "A Novel Means for ISRU Oxygen Production," Space Resources Roundtable IX, Golden, CO, 25-27 Oct 2007.

[16] Lewis, J., et. al, Eds. Resources of Near-Earth Space, University of Arizona Press, Tuscon, AZ, 1993.

[17] Kravetskii, G.A., et. al., "Increase in the Refractoriness of Carbon Composite Materials with the use of Heat-Resistant Ceramic Coatings," Refractories and Industrial Ceramics, v.49, no. 4, July 2008, pp 309-313.

[18] U.S. Nuclear Regulatory Commission statutes 10 CFR 30, 32, 33, 34, 35, 36, 39, and 40.

[19] Schubert, P.J., "A Novel Method for Element Beneficiation Applied to Solar Panel Production," Space Exploration 2005, Albuquerque, NM, 3-8 April 2005.

[20] Schubert, P.J., Beatty, M., "Harvesting of Lunar Iron: Competitive Hands-on Learning," Am. Soc. of Engineering Educators, Annual Conference, Pittsburgh, PA, 23-26 June 2008.

[21] Schubert, Peter J., Simpson, N., Lin, E., "Technical Feasibility of a Novel Method for Station Keeping," AIAA Space '09, Pasadena, CA, 14-17 Sept 2009.

[22] Duncan, S.A., Schubert, P.J., Delaurentis, D.D., "System for Electromagnetic Capture of Lunar-launched Payloads into GEO," AIAA/ASME/SAE/ASEE Joint Propulsion Conference, Cincinnati, OH 8-12 July 2007.

[23] Schubert, P.J., "Apparatus and method for maneuvering objects in a low/zero gravity environment," US 6,994,296, 7 Feb 2006.

[24] Schubert, P.J., "System and method for attitude control and station keeping," US 7,118,075, 10 Oct 2006.

CU$_2$ZNSNSE$_4$ THIN FILMS PRODUCED BY SELENIZATION OF CU-ZN-SN COMPOSITION PRECURSOR FILMS

O. Volobujeva, E. Mellikov, S. Bereznev, J. Raudoja, A. Öpik, T. Raadik
Dept. Mat. Science, Tallinn University of Technology, Ehitajate tee 5
Tallinn 19086, Estonia, v.olga@staff.ttu.ee, ph: +372620-3368, fax. -3367

ABSTRACT
    The influence of different constituent metallic and binary selenides precursor layers on the formation of Cu$_2$ZnSnSe$_4$ films with controlled morphology and phase composition is studied. It is shown that the selenization of metallic stacked precursor layers at temperatures higher than 420^{0}C leads to high quality films, but the films are multi-phased and contain additionally to Cu$_2$ZnSnSe$_4$ a separate ZnSe phase. The results indicate that stacked binary films allow the selenization step to be replaced with thermal treatment in an inert atmosphere, however, the formed Cu$_2$ZnSnSe$_4$ films are in Se deficit. It is only the sequence of constituent layers ZnSe/SnSe/CuSe in the precursor that leads to satisfactory adhesion of the formed Cu$_2$ZnSnSe$_4$ film to the Mo substrate. It was found that the phase composition and morphology of Cu2ZnSnSe4 films after the selenization of ZnSe/SnSe/CuSe precursors are strongly influenced by the morphology of the CuSe constituent layer.

INTRODUCTION
    Cu(In,Ga)(S,Se)$_2$ (CIGS) thin film technologies have shown good perspectives of complicated semiconductor compounds for solar cells production [1]. Despite the high promises of Cu(In,Ga)(S,Se)$_2$ (CIGS) technologies, limitations in supply for In and Ga require that new adsorber materials that consist of more earth-abundant constituents be looked for and studied [2]. Among these copper-zinc-tin-chalcogenide materials, Cu$_2$ZnSnS$_4$ and Cu$_2$ZnSnSe$_4$, are prospective for wide-scale energy production in the future, reaching efficiencies of up to 6.77% at the moment [3]. These new materials have as yet been investigated only by few groups and produced mainly as thin films [4–7] by methods that, in principle, are similar to those used so far to grow CIS and CIGS layers. These are mainly vacuum methods based on the deposition of metallic (Cu, Zn, Sn) constituent films with different sequences with the following selenization of precursor films [8-10]. Several authors have used precursor films from mixed binary and metallic layers with the following selenization [11-14], resulting in the highest effenciencies (6.77% [3]) of solar cells structures. Efficiencies of up to 6% have been achieved also by monograin powders and a monograin layer design of structures [15-16]. In a recent paper Todorov, Reuter and Mitzi achieved 9.6 % solar efficiency confirmed by NREL using soft chemistry methods [17]. All this is well below the reported best efficiencies for CIGS solar cells (up to 20 % in the lab [13]) but it is encouraging as a starting point.
    At the same time the use of only binaries (ZnSe, SnSe, CuSe) for the deposition of precursor films presents a challenge of avoiding the selenization step and making the solar cells technology simple. This paper deals with the influence of different constituent metallic and binary selenide layers on the formation of Cu$_2$ZnSnSe$_4$ films with controlled morphology and phase composition.

EXPERIMENTAL
    The precursor films with a different sequence of consistent elements Cu, Zn and Sn or binary chalcogenides (CuSe, ZnSe, SnSe) were deposited onto molybdenum covered soda-lime glass substrates by high temperature vacuum evaporation. Different parameters of deposition (sequence of layers, substrate temperature and speed of deposition) and the post deposition temperature were used to determine optimal parameters for the precursor films formation. The thickness of the precursor films was adjusted by the use of oscillatory quartz crystal microbalance, and controlled by the evaporation time and/or by SEM measurements.

The precursor films were annealed with elemental selenium in a temperature range between 270 and 520°C in isothermal sealed quartz ampoules, in which the Se pressure was controlled by the amount of elemental Se in the ampoule and by the temperature of the selenization. The duration of the selenization was between 0.33 and 1 hours.

Evolution of the surface morphology and the crystalline structure of the studied precursor and selenized films were analyzed by the high resolution scanning electron microscope (HR SEM) Zeiss ULTRA 55 equipped with the In-Lens SE detector for topographic imaging and energy and an angle selective backscattered detector (EsB) for compositional contrast.

The chemical composition and stoichiometry of the prepared thin film precursor and chalcopyrite materials and the distribution of the components in the investigated films were determined using an energy dispersive x-ray analysis (EDX) system (Röntec EDX-XFlash 3001 detector) and XRF (iXRF systems). Additionally, EDS was used to determine lateral and in-depth compositional uniformity of the developed films.

Bulk structure and phase compositions were additionally studied by X-ray diffraction (XRD) and Raman spectroscopy. Rikaku UltimaIV diffractometer was used to identify the phases of the films. Cu K_α served as an X ray source (40 kV, 40 mA) in the Bragg-Brentano geometry. The XRD peaks were identified using JCPDS fails. The room temperature micro-Raman spectra were recorded by a Horiba`s LabRam HR high resolution spectrometer. The incident laser light with the wavelength of 532 nm was focused on samples within a spot of 1 μm in diameter and the spectral resolution of the spectrometer was about 0.5 cm^{-1}.

RESULTS

Films from metallic precursor layers.

The morphology of precursor films consisting of metallic layers depends on the sequence of the consistent layers (Fig. 1). Precursor Sn-Zn-Cu films exhibit a well-formed "mesa-like" structure of the surface in which larger crystals are located on a "small-crystalline" valley (Fig. 1a) analogously to the In-Cu films [18, 19]. The size of "mesa-forming" crystals is about 1 μm and these "mesa–forming" crystals extend through the layer down to the Mo layer. The analysis of the cross-sectional SEM images and EDS data allows us to conclude that these crystals originate from the deposited Sn layer. During the Sn deposition it forms a discontinuous layer of semispheral crystals covered with several facets onto the surface of the Mo substrate, leading already to the "mesa-like" structure of the Sn layer. The following depositions of Zn and Cu lead to the formation layers of these metals quite uniform in thickness, on the whole surface of samples that result in a "mesa-like" structure of the surface of the final Sn-Zn-Cu films. For films with another sequence of metallic layers the mesa-like structure is not so well exposed, and well-formed flat precursor layers were produced replacing metallic layers with the Cu/Sn alloy layer (Fig. 1b). The analysis of XRD patterns gives evidence of strong interaction of metals in the deposition process and the formation of separate phases of Cu$_5$Zn$_8$ and Cu$_6$Sn$_5$ [19].

Regularities in the selenization of metallic precursor films of different sequences appeared quite similar. Our results indicate that the selenization of Cu in precursor films is dominating in any sequence of layers and binary Cu selenides in different compositions are formed at low temperatures on the surface of films [19]. The selenization of Cu dominates even for precursors where the Cu layer was the lowest (Cu-Zn-Sn). XRD investigations and EDS analyses refer to the formation of the CuSe$_2$ phase in the films selenized at 250°C and to the formation of the CuSe phase in the films selenized at 300°C. In addition to the CuSe, the films selenized at 300°C contain agglomerated particles of a mixture of ternary Cu$_2$SnSe$_3$ and quaternary Cu$_2$ZnSnSe$_4$. Selenization at temperatures higher than 420°C results in multi-phase films consisting of high quality Cu$_2$ZnSnSe$_4$ crystals with sizes of up to 2 μm (Fig 2) and of a separate phase of ZnSe. The content of ZnSe diminishes with the rise of the selenization temperature, but the selenized films stayed always multi-phased. Cu-rich precursors result in films with larger crystals and more dense films but contain a separate CuSe phase, as detected by Raman and SEM investigations (Fig. 2b)

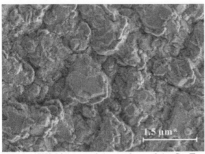 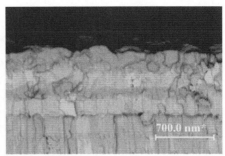

Fig. 1a. SEM image of the surface of the Sn-Zn-Cu sequential film [19]

Fig. 1b. SE SEM image of the cross section of the 2x(Cu/Sn-Zn) sequential film [18]

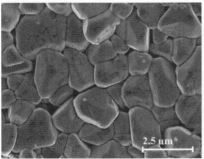 

Fig. 2a. SEM image of the surface of the stoichiometric Sn-Zn-Cu precursor film selenized at 470^{0}C for 1h [19]

Fig. 2b. SEM image of the cross section of the Cu-rich Sn-Zn-Cu precursor film selenized at 470^{0}C for 1h

Films from binary precursor layers

The binary constituent stacked layers have been used for CZTS formation by different authors [11-14, 20, 21]. Katagiri et al. [11] used ZnS-Sn-Cu stacked precursor layers for Cu$_2$ZnSnS$_4$ layer formation and noticed an improvement in the surface morphology of the obtained films. Solar cells on the basis of the absorber layers formed had an efficiency of up to 4.53%. The replacement of the Sn layer with the SnS layer in [13] resulted in higher efficiency of up to 5.74%. In their study of the formation of Cu$_2$ZnSnSe$_4$ layer from the stacked Cu-ZnSe-Sn-Se precursors, Babu et al. [12] point out the improved morphology of selenized films and the possibility of achieving single-phase films. In [14] thin films of Cu$_2$ZnSnS$_4$ films were produced by the sulphurization of stacked precursor SnS$_x$-CuS-ZnS films. The authors report a possible existence of a self–controlled process for the formation of Cu$_2$ZnSnS$_4$ from the stacked precursor of this composition. In-situ formation of Cu$_2$ZnSnS$_4$ from precursors of Mo-CuS-ZnS-SnS and Mo-CuS-SnS-ZnS compositions was studied in their following papers [20, 21]. The authors assume that the stacking sequence CuS-ZnS-SnS provides the best absorbers. The formation of Cu$_2$ZnSnS$_4$ starts just below 300°C and only the temperatures above 450°C lead to the formation of single-phase Cu$_2$ZnSnS$_4$ films. At the same time all the films studied by them had the tendency of delamination from the surface that was explained by the high diffusivity of copper [21].

The sequence of binary layers ZnSe-SnSe-CuSe was used in our experiments; other sequences bring to the peel off of films already during the process of evaporation. Above-given binaries in the precursor present a challenge of avoiding Se treatment or to replace Se atmosphere with an inert one as the molecular ratio 1:1:2 of the precursor layers has to result in the formation of stoichiometric Cu$_2$ZnSnSe$_4$ without any Se treatment if the loss of Se or Sn in the deposition and inert atmosphere in the thermal treatment processes is avoided. So far nobody has used CuSe as a layer in the stacked precursor for the Cu$_2$ZnSnSe$_4$ adsorber layer production. Our results indicate that the morphology of the deposited CuSe constituent films depends strongly on the speed of the deposition of CuSe and the substrate temperature. High deposition speed of the CuSe layer and low temperature of the substrate (100^{0}C) lead to the plate-like structure of CuSe crystals in the CuSe layer (Fig. 3a). The micro-Raman (Fig.4) and XRD investigations of films showed that the thermal treatment of ZnSe-SnSe-CuSe precursor films in the inert atmosphere already leads to the formation of homogeneous Cu$_2$ZnSnSe$_4$ films (Fig.3b), but EDS analysis showed that the films are in Se deficit ([Se]=43-45 at %). The selenization of these films results in single-phase Cu$_2$ZnSnSe$_4$ films with nearly stoichiometric composition ([Se]= 49.5-50.5at%), but crystals are in very different sizes.

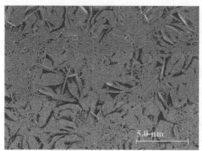 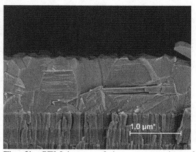

Fig. 3a. SEM image of the surface of the CuSe layer deposited onto ZnSe-SnSe layers (high speed of CuSe deposition, substrate temperature 100^{0}C)

Fig. 3b. SEM image of the cross section of the ZnSe-SnSe-CuSe precursor thermally treated in Ar atmosphere at 450^{0}C (high speed of CuSe deposition)

Low deposition speed of the CuSe layer and a higher temperature of the substrate (200^{0}C) result in a wide plate-like structure of CuSe crystals in the CuSe layer or in dense CuSe layers (Fig.5a). The SEM and micro-Raman investigations and EDS analysis of the films indicated to the inhomogeneous structure (Fig.5b) and the phase composition of thermally annealed films in an inert atmosphere. The Raman investigations confirm that the films consist of separate phases of Cu$_2$SnSe$_3$, SnSe and ZnSe. The multiphase composition of films is in good correlation of results of thermodynamic calculations by Shiyou Chen [22]. Selenization of these films preliminarily treated in an inert atmosphere leads to the formation of dense large-crystalline Cu$_2$ZnSnSe$_4$ films (Fig.5c, d).

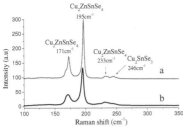

Fig.4. Raman spectra of Cu$_2$ZnSnSe$_4$ films from precursor with a slowly deposited CuSe layer: a) after Ar treatment at 450^{0}C for 1 h, b) after selenization at 470^{0}C for 15 min

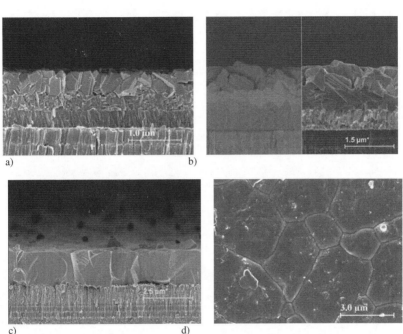

a)   b)

c)   d)

Fig. 5. SEM images of: a) the cross section of the precursor layer, CuSe deposited onto ZnSe-SnSe layers (low CuSe deposition speed, substrate temperature 200^{0}C), b) the ZnSe-SnSe-CuSe precursor thermally treated in Ar atmosphere at 450^{0}C, c,d) the cross section and surface of the ZnSe-SnSe-CuSe precursor selenized at 470^{0}C for 20 min

CONCLUSIONS

High quality $Cu_2ZnSnSe_4$ films are achieved at Se temperatures higher than 400°C, by the use of metallic stacked precursor layers, however, these films are multi-phased. The content of ZnSe diminishes with the rise of the selenization temperature, but the selenized films stayed always multi-phased.

Stacked binary films enable the selenization step to be replaced with thermal treatment in an inert atmosphere, but the formed films are in Se deficit. Only the sequence of constituent layers ZnSe-SnSe-CuSe leads to films with satisfactory adhesion to the Mo substrate. The phase composition of selenized films depends strongly on the morphology of the precursor's constituent CuSe layer.

ACKNOWLEDGEMENT

This work was supported by the target financing by HTM (Estonia) project No. SF0140099s08 and by the Estonian Science Foundation grants G-8147, G-7595.

REFERENCES

1. A. Goetzberger, C. Hebling, H.-W. Schock. Photovoltaic materials, history, status and outlook.- *Materials Science and Engineering*, **R 40**, 1-46 (2003).
2. A. Feltrin, A. Freindlich, Material considerations for terawatt level deployment of photovoltaics, *Renewable Energy*, **33**, 180 – 185 (2008).
3. H. Katagiri, K. Jimbo, W.S. Maw, K. Oishi, M. Yamazaki, H. Araki, A. Takeuchi, Development of CZTS-based thin film solar cells, *TSF*, **517**, 2455–2460 (2009).
4. J. J. Scragg, P. J. Dale, L. M. Peter, G. Zoppi, I. Forbes, New routes to sustainable photovoltaics: evaluation of Cu_2ZnSnS_4 as alternative absorber material, *Phys. Stat. Sol. (b)*, **245,** 1772 – 1778 (2008).
5. A. Ennaoui, M. Lux-Steiner, A. Weber, D. Abou-Ras, I. Kötschau, H.-W. Schock, R. Schurr, A. Hölzing, S. Jost, R. Hock, T. Voß, J. Schulze, A. Kirbs, Cu_2ZnSnS_4 thin film solar cells from electroplated precursors: novel low-cost perspective, *TSF*, **517**, 2511 – 2514 (2009).
6. N. Kamoun, H. Bouzouita, B. Rezig, Fabrication and characterization of Cu_2ZnSnS_4 thin films deposited by spray pyrolysis technique, *TSF*, **515**, 5949 – 5952 (2007).
7. T. Todorov, M. Kita, J. Carda, P. Escribano, Cu_2ZnSnS_4 films deposited by soft-chemistry method, *TSF*, **517**, 2541-2544 (2009).
8. G. Zoppi, I. Forbes, R.W. Miles, P. J. Dale, J. J. Scragg, L. M. Peter, $Cu_2ZnSnSe_4$ thin film solar cells produced by selenization of magnetron sputtered precursors, *Prog. Photovolt: Res. Appl.* **17,** 315-319 (2009).
9. H. Araki A. Mikaduki, Y. Kubo, T. Sato, K. Jimbo, W. S. Maw, H. Katagiri, M. Yamazaki, K. Oishi, A. Takeuchi, Preparation of Cu2ZnSnS4 thin films by sulfurization of stacked metallic layers, *TSF*, 517, 1457-1460 (2008)
10. T. Tanaka, D. Kawasaki , M Nishio ,Q. Guo, H.Ogawa, Fabrication of $Cu_2ZnSnSe_4$ thin films by co-evaporation, *Phys. Status Solidi c,* **3**, 2844– 2847 (2006).
11. H. Katagiri, $Cu_2ZnSnSe_4$ thin film solar cells, *TSF*, **480-481**, 426-432 (2005).
12. G. S. Babu, Y . B. Kumar. P. Uday Bhaskar , Effect of post-deposition annealing on the growth of $Cu_2ZnSnSe_4$ thin films for a solar cells adsorber layer, *Semicond. Sci. Technol.* 23 (2008) 085023
13. K. Jimbo, R. Kimura, T. Kamimura, S. Yamada, W. S. Maw, H. Araki K. Oishi, , Cu2ZnSnS4-type thin film solar cells using abundant materials, *TSF*, **515**, 5997-5999 (2007).
14. A. Weber , H. Krauth, S. Perlt, B. Schubert, I. Kötschau, S. Schorr, H.W. Schock, Multi-stage evaporation of Cu_2ZnSnS_4 thin films, *TSF,* **517**, 2524-2526 (2009)

15. M. Altosaar, J. Raudoja, K. Timmo, M. Danilson, M. Grossberg, J. Krustok, E. Mellikov, Cu$_2$Zn$_{1-x}$Cd$_x$Sn(Se$_{1-y}$S$_y$)$_4$ solid solutions as absorber materials for solar cells, *Phys. Status Solidi A,* **205**, 167 – 170 (2008).
16. E. Mellikov, D. Meissner, T. Varema, M. Altosaar, M. Kauk, O. Volubujeva, J. Raudoja, K. Timmo, M. Danilson, Monograin materials for solar cells, *Solar Energy Mat. Solar Cells,* **93**, 65 – 68 (2008).
17. T. K. Todorov, K. B. Reuter, D. B. Mitzi, High-Efficiency Solar Cell with Earth-Abundant Liquid-Processed Absorber, *Adv. Mater.,* **22**, 1 – 4 (2010).
18. O. Volobujeva, E. Mellikov, J. Raudoja, M. Grossberg, S. Bereznev, M. Altosaar, R. Traksmaa, SEM analysis and selenization of Cu–Zn-Sn sequential films produced by evaporation of metals, in: *Proceedings: Conference on Optoelectronic and Microelectronic Materials and Devices,* IEEE Publishing , 257 – 260 (2009).
19. O. Volobujeva, J. Raudoja, E. Mellikov, M. Grossberg, S. Bereznev, R. Traksmaa, Cu$_2$ZnSnSe$_4$ films by selenization of Sn–Zn-Cu sequential films, *Journal of Phys. and Chem. of Solids,* **93**, 11-14 (2009).
20. A. Weber, I. Kötschau, S. Schorr, H.-W. Schock, Formation of Cu$_2$ZnSnS$_4$ and Cu$_2$ZnSnS$_4$-CuInS$_2$ thin films investigated by in-situ energy dispersive X-ray diffraction, *Mater. Res. Soc. Symp. Proc.* **1012**, 1012-Y03-35 (2007).
21. S. Schorr, A. Weber, V. Honkimäki H.-W. Schock, In-situ investigation of kesterite formation from binary and ternary sulphides, *TSF,* **517**, 2461-2464 (2009).
22. Shiyou Chen, X. G. Gong, Aron Walsh, and Su-Huai Wei, Defect physics of the kesterite thin-film solar cell absorber Cu$_2$ZnSnS$_4$, Applied Physics Letters, **96**, 021902 (2010)

Hydropower

MARTENSITIC STAINLESS STEEL 0Cr13Ni4Mo FOR HYDRAULIC RUNNER

D.Z. Li, Y.Y. Li, P. Wang, S.P. Lu*
Institute of Metal Research, Chinese Academy of Sciences
Shenyang, Liaoning, China
*e-mail: shplu@imr.ac.cn

ABSTRACT

Three Gorges hydraulic runner is the largest one in the world with the weight of nearly 450 ton and the diameter of 10 meter. It consists of the top crown and the lower band through the blades which are assembled by welding. All of the crown, band and blade are made of 0Cr13Ni4Mo low carbon martensitic stainless steel.

The chemical composition optimization, heat treatment specification and microstructure controlling to the martensitic stainless steel 0Cr13Ni4Mo were systematically studied in this paper to meet the properties' requirements for the hydraulic turbine runner. By controlling the Ni/Cr equivalent proportion, the δ-Fe can be avoided in the microstructure and improve the impact performance. Single normalizing with double tempering is recommended for the heat treatment system. The normalizing microstructure consists of 100% lath martensite. The final microstructure in 0Cr13Ni4Mo stainless steel is tempered martensite and a certain amount of inversed austenite which has an excellent mechanical properties and good weldability.

The three gorges is the biggest hydropower facility in the world. It is of great importance for China to harness and develop the Yangtze River. By 2020, the gross installed capacity of hydropower will reach 300GW, 225GW of which will be from the large and middle-scale hydropower, while 75GW from small hydropower.

INTRODUCTION

Hydraulic power is one of the oldest resources and has been used to produce electricity for about a hundred years. Along with economic development and emphasis on environmental protection, more and more attention is paid to hydroelectric power around the world because of its renewability and cleanness. China plans to raise its hydro utilization ratio in the following decades and Three Gorges Power Plant is one of the important parts of the plan. There are thirty-two 700WM hydraulic turbines installed in Three Gorges Power Plant, which can produce over 100TWh electricity power per year. The hydraulic turbine runner, which transforms energy from moving water to generator, is the key equipment in the hydraulic power station. The 700WM Francis hydraulic turbine runner used in Three Gorges Power Plant comprises one crown, one band and 13~ 17 blades. The three components of the runner are cast separately and then welded together [1].

The crown, band and blade of the turbine runner are made of a kind of martensitic stainless steel 0Cr13Ni4Mo (similar to the ASTM A743-CA6NM). The 0Cr13Ni4Mo steel is famous for its combination of strength, ductility, toughness and excellent weldability and is widely used for hydraulic turbines, valve bodies, pump bowls, compressor cones, impellers and high-pressure pipes in power generation [2]. Although many researchers around the world paid attention to this famous steel [2-6], the manufacture of the 700WM hydraulic turbine runner castings is still very difficult because of their strict mechanical properties requirements (as shown in Table I) and huge size (over 10 meter diameter and 450 ton weight). As a result of higher power output and security of the turbine runner the material is required higher strength, ductility and toughness at the same time compared with the ASTM standard. Therefore in order to fulfill the new higher requirement systemic research of the 0Cr13Ni4Mo should be carried out.

Table I. Mechanical properties requirements of 0Cr13Ni4Mo

Standard	Tensile stress (MPa)	Yield stress (MPa)	Elongation (%)	Reduction of area (%)	Hardness (HB)	$0°C$ Charpy V-notch Impact (J)
ASTM A743 [7]	≥755	≥550	$\delta_4 \geq 15$	≥35	≤285	____
Three Gorges	≥780	≥580	$\delta_5 \geq 20$	≥55	220~285	≥100

It is well known that the microstructure of the material determines its mechanical properties. The relative researches indicate that in order to improve the comprehensive mechanical properties of 0Cr13Ni4Mo two important microstructures, delta ferrite and reversed austenite, should be controlled accurately [2,3,8,9]. The delta ferrite is mainly influenced by the chemical composition and the reversed austenite is affected by the heat treatment. Investigations on the composition optimization and heat treatment design are carried out to control the two key phases.

COMPOSITION OPTIMIZATION

According to already existing standard ASTM A743, the chemical composition of 0Cr13Ni4Mo steel is listed in Table II. Under this standard, two different microstructures in the as-cast station can be got as shown in Fig. 1 (a) and (b). Fig. 1(a) is full of low carbon lath martensite, however, Fig. 1(b) is a mixture of low carbon martensite and delta ferrite (the white strip phase shown in Fig 1 (b)). The delta phase region is the first solid phase region after solidification, and then it will transform to gamma phase by peritectic reaction with the temperature decreasing. In 0Cr13Ni4Mo martensitic stainless steel, if the chemical composition and cooling rate are not proper, the delta ferrite will remain to room temperature.

Table II The chemical composition limitation according to ASTM A743[7](wt.%)

	C	Si	Mn	S	P	Cr	Ni	Mo
CA6NM	≤0.06	≤1.0	≤1.0	≤0.03	≤0.04	11.5~14.0	3.5~4.5	0.4~1.0

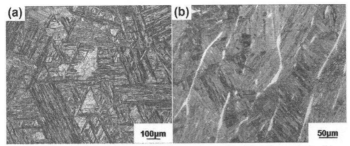

Fig. 1 The microstructure of 0Cr13Ni4Mo steel in the as-cast station (a) full martensite (b) martensite and delta ferrite

The delta ferrite always lies at the prior austenite boundaries, and is rich of ferrite stabilizing elements such as chromium and molybdenum. Since it does not experience any solid phase transformation during the cooling course, the defects in it is low, which makes its strength lower than the matrix martensite and tempered martensite. In addition, once the delta ferrite appears at room temperature, it is difficult to be removed by conventional heat treatments besides high temperature diffusion annealing. Although the effect of delta ferrite on martensitic steel were investigated by many researchers [10,11], controversial results still exist. Recently research on the delta effect on the impact properties of 0Cr13Ni4Mo martensitic stainless steel reveals that the presence of delta ferrite did not change the upper and lower shelf energy of the steel apparently, while lowering the impact energy remarkably in the transition temperature range and raised the ductile to brittle transition temperature of the material [9]. The detrimental effect of delta ferrite is caused by the differences of strength and ductility between delta ferrite and matrix tempered martensite.

As the delta ferrite has harmful effect on the material's toughness, it should be eliminated from the microstructure. However, the delta ferrite can't be removed easily by traditional heat treatment but only by high termperature diffusion annealing treatment. It is an easy and economic way to eliminate the delta ferrite by optimizing the material chemical compositions reasonably under the limitation of relative standard. Scheaffler[12] diagram illustrtates that the room temperature weld metal microstructure is contolled by the content of ferrite stablizing elements (Cr equivalent content) and austenite stablizing elements (Ni equivalent content). Howevwer, besides the chemical composition the cooling rate during the delta phase to austenite phase transformation also influences the final microstructure. Under a certain chemical compostion, the delta ferrite transform to austenite by peritectic reaction $L + \delta \rightarrow \gamma$ or diffusional phase transformation $\delta \rightarrow \gamma$. If the cooling rate is high during these transitions, the delta phase may retain to low temperature as a non-equilibrium phase (if the cooling rate is high enough, these transitions will not occur, additionally no $L \rightarrow \delta$ reaction will occur. However, for the large runner castings the cooling rate could not very high). Additionally, the calculation method of the Nieq and Creq of the Scheaffler diagram does not suit the 0Cr13Ni4Mo martensitic stainless steel perfectly.

Fig. 2 is the statistical results of the chemical compositions and the microstructures of some the runner castings in China, which illustrates that the delta ferrite can be eliminated in the microstructure of the runner castings when the Nieq/Creq≥0.42. In Fig. 3 the Nieq and Creq are calculated by:

$$Ni_{eq}=Ni+30(C+N)+0.5Mn$$

$$Cr_{eq}=Cr+Mo+1.5Si \tag{1}$$

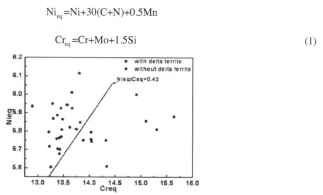

Fig. 2 Nieq/Creq statistic diagram of some runner castings in China

As mentioned above, the appearance of delta ferrite is influenced by the cooling rate during the delta to austenite phase transformation (which always happens at temperatures higher than 1300°C in the 0Cr13Ni4Mo steel). In order to research the cooling rate effects, two kinds of coupons were designed, one is cast separately with relative high cooling rate, and the other is attached to the bigger castings with lower cooling rate. Each kind of the coupon has a series of samples with different dimensions. The cooling rates of different samples were calculated by Procast software. The experiments indicate that with the same chemical compositions higher cooing rate between the pouring temperature to 1300°C induces higher delta ferrite content. Fig. 3 listed the microstructure of two different test coupons with the same chemical compositions but different cooling conditions: (a) attached on a blade, (b) cast separately. The Nieq/Creq of the two coupons is 0.421, and the cooling rates of the coupon attached on a blade and the coupon cast separately are about 300°C/h and 500°C/h. There is no delta ferrite in the coupon (a) but some delta ferrite in coupon (b). Based on the experimental results, it can be concluded that when the Nieq/Creq≥0.42 and the cooling rate between the pouring temperature to 1300°C is lower than 500°C/h, the delta ferrite can be eliminated in the castings.

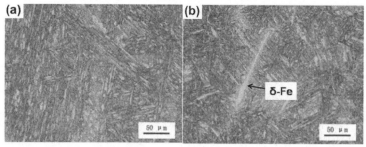

Fig.3 The microstructure of the different coupons (a) attached to the runner castings (b) cast separately

HEAT TREATMENT DESIGN

When the chemical composition of the 0Cr13Ni4Mo is decided, the heat treatment is the main way to adjust the microstructures. The martensitic steel is always used under the normalized and tempered condition, therefore the 0Cr13Ni4Mo used in Three Gorges hydraulic turbine runners is also treated by normalizing and tempering. However, the tempering is carried out in the intercritical temperature range in order to produce a key phase, reversed austenite.

The reversed austenite is the austenite appearing in the intercritical tempering and that remains to the room temperature [13]. Compared with the normal retained austenite, the reversed austenite is rich of austenite stabilizing elements, such as Ni [8] and has a very good thermal stability, no martensite transformation even at -196°C. The dimension of the reversed austenite is very small, about 100nm in width and 1μm in length, lying along the martensite interlath boundaries and prior austenite grain boundaries. Because of the small dimension, the reversed austenite can't be observed by optical microscopy, however, it can be detected by XRD or TEM [8]. The reversed austenite can transform to martensite under some deformation conditions [2]. Fig. 4 displays the XRD results of the sample with about 10% reversed austenite before and after bend test. It illustrates that the amount of reversed austenite decreases significantly after the bend test. The reversed austenite to martensite transformation during deformation can improve the ductility and toughness of the 0Cr13Ni4Mo [2].

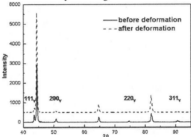

Fig. 4 The XRD spectrums of the sample with 10% reversed austenite before and after bend deformation

Normalizing

For the 0Cr13Ni4Mo martensitic stainless steel, the aim of the normalizing heat treatment is refining the prior austenite grain size, dissolving the precipitated phase and getting a full martensite microstructure. The variations of the prior austenite grain size with the normalizing temperature are shown in Fig. 5. The prior austenite grain grows with the normalizing temperature increasing and a rapid growth appears above the normalizing temperature of 1100°C. In order to reduce the micro-segregation and dissolve the precipitated phases in the big castings, the normalizing temperature of the runner castings can't be too low. After mechanical properties tests, 1050±10°C was chosen as the normalizing temperature.

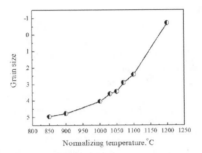

Fig. 5 The variations of the prior austenite grain size with the normalizing temperature

First-stage tempering

The aim of the tempering is changing the martensite to tempered martensite and adjusting the amount of reversed austenite into a proper range. Firstly, the phase transformation temperature of the 0Cr13Ni4Mo was measured by dilatometric method. The A_s, A_f, M_s and M_f of this steel are 578°C, 807°C, 318°C and 135°C, respectively[8]. As mentioned above, the reversed austenite appears in the intercritical temperature range, therefore the tempering temperature was chosen above the A_s point. The variations of reversed austenite content with the tempering temperature are listed in Fig. 6. The amount of the reversed austenite increases with the tempering temperature firstly, exhibiting a maximum amount of the reversed austenite, 6%, at 620°C, and then decreases with higher tempering temperatures. The decreas of reversed austenite content at higher tempering temperature is caused by the poorer stability of the austenite and some of the austenite returning to new martensite in the cooling process [2,3,8]. The mechanical properties of the 0Cr13Ni4Mo at different tempering conditions are listed in Table III. It indicates that the yield and tensile stress are inversely proportional to the inversed austenite contents, while the elongation is proportional to the inversed austenite contents. The elongation reaches its maximum 19.0% for the 620°C tempered sample, which can't satisfy the requirements. Therefore, the amount of reversed austenite should be raised in order to improve the material's ductility.

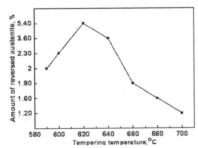

Fig. 6 The variations of the reversed austenite content with the tempering temperature [8]

Table III The mechanical properties of the 0Cr13Ni4Mo steel at different tempering conditions

Tempering temperature(°C)	$\sigma_{0.2}$ (MPa)	σ_m (MPa)	δ_5 (%)	ψ (%)	KV2 (0°C) (J)
600	620	847	17.0	65.5	136 , 138 , 142
620	655	840	19.0	69.0	128 , 131 , 140
640	770	880	16.5	73.5	115 , 117 , 124
660	790	920	16.5	69.0	72 , 101 , 104
680	820	945	16.0	65.5	

Second-stage tempering

Second-stage tempering is designed to increase the amount of reversed austenite. The variations of reversed austenite content with the second-stage tempering temperature are shown in Fig. 7. It can be observed that the reversed austenite content increases dramatically compared with the first-stage tempered samples. And the amount of the reversed austenite increases firstly and then decreases as the tempering temperature increases, which is the same with the first-stage tempering. The mechanism of the reversed austenite content increasing after second-stage tempering is that the reversed austenite grows in the second-stage tempering, and after first-stage tempering there are tempered martensite, reversed austenite and new martensite co-existing, which offers some new nucleation sites for the new reversed austenite at second-stage tempering [8].

The mechanical properties of the 0Cr13Ni4Mo martensitic stainless steel after second-stage tempering are listed in Table IV. The results show after second-stage tempering the ductility of the material increases, however, a higher tempering temperature will cause a lower strength. Therefore, the optimized tempering technology is 630°C for first-stage tempering and 590°C for second-stage tempering.

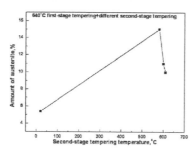

Fig. 7 The variations of the reversed austenite content with the second-stage tempering temperature

Table IV The mechanical properties of the 0Cr13Ni4Mo steel after second-stage tempering

Fist-stage tempering(°C)	second-stage tempering(°C)	$\sigma_{0.2}$ (MPa)	σ_m (MPa)	δ_5 (%)	ψ (%)	KV2 (0 °C) (J)
610	580	665	800	23.0	72.0	137,140,142
610	600	615	800	20.0	69.0	146,134,142
630	580	670	790	21.5	69.0	150,143,146
630	600	625	790	21.0	71.0	148,152,154
650	580	650	780	21.0	70.5	144,137,145
650	600	605	775	22.0	70.5	141,145,147

TYPICAL MICROSTRUCTURE

The typical microstructures of the 0Cr13Ni4Mo at different stages are listed as follows:

In the as-cast stage the microstructure of the 0Cr13Ni4Mo is the low carbon lath martensite with coarse prior austenite grain size and band size. The optical microscopy picture of the as-cast material is shown in Fig. 8 (a). The XRD spectrum of this sample is shown in Fig. 8 (b), which clearly shows there is no austenite existing under this condition.

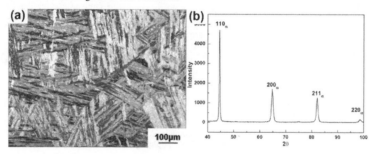

Fig. 8 The optical microscopy picture (a) and XRD spectrum (b) of sample in as-cast condition

The sample after normalizing consists of low carbon lath martensite with fine prior austenite grain size and band size (as shown in Fig. 9 (a)). Because of its good hardenability, there is no retained austenite after normalizing, which is demonstrated by the XRD spectrum as shown in Fig. 9(b).

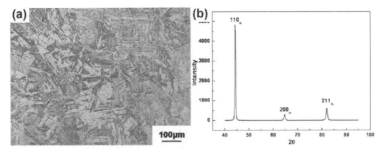

Fig. 9 The optical microscopy picture (a) and XRD spectrum (b) of sample in normalized condition

The first-stage tempering can change the microstructures from full martensite to tempered martensite and reversed austenite, and/or some new martensite. Fig. 10(a) is the optical microscopy picture of the sample after 620°C tempering, which consists of tempered martensite, about 5% reversed austenite and little new martensite. However, the reversed austenite and new martensite is too small to see in the optical microscopy. Fig. 10(b) shows the XRD spectrum of this sample, where the diffraction peaks of reversed austenite appears. Fig. 10(c) is the morphology of the reversed austenite in this sample under transition electron microscopy (TEM), which is platelet-like [8].

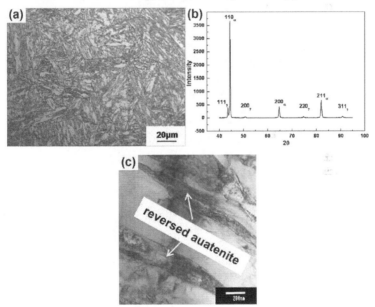

Fig. 10 The optical microscopy picture (a) and XRD spectrum (b) and the morphology of reversed austenite in TEM (c) of the sample with one-stage tempering

After second-stage tempering the microstructures of the sample consists of tempered martensite and reversed austenite. Fig. 11(a) is the optical microscopy picture of the sample after 620°C and 600 tempered, which consists of tempered martensite, about 10% reversed austenite. Fig. 11(b) shows the XRD spectrum of this sample. Fig. 11(c) is the morphology of the reversed austenite in this sample under TEM [8]. After second-stage tempering, there is not only platelet-like but also blocky reversed austenite existing.

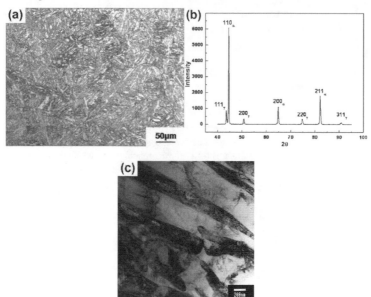

Fig. 11 The optical microscopy picture (a) and XRD spectrum (b) and the morphology of reversed austenite in TEM (c) of the sample with two-stage tempered

CONCLUSION

1. The δ ferrite which has a deleterious effect on impact properties of the 13Cr4Ni steel should be eliminated by optimizing the chemical composition;

2. The δ ferrite can be eliminated by controlling the Nieq/Creq≥0.42 and proper cooling rate in the hydraulic turbine runner castings;

3. The reversed austenite has a beneficial effect on improving the plasticity and ductility of the 13Cr4Ni steel;

4. A proper two-stage tempering heat treatment can adjust the amount of reversed austenite to satisfy the requirement of the turbine runner castings;

5. The Three Gorges Project requirements can be satisfied by chemical optimization and "one normalizing and two-stage tempering" heat treatment.

REFERENCES

[1]Y.F. Huang, W.X. Li Three Gorgoes hydraulic turbine properties and structure analyses, China Three Gorges Construction, 7, 23-6 (2000)

[2]P.D.Bilmes, M.Solari, C.L.Llorente. Characteristics and effects of austenite resulting from tempering of 13Cr-NiMo martensitic steel weld metals. Mater. Charact. , 46, 285-96(2001)

[3]Y.Iwabuchi Factors affecting on mechanical properties of soft martensitic stainless steel castings. JSME Inter. J. ,46, 441-46(2003)

[4]Y.Iwabuchi Intergranular failure along prior austenite grain boundary of type CA6NM stainless cast steel. Trans. Jap. Foundrymen's Soc. , 13,1-12(1994)

[5]Y.Iwabuchi Temper embrittlement of Type 13Cr-4Ni cast steel. ISIJ international,27, 211-17 (1987)

[6]P.D.Bilmes, C.Llorente, J.P.Ipina. Toughness and microstructure of 13Cr4NiMo high-strength steel welds. J. Mate. Eng. Perf. , 9, 609-15(2000)

[7]ASTM ASTM A743-2006

[8]P. Wang, S.P. Lu, D.Z. Li, X.H.Kang, Y.Y. Li, Investigation on phase transformation of low carbon martensitic stainless steel ZG06Cr13Ni4Mo in tempering process with low heating rate, Acta Metall. Sin. , 44,681-85 (2008)

[9]P. Wang, S.P. Lu, N.M.Xiao,D.Z. Li,Y.Y. Li, Effect of delta ferrite on impact properties of low carbon 13Cr-4Ni martensitic stainless steel. Mater. Sci. Eng. A, In press (2010)

[10]D.Carrouge, H.K.D.H. Bhadeshia, P.Woollin, Effect of delta-ferrite on impact properties of supermartensitic stainless steeel heat affected zones, Sci. Technol. Weld. Join. , 9,377-89 (2004)

[11]L. Schäfer, Influence of delta ferrite and dendritic carbides on the impact and tensile properties of a martensitic chromium steel, J. Nucl. Mater. , 262, 1336-39 (1998)

[12] A.L. Schaeffler Constitute diagram for stainless steel weld metal, Met. Prog. 56, 680-680 (1949)

[13]W.L. Friis, T.M.I. Noren Weldable, corrosion-resisting steel, US Patent 3378367 (1968)

ADVANCED COMPOSITE MATERIALS FOR TIDAL TURBINE BLADES

Mike Hulse, John Cronin, and Mike Tupper
Composite Technology Development, Inc.
Lafayette, Colorado USA

ABSTRACT
The energy potential of harnessing the kinetic energy of the world's ocean and tidal currents is part of the US's renewable energy plan. Technologies that improve the performance and reliability of marine current energy systems are needed. Marine current energy systems convert kinetic energy of moving tides and currents into electrical energy via underwater turbines. These turbine blades are designed to withstand the challenges of the marine environment including high hydrodynamic forces, corrosion due to salt water, and erosion due to cavitation, and impact from suspended particle. Advanced composites are attractive for constructing tidal turbine blades. However performance improvements are needed regarding impact-damage and cavitation-erosion resistance. If not properly addressed, this could dramatically reduce the size, efficiency, reliability and life span of tidal turbines thereby reducing the viability of marine current energy systems. Recent work focused on composite ship propellers have shown substantially improved cavitation-erosion and impact-damage resistance in comparison to industry-standard materials while also possessing necessary mechanical properties and processing characteristics. Adaptation of these cavitation resistant composite materials can significantly improve the performance and economic feasibility of ocean tidal systems.

INTRODUCTION
The energy potential of harnessing the kinetic energy of the world's ocean and tidal currents is currently being considered as a key part of the United State's renewable energy development plan. Similar to wind power energy systems, marine current energy systems convert the kinetic energy of moving tides and currents into electrical energy via underwater turbines. These turbines include blades that are designed to withstand the unique challenges of the marine environment including high hydrodynamic forces, corrosion due to saltwater and erosion due to cavitation and suspended particle impacts.

Composites are increasingly being considered in constructing tidal turbine blades due to their high specific properties and inherent corrosion resistance. Additionally, the complex curvature found in tidal turbine blades are more amenable to composite manufacturing techniques than metals, leading to lower fabrication costs. Unfortunately, commercially available composites possess questionable durability characteristics including poor impact damage and cavitation erosion resistance. Other durability concerns include long-term water immersion. If not properly addressed, these material deficiencies could dramatically reduce the efficiency, reliability and life-span of tidal turbines thereby reducing the viability of marine current energy systems.

To address these shortcomings in the performance of composite materials relevant to marine rotors, the application of the highly toughened TEMBO® family of composite materials has shown promising results for ship composite propeller blades for the U.S. Navy. In particular, TEMBO® materials possess substantially higher cavitation erosion and impact damage resistance in comparison to industry-standard materials while also possessing mechanical properties and processing characteristics consistent with requirements for making large composite turbine blades.

For the benefits of CTD's durable composites to be fully realized in the tidal turbine blade market, full-scale components must be mass-produced with high quality using cost effective techniques. Many successful composite marine structures including boat hulls are manufactured using resin infusion processes, such as Vacuum Assisted Resin Transfer Molding (VARTM). In fact, the wind and aerospace industries have begun to migrate toward this process due to lower cost with higher resultant quality and consistency.

TIDAL TURBINE RENEWABLE ENERGY SYSTEMS

Marine current energy is one of the most recent forms of renewable energy to be considered for development. Similar to wind energy, marine current energy systems convert the kinetic energy of ocean tides and river currents into electrical energy via underwater turbines. According to some estimates merely harnessing 1/1000th of available kinetic energy from the Gulf Stream would supply Florida with 35% of its electrical needs.[i] It also has the potential to provide continuous power unlike wind and solar, which are only available part of the time.[ii] Based on this potential, it is imperative that marine current energy be explored as a potentially critical part of the U.S.'s renewable energy portfolio and to that end, relevant technologies be developed and qualified that enable efficient, reliable, and cost effective marine current energy systems.

To explore the feasibility of marine current energy several prototype systems have been installed in high tidal flow areas in the United States and Europe. For example, Verdant Power currently operates a tidal turbine system in New York's East River while Marine Current Turbines operates two systems off the coast of the United Kingdom. These systems use a horizontal axis turbine and a yaw or blade pitch control system that permits the turbine to reverse its direction of rotation allowing capture of energy during both tidal ebb and flow.[iii] Verdant Power recently installed a system that consists of six 35-kW turbines with 16-foot-diameter, 3-blade rotors (see Figure 1(a)). Marine Current Turbine's first system, referred to as Seaflow, is a 300-kW system installed in 2003 and features a 36-foot diameter two-blade rotor. More recently, Marine Current Turbine installed the 1.2MW Seagen system that incorporates twin, two-bladed 52.5-foot diameter rotors (see Figure 1(b)).

Although perhaps not initially obvious, tidal turbine blades differ considerably from wind turbine blades primarily due to the difference in density between their respective energized fluids. Centrifugal loads typically dominate in wind turbine rotors, whereas bending caused by the high hydrodynamic loads experienced by the blades dominate in tidal turbine rotors.[iv] Indeed, a typical horizontal axis tidal turbine experiences nearly 3 times the thrust load that is experienced by a horizontal axis wind turbine of equivalent rated power, despite the tidal turbine rotor possessing a significantly smaller swept area.[v] For this reason, tidal turbine blades are typically thicker and shorter in span than similarly power-rated wind turbine blades.

(a) Verdant's rotors. (b) Seagen system.
Figure 1. Examples of currently operated tidal turbine systems.

In addition to the high hydrodynamic loads, tidal turbines must be resistive to the highly corrosive and erosive marine environment (e.g., salt water, marine growth, abrasive suspended particles and debris, and erosion due to cavitation near the blade). Furthermore, the blade designs feature significant compound curvature that makes manufacturing from metals both difficult and costly. For these reasons, the tidal turbine industry has migrated towards the use of fiber reinforced polymer (FRP) composites as the rotor blade material.[vi] Composites also provide the added benefit of affording substantially greater design tailoring to meet the exceptionally high hydrodynamic load requirements of tidal turbine blades. In addition, composite structures enable unique capabilities such as passive pitch control due to bend-twist coupling unique to certain anisotropic laminate architectures.[vii] Furthermore, composites materials are lighter than metals and therefore easier and cheaper to transport and install. Finally, the damping properties and the potential for neutral buoyancy of composite rotors can increase the operating life and reduce maintenance of the entire system by reducing wear on the moving components.

DURABILITY CONCERNS FOR COMPOSITE MATERIALS

Cavitation, characterized by the rapid formation and collapse of bubbles, is a phenomenon that has plagued the recreational and defense marine propeller industry for decades.[viii] Cavitation often occurs at the tips of rotating blades where the higher velocity reduces the local pressure to the vapor pressure of the surrounding water and can severely erode the blade over time, thereby reducing its efficiency and life-cycle (see Figure 2(a)). Although tip speeds in marine turbine rotors are relatively low, cavitation is often observed (see Figure 3(b)). Furthermore, cavitation-induced erosion has been identified as a concern for future tidal turbine systems.[ix] While studies have shown that the occurrence of cavitation can be reduced in carefully controlled environments[x], it will no doubt be an issue for current and future tidal turbine systems deployed in less controlled environments (i.e., locally varying tidal current speeds) ultimately reducing turbine reliability.

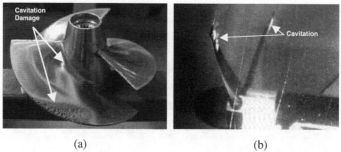

(a)   (b)

Figure 2.(a) Cavitation damage on a propeller and (b)cavitation in marine rotor systems.[xi]

There exists a substantial amount of research in the area of cavitation erosion of metals, but very few studies have focused on composites. However, the few existing studies have consistently shown that composites are highly susceptible to cavitation erosion.[xii] One such study performed at the University of Maine per ASTM G32,[xiii] involved an extensive material survey of the cavitation erosion resistance of numerous materials (e.g., metals, polymers, composites, etc.).[xiv] The results indicated that typical composites (e.g., carbon fiber/vinyl ester) exhibit more than 100 times the erosion rate (i.e., rate of material loss) than commonly used metals including stainless steel and nickel-aluminum-bronze (NAB).

In addition to high cavitation erosion, composites also offer relatively low impact damage resistance when compared with metals. Metals also possess the ability to absorb impact energy beyond elastic limits through plastic deformation, while maintaining structural integrity. It is therefore relatively easy to assess damage in metal components. Composites, however, often show little or no visible damage on the surface after impact beyond elastic limits, despite considerable internal damage and a reduction in structural integrity.[xv]

When exposed to an underwater environment, the internal moisture content of an organic matrix composite can increase over time. The increase in moisture content often degrades the mechanical properties of the laminate.[xvi] Structural designers must be cognizant of the effects of long-term exposure to the underwater environment. While many fiber-dominated properties are largely unaffected, resin dominated properties such as interlaminar shear strength can be reduced as much as 60%. Future viability of marine current energy systems will require system lifecycle of 10-20 years and the long term exposure to underwater environments may become a critical consideration.

The susceptibility of composites to impact damage, cavitation erosion and moisture absorption produces substantial uncertainty in the reliability and lifespan of tidal turbines. The offshore location of these systems only heightens this concern as servicing and/or replacing turbine blades may be both problematic and costly. Therefore, composite materials that are inherently durable (e.g., high cavitation erosion, impact damage, and corrosion resistance) while also possessing mechanical properties consistent with design requirements are of great interest. Furthermore, these materials must be cost-effective and have processing characteristics consistent with industry standard techniques.

CTD'S TEMBO® COMPOSITE MATERIALS FOR IMPROVED DURABILITY

To address these shortcomings in the performance of composite materials for marine rotor applications, CTD is leading an effort to advance the state-of-the-art of composite materials for the naval marine community.[xvii] This work is to enable the next-generation of composite propellers.[xviii] In addition, CTD is working to optimize composite material performance, structural design, and the manufacturing processes for tidal turbine blades. More specifically, CTD is developing materials that have substantially higher marine-relevant durability characteristics (e.g., cavitation-erosion, impact damage resistance, and water immersion) while also meeting material-level system design requirements (i.e.: mechanical performance). Furthermore, materials are being designed for compatibility with industry standard, large scale manufacturing processes such as VARTM. These materials, marketed under the trade name TEMBO®, are a family of highly toughened thermosetting epoxies. CTD's numerous variants of TEMBO® materials possess highly varying mechanical properties (e.g., elastic and strength) and toughness characteristics each developed for specific product needs. Demonstrated material characteristics relevant to the marine turbine industry as compared to industry-standard composite materials include:

- Higher cavitation erosion resistance.
- Higher impact damage tolerance.
- Similar mechanical properties.

The high durability of TEMBO® composite materials in comparison to industry standard composite materials is illustrated in Figure 3 and Figure 4. In particular, both the cavitation erosion resistance and impact durability of a TEMBO® composite material were evaluated in comparison to composite materials manufactured with the vinyl ester resin Derakane® 8084, which is commonly used in the naval marine industry. Figure 3(a) compares the cavitation erosion rates for TEMBO® composite, vinyl ester composite and nickel-aluminum-bronze measured via ASTM G32.[14] TEMBO® composites exhibited nearly 50-fold reduction in the cavitation erosion rate over vinyl ester composites while also approaching the performance of NAB. The high cavitation erosion resistance of TEMBO® composites is further illustrated by the photographs shown in Figure 3(b), which show the surface of test samples following 10 minutes of exposure to induced cavitation via cavitating jets test protocol (ASTM G134), which produces much higher cavitation intensity relative to the ASTM G32 protocol.[xix] The TEMBO® composite sample exhibited virtually no erosion whereas the vinyl ester sample allowed the highly localized cavitation to erode through nearly 3-plies of the laminate.

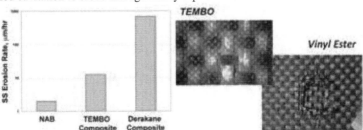

(a) Cavitation erosion resistance comparison.   (b) Samples after 10-minutes exposure.
Figure 3. High cavitation erosion resistance of CTD's TEMBO® composites.[18]

The high toughness characteristic of TEMBO® composite materials also provides substantial impact durability, demonstrated by the results of a recent study summarized in Figure 4. TEMBO® and vinyl ester composite materials were subjected to a drop-weight impact test while a three-point bend method was used to quantify impact-induced damage. Figure 4 presents photographs of the microsections obtained from samples subjected to impact energies of 144 in-lb. The images reveal no fiber damage or delamination in the TEMBO® sample whereas the vinyl ester sample exhibits substantial failure including underlying delamination and fiber and matrix fracture.

TEMBO Post Impact   **Derakane Post Impact**

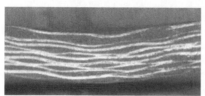 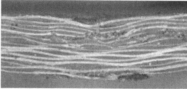

Figure 4. Impact damage tolerance of CTD's TEMBO® composite materials.[18]

COMPOSITE MANUFACTURING CONSTRAINTS FOR DURABLE TEMBO® MATERIALS

CTD recognizes that no advanced material solution is viable if it cannot be manufactured using cost-effective industry-standard techniques. Many successful composite marine structures, such as boat hulls, are manufactured using resin infusion processes such as VARTM. In this process, dry fiber reinforcement is placed into a closed mold or onto an open mold and enclosed in a vacuum bag. Inlet tubes are fed into the mold or vacuum bag at selected points (see Figure 5), the mold is evacuated of air and the resin is introduced into the evacuated volume. Pressure differential created by the vacuum pulls the resin through the reinforcement to impregnate it fully. The resin's constant exposure to vacuum during infusion results in a composite with a very low void content. The component is then cured at elevated temperatures in an oven or using heaters placed directly on the mold. Previously, high performance composites were manufactured using pre-preg (pre-impregnated) reinforcements. As the name implies, these materials are fiber reinforced materials that have a partially cured resin pre-impregnated into the reinforcement. The state of partial cure of the resin is often referred to as the b-stage form, which allows for handling of the material and application to molds to form the desired shape of the composite, such as a tidal turbine blade. Most often composites manufactured using pre-preg are cured in an autoclave, which is used to provide extremely high consolidation pressures to the composite during cure. However, use of an autoclave adds significant time and cost to the manufacturing process, particularly for the manufacture of large components. The pre-preg is placed on a mold, vacuum bagged and cured without a need for resin infusion. Pre-preg components can be made without an autoclave, but this usually results in a higher void content, reducing mechanical properties and overall component quality. De-bulking techniques, which are used to remove entrapped air and better consolidate the layers of reinforcement, have been developed. However, these additional steps add time and cost to the manufacturing process.

The VARTM process presents a significant opportunity for reduced cost and improved quality with shorter manufacturing time. In fact, the wind energy and aerospace industries have begun to migrate toward resin infusion processes and away from more traditional pre-preg and hand layup technologies.[xx,xxi] Previously, reduced mechanical properties due to low fiber volume fraction and fiber waviness in resin infused parts has been cited as a limiting factor.[xxii] However, recent advancements in VARTM processes made for next-generation commercial aircraft (i.e.: the Boeing 787 Dreamliner) have shown promise for increasing fiber volume fraction, and thus mechanical properties (i.e.: elastic modulus and strength).

Vacuum bagged laminate

Resin infusion tubes

Figure 5. Composite laminates being infused with the VARTM process.

The successful manufacturing of composite components, particularly one as large as a tidal turbine blade, depends on the resin processing parameters. Additionally, the fiber selection, fiber architecture, tooling design, processing temperature, vacuum and leak integrity of the molding tool, and resin flow pattern within the mold are all important to the final performance of the composite component and the cost of manufacturing. CTD is currently optimizing these design aspects to provide a robust composite tidal turbine blade, which will enable CTD to provide more cost-effective, more durable, and higher performance composite tidal turbine blades with shorter manufacturing lead times to the emerging tidal current energy industry.

FUTURE PLANS

CTD plans to become a supplier of composite tidal turbine blades. CTD is in the process of designing highly robust composite tidal turbine blades based on our cavitation resistance, impact tolerant TEMBO® materials. Sub-scale blades will be fabricated and tested in relevant laboratory tests in air and in water. This will be followed by design and manufacture of full-scale blades that will be tested both in the laboratory and in the field.

SUMMARY

The energy potential of harnessing the kinetic energy of the world's ocean and tidal currents is currently being considered as a key part of the United State's renewable energy development plan. Similar to wind power energy systems, marine current energy systems convert the kinetic energy of moving tides and currents into electrical energy via underwater turbines. Potential also exists to provide continuous power, 24 hours per day, unlike wind and solar which are available only when the wind blows or the sun shines respectively. Composites are increasingly being considered in constructing tidal turbine blades due to their high specific properties and inherent corrosion resistance.

Unfortunately, commercially available composites possess questionable durability characteristics including poor impact damage and cavitation erosion resistance. If not properly addressed, these material deficiencies could dramatically reduce the efficiency, reliability and life-span of tidal turbines thereby reducing the viability of marine current energy systems. CTD is developing materials that possess substantially higher marine-relevant durability characteristics while also meeting material-level system design requirements. Finally, materials are being designed for compatibility with industry standard, well proven, large scale manufacturing processes to ensure a new source of cheap and sustainable renewable energy.

[i] U.S. Department of the Interior, Renewable Energy and Alternate Use Program, Technology White Paper on Ocean Current Energy Potential on the U.S. Outer Continental Shelf, http://ocsenergy.anl.gov.

[ii] Freeman, Kris. "Tidal Turbines: Wave of the Future?" *Environmental Health Perspectives*. January 2004.

[iii] Gardiner, G., "Tidal Turbines to Mine Marine Megawatts," *Composites Technology*, June, 2007.

[iv] "Windmills Below the Sea: A Commercial Reality Soon?", *ReFOCUS*, March/April, 2004, pp 46-50.

[v] Nichols-Lee, et al, "Simulation Based Optimisation of Marine Current Turbine Blades," In Proc. of 7th International Conference on Computer and IT Application in the Maritime Industries, 21-23 April, 2008, Liege, Belguim.

[vi] "Tidal turbines harness the power of the sea," *Reinforced Plastics*, June, 2004, pp44-47.

[vii] Nicholls-Lee, R.F., et al, "Enhancing Performance of a Horizontal Axis Tidal Turbine using Adaptive Blades," Ocean 2007 – Europe, 18-21 June, 2007.

[viii] The Specialist Committee on Cavitation Erosion on Propellers and Appendages on High Powered/High Speed Ships, Final Report and Recommendations to the 24th International Towing Tank Conference (ITTC), Proceeding of the 24th ITTC – Volume II, 2005.

[ix] Mason, K., "Composite Tidal Turbine to Harness Ocean Energy," *Composites Technology*, December, 2005.

[x] Batten, W.M.J., et al, "The Prediction of the Hydrodynamic Performance of Marine Current Turbines," *Journal of Renewable Energy*, Vol. 33, pp1085-1096, 2008.

[xi] Bahaj, A.S., et al, "Power and Thrust Measurements of Marine Current Turbines Under Various Hydrodynamic Flow Conditions in a Cavitation Tunnel and Towing Tank, *Journal of Renewable Energy*, Vol. 32, No. 3, pp407-426, 2007.

[xii] Hammond, D.A., et al, "Cavitation Erosion Performance of Fiber Reinforced Composites," *Journal of Composite Materials*, Vol. 27, No. 16, pp1522-1544, 1993.

[xiii] ASTM G32 – 06, Standard Test Method for Cavitation Erosion Using Vibratory Apparatus.

[xiv] Caccese, V., et al, "Cavitation Erosion Resistance of Various Material Systems," *Journal of Ships and Offshore Structures*, Vol. 1, Issue 4, June, 2006, pp309-322.

[xv] Reid, S.R., and Zhou, G., Impact Behaviour of Fibre-Reinforced Composite Materials and Structures, World Publishing Limited, 2000.

[xvi] Greene, Eric. "Chapter Four: Water Absorption." *Marine Composites*.

[xvii] Campbell, D., Caccese, V., et al, "Improvements in Composite Durability for Naval Marine Applications," Presented at the ShipTech 2008 Conference, Biloxi, MS, 12-13 February, 2008.

[xviii] Phase II SBIR, Office of Naval Research, "Propulsor Blade Advanced Composite Materials," Contract No. N00014-08-C-0763, Dr. Ki-Han Kim ONR PM.

[xix] Website: http://www.dynaflow-inc.com/Products/Jets/Jets.htm.

[xx] "Vacuum Assisted Resin Transfer Molding for Aerospace." U. S. Department of Defense Mantech, Manufacturing Technology Program: Success Stories. Accessed August 31, 2009: <https://www.dodmantech.com/successes/AirForce/04-08/CAI_VARTM.pdf>

[xxi] Griffin, D.A. "Infused Carbon-Fiber Spar Demonstration in Megawatt-Scale Wind Turbine Blades." SAMPE Journal. Vol. 45, No. 5, Sept./Oct. 2009. pp. 6-16.

[xxii] Gilbert, E. N., Hayes, B. S. and Seferis, J. C. "Characterization of VARTM Resin Systems for Commercial Airplane Applications." Proceedings of the 33rd SAMPE ISTC Conference. Seattle, WA, Novemeber 5-8, 2001.

Nuclear

IMMOBILIZATION OF Tc IN A METALLIC WASTE FORM

W.L. Ebert[a], J.C. Cunnane[a], S.M. Frank[b], and M.J. Williamson[c]
[a]Argonne National Laboratory, Argonne, IL USA
[b]Idaho National Laboratory, Idaho Falls, ID USA
[c]Savannah River National Laboratory, Aiken, SC USA

ABSTRACT

A multi-phase iron-based metallic waste form is being developed to immobilize the metallic and Tc-bearing waste streams that are generated during the reprocessing of used nuclear fuels with either aqueous or electrochemical methods. A metallic waste form provides for the efficient processing and immobilization of metallic wastes and other components that can be readily reduced to metals prior to or during immobilization. These waste streams can be processed with added iron (or stainless steels) at about 1600 °C to incorporate transition metal fission products and other waste components into durable iron solid solution and intermetallic phases. Work is in progress to (1) formulate and produce an alloy composition to immobilize the anticipated range of waste compositions within a small number of phases, (2) identify processing conditions for producing waste forms with high waste loadings and consistent chemical, physical, and radiological properties, and (3) develop a mechanistically-based corrosion and radionuclide release model for calculating long-term waste form performance under the range of possible disposal conditions. The experimental and modeling approaches are presented with some representative results.

INTRODUCTION

The US Department of Energy is evaluating both aqueous and electrochemical methods for reprocessing used nuclear fuel to recover and recycle actinides from existing fuels in advanced fuels being developed for possible closed fuel cycle in the future. Both reprocessing methods generate high-level radioactive waste streams containing transition metal, lanthanide, alkaline earth and alkali metal fission products that must be immobilized and disposed in a geologic facility. Work is in progress within the DOE fuel cycle research and development (FCRD) program to develop durable waste forms for the separate or combined waste streams and captured volatile radionuclides from reprocessing operations. Separate waste forms are being developed and evaluated for metallic and oxide waste streams, and for waste streams not amenable to high-temperature processing methods.[1]

Management of Tc during the reprocessing of used nuclear fuel, production of waste forms, and long-term waste form disposal is important due to the radiotoxicity of ^{99}Tc and the high mobility of the pertechnetate ion (TcO_4^-) in the environment. For example, about 1 kg Tc will be recovered from each metric ton of 51 GWd/MTIHM oxide fuel that is processed 20 years after being removed from the reactor.[1] The most environmentally important Tc isotope is ^{99}Tc, which is a long-lived fission product with a half-life of 0.21 Myr and a specific activity of 1.69 x 10^{-2} Ci/g (0.292 MeV β^-). Other isotopes are either generated in negligible quantities (^{97}Tc and ^{98}Tc) or short-lived (^{95m}Tc, ^{96}Tc, ^{97m}Tc, ^{99m}Tc, $^{101-107}$Tc). The fission yield of ^{99}Tc from ^{235}U is 6.06%; ^{99}Tc is also generated by thermal neutron fission of ^{233}U and ^{239}Pu, and by the fast neutron fission of ^{239}Pu, ^{238}U, and ^{232}Th.[2] As the fuel burns, Tc combines with the other fission products Mo, Ru, Rh, and Pd to form micrometer-sized particles of a 5-metal alloy referred to as the ε-phase (due to its similarity to the ε-phase of metallic Ru) that are distributed throughout the fuel. The relative amounts of Tc, Mo, Ru, Rh, and Pd in the ε-phase particles depend on the fission yield, the initial O/(U+Pu)-ratio of the fuel (the oxygen potential), the temperature gradients in the pin, and the fuel burn-up.[3] The masses of these elements calculated to be present in 1000 kg of 51 GWd/MTIHM fuel that has aged 20 years, which has been used as a reference case for initial waste form development,[1] are 5.11 kg Mo, 3.49 kg Ru, 2.35 kg Pd, 1.14 kg Tc, and 0.61 kg Rh.

The dissolution of used nuclear fuel in nitric acid prior to aqueous separation operations results in only partial dissolution of the ε-phase particles and generates pertechnetate ions in the oxidizing solution, as well as dissolved Mo, Ru, Rh, and Pd. The extent to which the ε-phase particles and other metal wastes dissolve depends on the conditions used to dissolve the fuel; for example, between about 30% and 90% of the Tc inventory in the fuel is dissolved under the range of conditions that have been used in laboratory experiments.[1] The remaining Tc is retained in the residual undissolved metal particles that are comingled with scraps of cladding material and Zr from the fuel in what is referred to as the undissolved solids (UDS) waste stream. In the first Coupled End-to-End (CETE) demonstration conducted at Oak Ridge National Laboratory with low burn-up fuel from the Dresden reactor to evaluate the UREX+ aqueous flow sheet, about 50% of the Tc in the fuel was dissolved; the retained Tc represented 10 mass% of the UDS.[2]

Both the UDS and recovered dissolved Tc waste streams from aqueous processing are to be disposed as high-level radioactive waste. The chemical composition of the UDS waste stream will be dominated by Mo, Zr, Ru, and Pd; ^{99}Tc is a minor component of that waste stream by mass, but has the greatest radiotoxicity. Although it will depend on the conditions used to dissolve the fuel (temperature, nitric acid concentration, time), the mass of Tc recovered from the dissolver solution (which is referred to as recovered Tc) will be a small fraction of the total mass of the Tc-bearing waste to be immobilized, which will be dominated by the UDS. Trace concentrations of other contaminants, such as ^{90}Sr, ^{137}Cs, and various transuranics will greatly increase the radiotoxicity of the UDS waste stream. The waste form must be designed to accommodate these contaminants. Although the recovered Tc and UDS waste streams could be immobilized in separate waste forms, combining the UDS and recovered Tc in a single waste form will be more economical than producing separate waste forms due to lower operational and waste form qualification costs. The current approach being studied within the FCRD Waste Form Campaign is to combine the UDS and recovered Tc in a single metallic waste form. Including the transition metal fission products (TMFP) recovered from the dissolved fuel solution in a later separation step within the same waste form is being considered as an option. Separation and reduction of the dissolved TMFP metals would probably be required prior to their incorporation in the waste form.

Waste Streams

In the aqueous UREX+ separation method that is being developed and studied in the FCRD program (uranium extraction followed by other extraction operations to separate TMFP, lanthanides, and actinides)[3], the technetium and uranium in the dissolver solution are separated from the other dissolved constituents into a nitric acid product solution. For a reference fuel being used for waste form development (e.g., 51 GWd/MTIHM fuel), the UREX product solution is expected to contain about 100 g U/L as uranyl nitrate (0.4 $\underline{M}$) and 0.13 g Tc/L as pertechnetate (1.3 $\underline{mM}$ TcO$_4^-$) in a ~0.01 $\underline{M}$ nitric acid solution. The pertechnetate ions can be recovered from this solution by using an anion exchange column to purify the uranium stream for recovery and reuse. The pertechnetate ion can be eluted from the column with an ammonium hydroxide solution, and several methods to recover the eluted Tc in a form suitable for immobilization are being investigated.[4] These include methods that provide small particles of metallic Tc or metallic Tc plated onto a steel wool substrate as feed streams for alloy waste form production. Either form can be alloyed directly.

In the electrorefining process, uranium and other fuel components that are oxidized under the refining conditions become dissolved in molten salt, such as the LiCl/KCl eutectic salt being used for EBR-II fuel.[5] Essentially all of the Tc remains in the anode basket with the metal cladding hull scrap and fuel components that are not oxidized during electrorefining; these are disposed as high-level radioactive waste. The salt is further processed to recover the uranium for storage, separate the actinides for recycle, and separate the lanthanides and remaining fission products (e.g., Cs and Sr) for disposal. The metallic waste stream from electrochemical processing is expected to contain essentially the same amount and composition of fuel wastes as the combined UDS, recovered Tc, and TMFP waste streams generated by aqueous processing. It is anticipated that the same waste form can be used to immobilize metallic wastes from aqueous and electrochemical processing, although different pre-conditioning operations will probably be required. However, the metallic electrochemical waste stream might include cladding hulls, which would dominate the waste stream and require a different waste form composition.

Formulating an appropriate waste form for a particular waste stream involves consideration of three key factors: the ability to produce the waste form from the material in the waste stream, the capacity to incorporate all waste components into a stable phase, and the ability to reliably calculate the release of the radionuclides as the waste form degrades over the regulated service life of the disposal system. High waste loadings are desirable to lower processing and disposal costs, but may be limited by processing limitations and waste form durability. The Tc in the UDS is alloyed in the metallic ε-particles and the recovered Tc is readily reduced to either pure Tc metal or metallic Tc deposited on iron; both waste streams are well-suited for immobilization in a tailored iron-based alloy waste form. A small number of phases with compositional flexibility is beneficial to modeling the waste form corrosion behavior and tracking the consistency of waste forms produced over time.

Waste Form

The technically-preferred waste form for Tc-bearing waste streams generated by either the aqueous or electrochemical processing operations is a multi-phase alloyed metallic waste form similar to that developed for metallic wastes from the electometallurgical treatment of used sodium-bonded fuel from the Argonne Experimental Breeder Reactor-II (EBR-II).[1,6] An alloy of stainless steel with 15 mass% zirconium (SS-15Zr) was developed to immobilize and dispose metallic wastes from the electrometallurgical treatment of steel-clad spent sodium-bonded nuclear fuel from the EBR-II. The target composition SS-15Zr was selected to maintain consistency in the phase compositions for a range of waste stream compositions (e.g., for the treatment of driver and blanket fuel rods) by adding various amounts of Zircaloy to the waste stream to achieve the desired ratio. This was done to simplify the qualification of the small number of waste packages that were (at that time) scheduled for disposal in the proposed Yucca Mountain repository. The waste form is comprised of two predominant phases: a steel-like iron solid solution and a $ZrFe_2$ intermetallic.[6] In the treatment of spent sodium-bonded fuels from the EBR-II, the steel cladding hulls from the fuel are processed along with the fuel and the hulls dominate the composition of the metallic waste stream. The cladding hulls are not included in the FCRD wastes being considered for immobilization in the alloy waste form.

Previous analyses indicated that the iron-dominant Fe-15Zr alloy composition would be suitable for the metallic fuel wastes.[7] This conclusion was based on consideration of the likely compositions of the UDS, the likely amount and chemical form of recovered Tc, the projected compositions of electrochemical metallic waste streams, expected processing limitations, and, primarily, the expected distributions of individual waste elements into the component phases of the alloyed waste form. The advantages of an iron-based alloy waste form are that the predominant waste components can be readily dissolved in molten iron at <1600 °C, a temperature that is well below the melting points of the individual waste components, and the waste components are either soluble in an iron solid solution (e.g., austenite or ferrite) or form an intermetallic phase with iron. Another advantage is that other iron-bearing waste streams can be incorporated into the waste form. For example, iron might be added to the TRUEX feed (as ferrous sulfamate) to increase the efficiency of neptunium recovery or could be added to the TRUEX raffinate waste stream to precipitate and recover transition metal fission products.[8] Iron may also be used as a substrate to capture the Tc eluted from an anion exchange resin used to recover pertechnetate from the UREX solution.[9]

EXPERIMENTAL APPROACH

The approach being taken to develop an alloy waste form for the FCRD processing wastes is to tailor the alloy to immobilize the waste stream components within a small number of durable alloy phases. The alloy waste form will be composed of a small number of phases that are aggregated by a stainless steel-like solid solution phase. The assemblage of phases that comprise the waste form is referred to collectively as the phase composition, which is used to denote both the identities and the compositions of the phases that comprise the waste form. Insights gained during development of the EBR-II alloy waste form and characterizations of the phase composition and distribution of radionuclides[5] in those alloys are being used as guidance for activities being conducted at several national laboratories and universities to address the formulation, perform material analyses and corrosion testing, and develop a corrosion behavior model for iron-based alloy waste forms for FCRD wastes.

The objective of the alloy waste form development activities is to formulate a waste form that can be produced by adding steel to the combined UDS, recovered Tc, and TMFP waste streams and processed at about 1600 °C to generate a small number of durable phases that, together, will accommodate all anticipated fuel wastes and processing chemicals over their expected concentration ranges. A small number of component phases (e.g., 5 or fewer) is desired in order to maintain a consistent phase assemblage in waste form products made with a range of waste compositions to facilitate waste form qualification (for eventual disposal in a federal repository) and corrosion modeling over long disposal times. The addition of steel will allow the waste streams to be processed at a reasonable temperature because the waste components with high individual melting points can be dissolved in molten iron at <1650 °C and all the waste components (including Tc) will form intermetallic phases or be incorporated in a steel-like solid solution as the mixture cools. Proper formulation of the alloy will ensure acceptable waste form corrosion resistance and retention of long-lived radionuclides. For example, trim metals such as Cr can be added to improve the corrosion resistance of the steel-like solid solution phase and perhaps also the intermetallics that incorporate Cr or other passivating components. It is expected that various contaminated steel wastes (from fuel assembly hardware, activated metal waste, etc.) can be utilized in making the alloy waste forms.

Formulation of Alloy Waste Form

Variances in fuel burn-up, storage time, dissolution and separation efficiencies can lead to a range of waste stream compositions that could impact the alloy phase composition. Therefore, understanding the flexibility of the phase composition to accommodate a range of waste compositions will be important for establishing processing control limits. The ranges and possible combinations of concentrations of the major components in the combined waste streams (Zr, Mo, Ru), several minor components (Pd, Rh, Te, Cr, Ni, and Sn), and Tc define a very large composition space. The approach that is being followed is to first define representative compositions for the UDS, recovered Tc, and recovered TMFP waste streams (see Table I), then study the effects of adding different amounts of Fe and Zr on the assemblage of phases that form.[5] Representative compositions of the UDS, recovered Tc, and recovered TMFP waste streams were estimated based on the Origen calculations for the used fuel and the estimated efficiencies for the separation operations.[6] Rhenium (Re) is being used as a surrogate for Tc in these studies. The relative amounts of Fe, Zr, Mo, and Ru in the alloy are expected to establish the assemblage of host phases that incorporate the other waste components. After the amount of Fe needed to dissolve the waste components and form a consistent set of host phases was determined, the capacity of the phases to accommodate a particular waste component was evaluated. The ability to accommodate wastes with higher Zr contents was measured first.

Trace amounts of actinide contaminants are expected to be present in the waste streams due to finite separation efficiencies; these have been neglected in these formulations. Based on the EBR-II

Table I. Masses of Major Components in Alloy Waste Form, kg/1000 kg processed fuel

Element	Fuel (Origen)	UDS waste	Recovered Tc	TMFP waste
Zr	5.64	1.86	0	3.78
Mo	5.11	4.75	0	0*
Tc	1.14	0.27	0.87	0
Ru	3.49	1.64	0	1.65*
Rh	0.61	0.24	0	0.37
Pd	2.35	0.68	0	1.29*
Sn	0.15	0	0	0.14*
total	18.49	9.44	0.87	7.23

*Based on estimated recovery of dissolved metal.

metal waste form,[ ] the actinides are expected to be sequestered in a $ZrFe_2$ intermetallic phase. Alloys Alloys can be made with actual wastes (e.g., wastes recovered during the demonstrations of separation operations), with simulated wastes, or doped with particular elements of interest after the optimum alloy composition has been determined in order to verify the distributions of trace contaminants.

The element concentrations in the combined UDS and recovered Tc waste streams and the combined UDS, recovered Tc, and TMFP waste streams were used to estimate the amount of Fe needed to (1) dissolve individual waste component at 1600 °C and (2) incorporate that component in a stable Fe-intermetallic phase or in a solid solution with Fe (α-Fe, γ-Fe, or δ-Fe, depending on the component) based on the binary phase diagrams.[ ] Reagent iron was used for this purpose in these initial studies,[10, 11] but stainless steel is being used for further development. Other alloys were formulated with added Zr. The sum of the amounts of Fe needed to process and immobilize the individual waste components and added Zr was then taken to be the stoichiometric amount of Fe to be added for each mixture. Other alloys were formulated by adding less Fe with the expectation that element substitution in the intermetallic and solid solution phases would lower the demand for Fe. Formulations for estimated stoichiometric and smallest amounts of Fe added to combined waste streams and formulations with added Zr are given in Table II. Because the alloys with the least added iron were sent to incorporate all of the waste components, formulations with intermediate amounts of Fe were not produced. Future studies will evaluate the effects of added Mo, Ru, and Cr.

Table II. Formulated Compositions for Composition Study, mass%

Alloy:	UFe-1	UFe-4	UTFe-1	UTFe-4	UZr-1	UZr-4	UTZr-1	UTZr-4
Waste Streams	UDS+Tc	UDS+Tc	UDS+Tc +TMFP	UDS+Tc +TMFP	UDS+Tc	UDS+Tc	UDS+Tc +TMFP	UDS+Tc +TMFP
Added Fe	17	7	32	15	10	10	20	20
Added Zr	0	0	0	0	0.5	2.5	0.5	2.0
Fe	59.97	38.16	62.12	43.46	45.78	41.94	49.98	48.17
Zr	6.56	10.14	10.95	16.34	10.80	18.28	15.34	18.40
Mo	16.76	25.89	9.91	14.80	21.74	19.92	12.76	12.30
Re*	7.67	11.85	4.22	6.30	9.95	9.11	5.43	5.23
Ru	5.79	8.94	6.77	10.11	7.51	6.88	8.72	8.41
Rh	0.84	1.30	1.19	1.77	1.09	1.00	1.53	1.47
Pd	2.41	3.73	4.57	6.82	3.13	2.87	5.88	5.67
Sn	0	0	0.27	0.41	0	0	0.35	0.34

*Surrogate mass Re representing equivalent atomic% of Tc.

Composition Study

Reagent metals were used to make 50-g samples according to the alloy compositions in Table II at about 1700 °C in a vacuum furnace, which were then analyzed with scanning electron microscopy and X-ray diffraction to characterize the phases that formed.[11] All mixtures resulted in a multi-phase alloy composed of intermetallic phases similar to $MoFe_2$ and $ZrFe_2$, a small amount of $(Zr,Fe)Pd_2$ intermetallic, and a small amount of an iron solid solution phase. A scanning electron microscopy photomicrograph of material UTFe-4 is shown in Fig. 1. The brightest regions are $(Zr,Fe)Pd_2$ phases (this region had an unusually large amount of this phase) and the darkest regions are iron solution phases. The predominant grey regions include similar amounts of the $ZrFe_2$ and $MoFe_2$ phases, but don't distinguish between the two phases (the $ZrFe_2$ phase surrounds the $MoFe_2$ phase and probably nucleated later). The other waste components were sequestered in these phases: most of the Ru was in the $ZrFe_2$ phase, most Re (used as a surrogate for Tc) was in the $MoFe_2$ and $ZrFe_2$ phases, with a small fraction in the iron solid solution, while Rh was fairly evenly distributed between all phases. Although the microstructures vary with the gross alloy composition, the same component phases are present in each material with similar domain sizes and the phases are well-mixed on the millimeter scale. The XRD analyses showed the presence of $MoFe_2$, both the C14 and C36 polytypes of $ZrFe_2$, and the iron solution phase, but did not detect the $(Zr,Fe)Pd_2$ phase.[11]

43Fe-16Zr-15Mo-10Ru-6.8Pd-6.3Re-1.8Rh

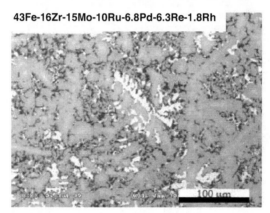

Figure 1. Photomicrograph of the microstructure of UTFe-4.

In the EBR-II alloy waste form, $ZrFe_2$ is the host phase for all of the actinides and about half of the Tc; the balance of the Tc is in the iron solid solution phase.[6] Tc is likewise expected to report to both the intermetallic and iron solid solution phases in the FCRD waste form. For that reason, initial corrosion studies are being conducted with separate Tc-bearing materials composed of the $MoFe_2$ phase (primarily) and the iron solid solution.[10] The intent is to characterize the corrosion behaviors of the separate phases in order to better interpret test and analytical results and better understand and model the corrosion behavior of the multi-phase waste form.

As stated above, the initial suite of alloys was made using reagent iron to simplify the system in order to estimate the amount of iron needed to process the waste elements and immobilize them in stable phases. It was not expected that the iron solid solution phase would be adequately durable as a waste form, and the intent is to use various stainless steels as the source of Fe to form the intermetallics and as a source of Cr to tailor superior corrosion resistance into iron solid solution phase within the alloy waste form. Based on the initial composition study, two additional alloys were made using Type 304L and Type 316L stainless steels as the source of added iron following the formulation for UTFe-4, but replacing the reagent iron with an equal mass of either Type 304L or 316L stainless steel. The resulting alloys had microstructures very similar to that of UTFe-4. Significant amounts of Cr and Ni were incorporated into the $MoFe_2$ and $ZrFe_2$ intermetallics, respectively, and the Cr and Ni concentrations in the iron solid solution phase were about half that in the added stainless steel.

Alloy Corrosion Behavior Studies

The corrosion resistance of the alloys made for the initial composition study and with stainless steels were evaluated using cyclic potentiometric polarization and linear polarization resistance methods. Figure 2 shows compilations of cyclic potentiometric polarization results in acidic and alkaline solutions.[11,12] The voltage represents the oxidizing strength of the environment and the current density represents the extent of alloy corrosion. Alloys made with reagent iron (scans labeled Fe) group fairly tightly and show active corrosion behavior with pitting. Alloys made with stainless steel (scans labeled SS) have some passive characteristics and are less corroded, but do show pitting behavior. It is hypothesized that pits and crevices will form as the iron solid solution phases dissolve in preference to the intermetallic phases, but the reacted surfaces have not yet been examined. These results reveal the combined responses of the intermetallic and iron solid solution phases, and it is not

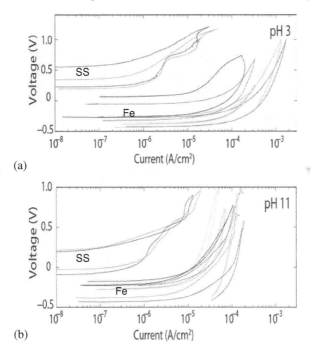

Figure 2. Cyclic potentiometric polarization of materials made using reagent iron (Fe) and stainless steels (SS) in (a) acidic (1 mM HNO_3) and (b) alkaline (1 mM NaOH) solutions.

clear if one or the other phase dominates the response or how the phases are coupled. It appears that the response is dominated by corrosion of the iron solid solution, but this remains to be verified.

Valuable insights into the likely corrosion behavior are provided by previous tests and analyses performed with EBR-II metal waste form materials. For example, Fig. 3 shows transmission electron microscopy images of a cross-sectioned corroded test specimen showing two or more oxide surface layers overlying both the iron solid solution and $ZrFe_2$ intermetallic phases.[13] The regions shown are for neighboring phases on the same reacted surface and indicate that the extents of corrosion of the two phases are similar. The iron phase was identified as ferrite with a nominal composition of 67Fe-25Cr-4Ni-2Mn-2Mo.[6] Coupled with the results of dissolution tests, these layers have been interpreted to slow the corrosion rate of EBR-II waste form, but it has not been demonstrate the surfaces are passivated.[6] For example, Fig. 4 shows the results of a dissolution test of an EBR-II metal waste form material in a simulated groundwater.[6, 14] The measured concentrations of U and Tc have been normalized to their mass fractions in the alloy and to the surface area of the test specimen. The preferential release of U relative to Tc may be due to differences in the extent of oxidation of the two metals or differences in the solubilities of the resulting oxides. The lower release of Tc suggests that it is not completely oxidized to Tc(VII), which is highly soluble, and the majority of Tc is being retained in the oxide layer as sparingly soluble Tc(IV). Tc was detected in the oxide layers of reacted EBR-II materials,[13] but was not quantified. Similar oxide layers are expected to form over these and the $MoFe_2$ intermetallic phases in the FCRD waste form and to have a similar effect on the corrosion behavior. Corrosion tests are being conducted to evaluate the release of Tc and formation of surface layers as various alloys corrode.

Figure 3. Reaction layers formed on iron and intermetallic phases of EBR-II metal waste form.

Work is in progress with materials representing the separate phases to distinguish and quantify the corrosion behaviors of the individual phases and their coupling. It is clear from the polarization results in Fig. 2 that using stainless steel as the source of iron will result in a more durable waste form, and the expected benefit of adding more Cr or other passivating metals when processing the waste will be investigated.

Tc Distribution and Release Behavior

Materials with compositions matching those of the individual phases in the multi-phase alloys have been made with Tc instead of Re to study the behaviors of individual Tc-sequestering phases. Two materials were made to match the composition of the $MoFe_2$ phase formed in the alloy made with

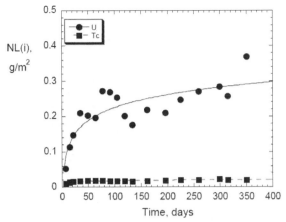

Figure 4. Results of Immersion Tests with a SS-15Zr-4(noble metals)-2U-1Tc Alloy.

Type 316L stainless steel; one was made with Re and the other with an equivalent amount of Tc, based on atomic%. Another series of materials was made to represent the iron solid solution phase formed in the alloy made with Type 316L stainless steel. The iron phase materials were doped Re or various amounts of Tc to measure the effect of the Tc concentration on the durability of the iron phase, as measured with LPR, and evaluate the solubility of Tc in the iron phase. Various tests and analyses are being performed with these Tc-bearing materials and some are summarized below.

Production of a Tc-Bearing Intermetallic Phase

Composition analysis with X-ray emission spectroscopy in the SEM revealed that most of the Re in alloys UFe-1 and UTFe-1 was sequestered in an intermetallic phase containing 64 atomic% Fe, 30 atomic% Mo, and 6 atomic% Re.[12] Approximately 10-g samples of an alloy with this composition and a companion alloy with Tc substituted for Re (on an equal atomic% basis) were made from reagent metals in a vacuum furnace with the objectives of (1) verifying the use of Re as a surrogate for Tc and (2) providing a Tc-bearing material for testing.[15] The as-batched compositions are 48Fe-38Mo-14Re and 51Fe-41Mo-8Tc on a mass% basis. Analysis indicates that both materials are multi-phase alloys dominated by the MoFe$_2$ intermetallic phase. Figure 4 shows a scanning electron microscopy photomicrograph of the microstructure. The intermetallic (light phase) occupies about 93% volume% and the iron (dark phase) occupies about 7 volume%. The few black spots are voids in the specimen. The nominal compositions (in atomic%) of the intermetallic and iron phases in the Tc-bearing alloy are 56Fe-36Mo-8Tc and 93Fe-42Mo-2Tc, respectively. Reference to binary phase diagrams for Fe-Mo, Fe-Tc, and Fe-Re indicates that the gross composition that was processed should not produce a single thermodynamically stable phase at equilibrium.[16] This suggests that the phases in alloy UTFe-4 and in the other alloys do no provide the equilibrium compositions, which is not unexpected considering the relatively rapid cool-down during processing. From the phase diagram and for Fe-33Mo, the solubility of Mo in α-Fe decreases as the temperature decreases from 1600 °C to below 900 °C and the thermodynamically stable intermetallic phase changes from σ to R to μ to λ; these intermetallic phases have different structures and composition ranges, although the phase changes below about 1000 °C were probably kinetically prohibitive as the materials cooled. The effects of processing time and heat treatments on the microstructure and phase compositions of the alloy are being studied as part of waste

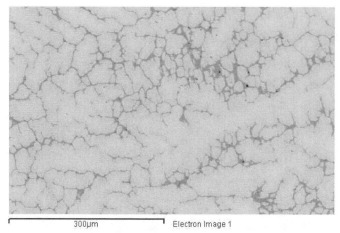

300μm   Electron Image 1

Figure 4. Scanning Electron Photomicrograph of 51Fe-41Mo-8Tc Alloy.

form development. Scaling up from 50-g laboratory samples to full-sized waste forms will greatly affect the cooling rate and probably the phase composition. Of course, the crucial aspect of the alloy phase composition is that Tc be retained in one or more durable phase within the alloy. Therefore, the distribution of Tc and other radionuclides is of interest during the entire cooling process.

Production of Tc-Bearing Iron Solid Solution Phases

A series of iron solid solutions was made to match the nominal composition of the iron solid solution phase present in the alloys made with reagent iron and the alloys made with Type 316 stainless steel. These materials are being used to study the effects of Tc and Cr on the durability of the metal phase. The nominal compositions of the metallic phases in the alloys made with reagent iron and with Type 316 stainless steel are given in Table III. Note that the Cr/Fe ratio in the alloy made with Type 316 stainless steel (14%) is significantly lower than the Cr/Fe ratio in Type 316 stainless steel (which is about 25 mass%). The lower Cr content may explain the behaviors of the alloys made with stainless steel seen in the cyclic polarization scans in Fig. 2. Since Cr is incorporated into the intermetallic phase, additional Cr may be needed to produce an iron solid solution phase as corrosion resistant as Type 316 stainless steel. Alloys were made with Tc instead of Re and the amount of Tc was varied to achieve loadings of about 1, 2, 6, and 8 atomic% Tc. The microstructures of these materials are currently being analyzed and the corrosion rates will be measured in electrochemical tests to assess the effect of the Tc content (and the Cr content) on the corrosion resistance.

Table III. Nominal Concentrations of Metallic Phases in Alloys Made with Reagent Iron and Type 316 Stainless Steel, atomic %.

	Al	Cr	Fe	Ni	Zr	Mo	Ru	Rh	Pd	Re
Iron	—	—	87.50	—	—	7.28	3.86	—	—	1.36
Type 316	0.32	8.73	56.64	23.19	0.02	5.48	4.03	0.32	0.69	0.58

CONCEPTUAL ALLOY WASTE FORM CORROSION MODEL

A conceptual model has been developed to guide the testing activities being conducted as part of waste form development.[17] It is a mechanistic model in the sense that it is based on process steps involved in the metal waste form corrosion and radionuclide release; that is, the charge transfer and mass transport processes that control the rate of alloy waste form corrosion and the oxidation and dissolution of Tc and other radionuclides. The model is being used to identify information needs to be addressed by laboratory tests and analyses for processes such as the redox behavior of the radionuclides at and near the corrosion interface, the retention of radionuclides in oxide layers that form on the surface, and the oxide dissolution and mass transport steps that may limit the release of radionuclides into the bulk water. The model will be further developed to link the results of electrochemical tests that highlight the oxidation steps and near-surface mass transport processes with laboratory corrosion tests highlighting the dissolution and release of radionuclides as the host phases corrode to calculate the radionuclide release rate used in reactive transport models for site assessments.

Environment

As neither a disposal system nor environment has been identified for high-level waste in the US, a full range of possible environmental conditions is being considered in the modeling and testing activities. This includes a variety of hydrologically saturated host rocks (salt, basalt, and granite) and unsaturated (tuff) rock. The groundwater is likely to have near-neutral pH and to be reducing in the saturated rocks and oxidizing in the unsaturated rock. Demineralized water, a silicate solution, and a brine will be used to span the range of deep groundwater compositions; preparation procedures are provided in ASTM standard C1220.[18] Because radiolysis, dissolution, and chemical reactions will alter the environment near the waste package relative to the unperturbed environment, those compositions will be modified to represent a wider range of pH, PO2, and component concentrations. In general, tests are expected to span the stability field of water over the range from about 4 to 10.

Anodic and Cathodic Reaction Currents

The alloy corrosion model is based on charge conservation between anodic reactions corroding the alloy and cathodic reactions of the available oxidizing agents. The sum of the cathodic currents $i_{R/O}$ that are generated by the set of n redox couples defined by the environment must be equal and opposite to the anodic current i_{corr} generated by metal corrosion as

$$\sum_{j=1}^{n} i_{R/O,j}\left(E, pH, T, [O_2], [Cl^-], ...\right) = -i_{corr}\left(E, pH, T, [O_2], [Cl^-], alloy\ composition, surface\ films\ ...\right). \quad (1)$$

Important cathodic reactions will include the reduction of dissolved oxygen, radiolytic species such as hydrogen peroxide, water, and hydrogen ions. Identifying and quantifying the contributing cathodic reactions is probably the most challenging aspect of the modeling effort. Modeling the corrosion rate of the alloy is equivalent to modeling the corrosion current under the steady state conditions that prevail when the charge conservation constraint in Eq. 1 is satisfied. As implied in Eq. 1, these currents will depend on the potential (E) at which the charge transfer processes occur, environmental variables, the composition of the alloy and the presence of surface films, and perhaps other factors. An important role of testing activities is to determine the functional dependencies of the anodic and cathodic currents on environmental conditions and the state of the alloy surface, and then to express those dependencies mathematically. The cathodic and anodic polarization curves give the dependencies $i_{R/O,j}(E)$ and $i_{corr}(E)$, respectively, under the particular test conditions. The shapes of these curves reveal important system properties, such as regions of active, passive, and transpassive behavior, that affect how the reactions are modeled. Although time is not listed as a variable for either anodic or cathodic reactions, it will affect the reactivity both through changes in concentrations of reactants (e.g., oxygen) in the solution at the alloy surface and the generation of surface layers covering the alloy surface, including oxides and passivating layers. Additional time dependencies may be imposed by advection and diffusion processes.

Mathematical models are available for calculating the dependence of the cathodic and anodic currents on the potential under particular environmental conditions. The dependence of the current of each cathodic reaction on the applied potential can be modeled using a Mixed Potential Model[19]

$$
i_{R/O} = \frac{\exp\left[\left(E - E^e_{R/O}\right)/b_a\right] - \exp\left[-\left(E - E^e_{R/O}\right)/b_c\right]}{\dfrac{1}{i_{0,R/O}} + \dfrac{1}{i_{i,f}}\exp\left[\left(E - E^e_{R/O}\right)/b_a\right] - \dfrac{1}{i_{i,f}}\exp\left[-\left(E - E^e_{R/O}\right)/b_c\right]} , \tag{2}
$$

where

$i_{0,R/O}$ = exchange current density
$i_{i,f}$ and $i_{i,r}$ = mass transfer limited current densities
$E^e_{R/O}$ = equilibrium potential from the Nernst equation
b_a and b_c = anodic and cathodic Tafel constants.

The anodic polarization curve in the active corrosion region can be described by the Butler-Volmer expression for charge transfer kinetics when the mass transport kinetics to and from the corroding surface are unimportant.

$$
i_{corr} = i_0 \times \left\{ \exp\left[\frac{(1-\alpha)\times n \times F}{R \times T} \times \left(E - E_{eq}\right)\right] - \exp\left[\frac{-\alpha \times n \times F}{R \times T} \times \left(E - E_{eq}\right)\right] \right\}, \tag{3}
$$

where

i_{corr} = corrosion current density (measured) α = symmetry factor
i_0 = exchange current density (measured)   F = Faraday constant
n = number of electrons involved in the reaction   R = gas constant
E_{eq} = equilibrium potential from the Nernst equation   T = absolute temperature.

The environmental conditions, alloy composition, and alloy surface condition are variables that affect both of these models and those effects must be determined experimentally. Mathematical expressions for these dependencies can be combined with the equations for the cathodic and anodic reactions and subjected to the conservation of charge constraint in Eq. 1 to calculate the corrosion current density over the relevant range of environmental conditions in a particular disposal system. The corrosion current density is related to the alloy corrosion rate (penetration rate) using the equivalent weight of the alloy (EW) and its density (ρ) as[20]

$$
\text{corrosion rate}, (\mu m/y) = = \frac{3.27 \times i_{corr} \times (EW)}{\rho} . \tag{4}
$$

The equivalent weight of an alloy is the reciprocal of the sum over components of the product of the mass fraction of the component times the number of electrons involved in the oxidation of that component divided by its atomic weight.[20] The equivalent weight of Type 316 stainless steel is 25.50 and the equivalent weight of the iron solution phase in the alloy made with Type 316 stainless steel (see Table III) is 24.45. The constant 3.27 in Eq. 4 is a unit conversion factor.

Although the general (uniform) corrosion processes described above are expected to represent the oxidation of most radionuclides, localized processes at grain boundaries and pits may be important. The susceptibility of the alloy waste form to localized corrosion processes must be evaluated in laboratory experiments to utilize the appropriate process model. It may be that corrosion of the small iron solid solution domains is better modeled as pit corrosion rather than uniform corrosion.

The calculated alloy corrosion rate provides the amounts of radionuclides that are oxidized, but the dissolution kinetics will control the rates at which they are released from the waste form and transportable. Therefore, the electrochemical studies conducted to measure corrosion reaction currents must be coupled with test methods to evaluate dissolution behavior and release rate to develop a mathematical radionuclide source term model. Test methods are being developed to study the coupled oxidation, dissolution, and mass transport processes. Many of the variables affecting the oxidation behavior will also affect the dissolution behavior.

CONCLUSION

Work is in progress to develop a metallic waste form that can be used to immobilize Tc and other metallic wastes from used fuel reprocessing operations in durable phases. Based on insights from the waste form developed for metal wastes from electrometallurgical treatment of EBR-II spent sodium-bonded fuel, an iron-based multi-phase alloy material has been formulated to meet the following goals: process the waste streams at about 1600 °C; use waste steel as an iron source; accommodate the expected range of waste stream compositions in a small number of phases that incorporate and immobilize all waste components. The alloy being studied is composed of two predominant intermetallic phases and a steel-like iron solid solution phase. Current activities are focused on characterizing the corrosion behaviors of the individual phases and measuring the effects of environmental variables. The measured chemical and electrochemical behaviors will be used to parameterize the process models. A conceptual model that couples oxidation/reduction processes with dissolution is being used to guide these laboratory experiments and analyses. The long-range intent is to develop a mechanistically-based source term model for radionuclide release that takes into account the effects of environmental variables and the evolution of the corroding alloy surfaces over time. That model can then be used in reactive-transport models to evaluate the long-term performance of a high-level waste disposal system.

ACKNOWLEDGEMENTS

The work summarized in this report is part of a collaborative effort by scientists at Argonne, Idaho, Los Alamos, Pacific Northwest, and Savannah River National Laboratories and at the University of Nevada-Las Vegas and the University of Western Ontario. The work of collaborators at each national laboratory in performing laboratory experiments and analysis is gratefully acknowledged: J.A. Fortner, and J.J. Jerden, Jr. (ANL), T.P. O'Holleran, M. Simpson and P. Hahn (INL), G. Jarvinen and D. Kolman (LANL), E. Buck (PNNL), R.L. Sindelar (SRNL), K.Czerwinski and E. Mausolf (UNLV), and D. Shoesmith (UWO). Programmatic guidance provided by T. Todd (INL), J. Vienna (PNNL), and S. Lesica (DOE) is also gratefully acknowledged.

Government License Notice

The submitted manuscript has been created by UChicago Argonne, LLC, Operator of Argonne National Laboratory ("Argonne"). Argonne, a U.S. Department of Energy Office of Science laboratory, is operated under Contract No. DE-AC02-06CH11357. The U.S. Government retains for itself, and others acting on its behalf, a paid-up nonexclusive, irrevocable worldwide license in said article to reproduce, prepare derivative works, distribute copies to the public, and perform publicly and display publicly, by or on behalf of the Government. This work was supported by the U.S. Department of Energy, Office of Nuclear Energy, under Contract DE-AC02-06CH11357.

REFERENCES

[1]Gombert, D. et al. 2007. *Global Nuclear Energy Partnership Integrated Waste Treatment Strategy Waste Treatment Baseline Strategy*, Idaho National Laboratory report GNEP-WAST-AI-RT-2007-000324.

[2]Till, J.E. 1984. "Source Terms for 99Tc from Nuclear Fuel Cycle Facilities," in *Technetium in the Environment*, eds. Desmet, G, and Myttenaere, C., Elsevier, New York, New York.

[3]Kleykamp, H. 1985. "The Chemical State of the Fission Products in Oxide Fuels." *Journal of Nuclear Materials, 131*, 221-246.

[4]Collins, E. (2009). Personal communication February 18, 2009 and May 13, 2009, presentations at AFCI Waste Form campaign videoconference workshops.

[5]Ebert, W.L., Fortner, J., Shkrob, I., and Jarvinen, G. (2009). *Options for Recovering and Immobilizing the Tc Dissolved During Oxide Fuel Dissolution*, Idaho National Laboratory report AFCI-SEPA-PMO-MI-DV-2009-000161.

[6]Ebert, W.L. (2005). *Testing to Evaluate the Suitability of Waste Forms Developed for Electrometallurgically Treated Spent Sodium-Bonded Nuclear Fuel for Disposal in the Yucca Mountain Repository*, Argonne National Laboratory report ANL-05/43.

[7]Ebert, W.L. 2008. *Immobilizing GNEP Wastes in Pyrochemical Process Waste Forms*. Idaho National Laboratory report GNEP-WAST-PMO-MI-DV-2008-000150.

[8]Ebert W. L., J. C. Cunnane, and J. L. Jerden Jr. (2009). *A Strategy for Conditioning TRUEX Raffinate for Immobilization,* Idaho National Laboratory report AFCI-SUI-WAST-WAST-MI-DV-2009-000001.

[9]Ebert, W.L. 2008. *Alloy Formulations*, GNEP-WAST-WAST-MI-RT-2009-000014.

[10]Ebert W., M. Williamson, and S. Frank (2009). *Immobilizing Tc-Bearing Waste Streams in an Iron-Based Alloy Waste Form,* Idaho National Laboratory report AFCI-WAST-PMO-MI-DV-2009-000160.

[11]Williamson, M.J. and Sindelar, R.L. (2009). *Development of an Fe-Based Alloy Waste Form for Spent Nuclear Fuel*, Idaho National Laboratory report AFCI-SEPA-WAST-MI-DV-2009-000143.

[12]Williamson, M.J. and Sindelar, R.L. (2009). Personal communication October 30, 2009 presentation made as part of Fuel Cycle Research & Development Program teleconference.

[13]Dietz, N.L. 2005. *Transmission Electron Microscopy Analysis of Corroded Metal Waste Forms*. Argonne National Laboratory report ANL-05/09.

[14]Johnson, S.G.; Noy, M.; DiSanto, T.; and Keiser, D.D., Jr. 2002. "Long-Term Immersion Test Results of the Metallic Waste Form from the EMT Process of EBR-II Spent Metallic Fuel." *Proceedings of the DOE Spent Nuclear Fuel and Fissile Materials Management Meeting held September 17–20, 2002*. Charleston, South Carolina. Waste Form Testing session. La Grange Park, Illinois: American Nuclear Society.

[15]Frank, S., O'Holleran, T., and Hahn, P. (2009). *Composition of Tc-Fe Alloy Produced for Testing,* Idaho National Laboratory report FCRD-WAST-2010-000011.

[16]Massalski, T.B. 1990. Binary Alloy Phase Diagrams. ASM International.

[17]Cunnane, J.C. (2010). *Initial Conceptual Model and Planned Testing Approach-Metal Waste Form,* Idaho National Laboratory report FCRD-WAST-2010-000014.

[18]ASTM (2009) *Annual Book of ASTM Standards, Vol. 12.01*, West Conshohocken, Pennsylvania: ASTM-International.

[19]Macdonald, D. D. (2001). "The Deterministic Prediction of General Corrosion Damage in HLNW Canisters," in Proceedings from an International Workshop on Long-Term Extrapolation of Passive Behavior, July 19-20, Arlington, VA, US Nuclear Waste Technical Review Board.

[20]Jones, D.A. 1992. *Principles and Prevention of Corrosion*. Prentice Hall, Englewood Cliffs, NJ.

DEVELOPMENT OF IODINE WASTE FORMS USING LOW-TEMPERATURE SINTERING GLASS

Terry J. Garino, Tina M. Nenoff, James L. Krumhansl, and David Rademacher
Sandia National Laboratories
Albuquerque, New Mexico, USA

ABSTRACT

Radioactive iodine, [129]I, a component of spent nuclear fuel, is of particular concern due to its extremely long half-life, its potential mobility in the environment and its effects on human health. In the spent fuel reprocessing scheme under consideration, the [129]I is released in gaseous form and collected using Ag-loaded zeolites such as Ag-mordenite. The [129]I can react with the Ag to form insoluble AgI. We have investigated the use of low temperature-sintering glass powders mixed with either AgI or AgI-zeolite to produce dense waste forms that can be processed at 500°C, where AgI volatility is low. These mixtures can contain up to 20 wt% crushed AgI-mordenite or up to 50 wt% AgI. Both types of waste forms were found to have the high iodine leach resistance in these initial studies.

INTRODUCTION

Radioactive iodine ([129]I, half-life of 1.6×10^7 years) is generated at relatively low concentration in the nuclear fuel cycle. However, due to its long half-life, the potential high mobility of I^- in the environment and its effects on human health, safe, long-term storage of [129]I is of particular concern. As part of the DOE/NE Fuel Cycle R&D, the separation of [129]I from spent fuel during fuel reprocessing, so it can be subsequently incorporated in a suitable waste form, is being studied. In the process under consideration, gas containing [129]I vapor is passed through a bed of silver-exchanged zeolite, such as Ag-mordenite (Ag-MOR) that captures the [129]I. The Ag in the mordenite can be in either ionic (Ag^+) or metallic (Ag) form. Depending on the specific zeolite used and the temperature, water content, etc of the vapor, the captured [129]I can then be either (1) converted to AgI in the zeolite or (2) later driven off the zeolite and directly reacted with silver to produce AgI. Silver iodide (melting point = 552 C and density = $5.675 g/cm^3$) has an extremely low solubility in water, 1.3×10^{-8} M at 20 C, which makes it a good candidate for long term storage. However, the AgI, whether in the zeolite or not, must be incorporated into a dense solid waste form prior to storage.

One potential problem with using AgI to contain [129]I is its relatively high vapor pressure at moderate temperatures[1], ~1 millitorr at 500 C and ~10 millitorr at 600 C. This and the melting of AgI at 552 C, limit the thermal processing that can be done on a AgI-containing waste form without excessive loss of [129]I (>0.1%) to ~550 C. Because of this, immobilization using molten borosilicate glass, for example, is not feasible due to excessive iodine loss at the required processing temperature.[2] Other, low-melting, glasses such as vanadium or lead oxide-containing glasses[3,4], do not meet chemical stability requirements. Another possibility that has recently been investigated is forming a glass using AgI and $Ag_4P_2O_7$.[5-7] In this case, the glass was melted at 500°C and was shown to have low solubility due to the formation a protective AgI surface layer after exposure to water. However, this approach is not applicable to AgI-containing zeolites. The use of grout to contain AgI-MOR has also been investigated, but possibilities of carbonate release with time limit its effectiveness as a long term storage material.[8] Hot isostatic pressing with Ag-MOR and metal[9] or Pb-Fe-phosphate glass[4], or of AgI-zeolite to form sodalite[10] have also been investigated. Hot isostatic pressing has also been investigated for AgI without zeolite.[11]

In this work, encapsulation of either AgI or AgI-zeolite using a low temperature sintering glass to form dense, stable waste forms was further investigated.[12] In this approach, the AgI or AgI-MOR is mixed with a glass powder that can be sintered to high density at 500˚C. The mixture is first formed into the desired shape by pressing and then heated to sinter the glass by the viscous sintering process. Since the glass is not melted as in other methods, a more refractory and therefore more chemically stable glass can be used. A commercially available bismuth oxide-based glass was chosen due to the low solubility of bismuth oxide in aqueous solution at pH>7.[13] In this work, the feasibility of such a process was demonstrated and the materials produced were analyzed using a variety of techniques. An aqueous leaching study was also performed at 90˚C on crushed samples of the AgI-zeolite and AgI mixed with the bismuth oxide-based glass. The behavior of I_2 vapor-loaded Ag-MOR was also studied.

EXPERIMENTAL PROCEDURE

The zeolite used was mordenite (LZM5, UOP) that was silver-exchanged and then either used in the ionic form or in the reduced, metallic silver form by heating in 3% hydrogen to 500 C. The Ag-MOR was exposed to I_2 vapor at 90 C in either ambient air or in air saturated with water vapor. The samples were then heated in an open container for 15 hr at 95 C to remove excess iodine. Simultaneous thermal gravimetric and differential thermal analyses were performed on the iodine-treated Ag-MOR and on silver iodide (Aldrich Chem. Co) in flowing air using a heating rate of 10˚C/min. A sample of the Ag -MOR that was loaded with dry I_2 vapor was heated in a closed Petri dish on a hotplate to 500 C to observe the evolution of I_2 vapor. The temperature was held every 50 C for 10 min and then the I_2 vapor was allowed to escape before replacing the cover and heating to the next hold temperature. Powder x-ray diffraction was performed on the I_2-loaded Ag-MOR samples before and after heating to 600°C in air.

The low sintering temperature glass used in this work was a bismuth-zinc-boron oxide glass that is available commercially (8 μm average particle size, a density of 5.71 g/cm^3, from Ferro Corp., Cleveland, OH). This glass sinters to high density after only 15 min at 500˚C and crystallizes during the sintering process.

Two types of samples were made using a mixture the glass and AgI. In the first type, the glass powder was simply mixed by mortar and pestle with AgI powder in equal parts by mass. The mixture was then dry pressed in a cylindrical steel die (2.5 cm diameter) at 70 MPa. The second type of sample contained the AgI-MOR that had been iodine-loaded either using an aqueous KI solution or using I_2 vapor. A mixture of 20 wt% AgI-MOR ground to less than 38 μm was mixed by mortar and pestle with the glass powder and then dry pressed at 70 MPa. The samples were then heated in air for up to 3 hr at 500˚C. After sintering, the samples were analyzed using powder X-ray diffraction, TGA/DTA, and SEM using a back-scattered electron detector and energy dispersive spectroscopy (EDS). Also, some of each sample was used for solubility testing. A sample was ground to a 150 to 250 μm fraction and placed in a PTFE container (Parr Instrument Co.) with deionized water. The mixtures were heated at 90˚C for 7 days, in accordance with the Product Consistency Test (PCT).[14] The amount of dissolved I⁻ was determined using an I⁻-specific electrode (Orion 4Star, Thermo Scientific). For comparison, a sample of AgI-Ag phosphate glass (53.8 wt% AgI heated to 500˚C for 3 hr) was prepared in accordance with reference 6 and was also leach tested.

RESULTS AND DISCUSSION

The results of the TGA on silver iodide powder are shown in Fig. 1. With the 10 C/min heating rate used, no mass loss is observed below 600 C. Since processing a sample that contains a glass powder typically requires hold times from tens of minutes to an hour to achieve full densification, a temperature no higher than 550 C should be used to ensure minimal loss of AgI. This is just below the melting point of AgI so that melting will not occur. This is also beneficial since liquid AgI could be squeezed out of the sample by the shrinking glass framework. Therefore, we chose to study the use of glass powders with sintering temperature below 550 C. After initial compatibility and sintering experiments with several bismuth oxide and lead oxide based glasses, we chose the best performing glass, one that sinters at 500 C, for the results reported here.

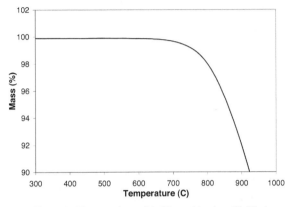

Figure 1. The mass loss of AgI heated in air at 10 C/min.

The mass loss during heating in air of both ionic and metallic Ag-MOR that was treated with either dry or wet I_2 vapor is shown in Fig. 2. All of the samples begin to lose mass just above 100 C and continue to do so up to 600 C, with a slower rate above 450 to 500 C, depending on the sample. Both water and iodine vapor are given off by the samples and from the TGA alone it is not possible to determine how much of each and over what temperature range each is evolved. However, from the magnitude of the mass loss and from the fact that a large amount I_2 vapor can be seen coming off of the vapor-loaded Ag-MOR as shown in Fig. 3 throughout the temperature range of 200 to 500 C, indicates that much of the iodine is chemically adsorbed and not reacted with the Ag to form AgI. This result indicates that if [129]I vapor-loaded Ag-MOR is loaded to capacity and is then to be incorporated into a waste form, a strategy to capture the chemisorbed iodine that evolves at temperatures as low as 200 C must be employed.

Even though much of the initially captured iodine is lost during heating of the I_2-loaded Ag-MOR, a significant amount of AgI is formed. Fig. 4 shows the x-ray diffraction pattern after loading with AgI crystal peaks prominently present. However, as shown in the bottom half of the figure, after the material has been heated to 600 C, the AgI peaks are no longer present in the diffraction pattern. Based on the vapor pressure of AgI and the results shown in Fig. 1, the AgI could not have sublimed or evaporated completely from the sample. This suggests that the AgI is still present after heating but that it is not in the normal crystalline form. This phenomenon was further investigated using differential thermal analysis (see Fig. 5). Fig. 5a shows that for all samples, a peak can be seen at low temperature (147 C) where AgI undergoes a crystalline phase change from the β phase where the silver ions are on fixed sites to the phase where the silver lattice has melted, consistent with the AgI still being crystalline at that temperature. However, as shown in Fig. 5b, when a sample first heated to 200 C is cooled and then reheated, the size of the peak associated with the phase change has decreased. With further heating, cooling and reheating, the peak continues to decrease in magnitude until it essentially is gone after heating to 350 C. Therefore, the silver iodide is becoming non-crystalline between 200

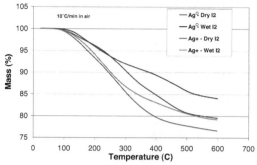

Figure 2. The mass loss of AgI-MOR samples heated in air at 10 C/min.

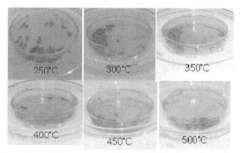

Figure 3. Images of dry I$_2$-vapor loaded Ag -MOR heated to a series of temperatures showing the evolution of pink I$_2$ vapor.

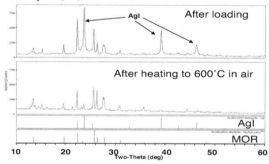

Figure 4. Powder x-ray diffraction patterns of Ag -MOR loaded with dry I$_2$ vapor after loading and after heating to 600 C in air.

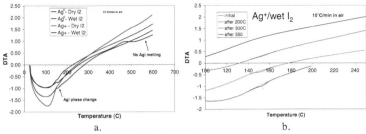

a.   b.

Figure 5. Differential thermal analysis of a) various iodided Ag-MOR samples heated to 600 C
and b) a wet I$_2$ loaded Ag$^+$-MOR heated successively to 200 , 250 , 350 and 400 C.

and 350 C. A similar result was previously reported by Vance and coworkers.[15] This structural
change is due to the interaction of the AgI with the pores in the mordenite crystal structure. Recent
structural studies[16] have determined that in fact the AgI is crystalline nanoparticles (<7Å) and confined
to the MOR pores. The size and discrete nature of the AgI result in an amorphous result in the powder
X-ray diffraction studies.

Sintered mixtures of the low-sintering glass and AgI-MOR or AgI that were sintered at 500 C
are shown in Fig. 6. In Fig. 6a, the glass was mixed with 20 wt% AgI-MOR that had been previously
heated to 500 C to drive off all the chemisorbed iodine. The sample in Fig. 6b contains 50 wt% AgI,
which is essentially also 50 vol% since the density of the glass is nearly the same as AgI. Both of the
samples in Fig. 6 are essentially non-porous and impervious to water. The results of thermal
gravimetric analysis on these samples after grinding are shown in Fig. 7. As expected, neither sample
lost mass until above 600°C where AgI becomes volatile. Since the AgI-MOR contains on the order of
10 wt% AgI and it was mixed at 20 wt% with the glass, only ~2 wt% AgI is present in that sample as
opposed to the 50 wt% AgI in the other sample. Obviously, a much smaller waste form volume can be
achieved when AgI is used instead of AgI-MOR. The microstructure and x-ray diffraction patterns of
AgI-MOR/glass and AgI/glass samples are shown in Figs. 8 and 9, respectively. In both cases, the
glass crystallized during the sintering process, as expected since this particular glass was identified by
the manufacturer as one that crystallizes upon sintering. The primary phase formed appears to be a
form of bismuth oxide, possibly stabilized by a small amount of zinc.[17] The crystallized glass phase
encapsulates the mordenite or the AgI particles. The sample that contained 50 wt% AgI had some
cracks in the crystallized glass phase, probably due to the anomalous thermal expansion behavior[18] of
AgI. This might limit the maximum amount of AgI that can be incorporated to below the 50% level.

a.   b.

Figure 6. Dense pellets of low-temperature sintering glass with a) 20 wt% AgI-MOR and b) 50
wt% AgI that were sintered at 500 C.

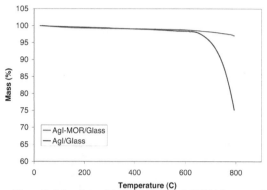

Figure 7. Mass loss of AgI/glass and AgI-MOR/glass samples.

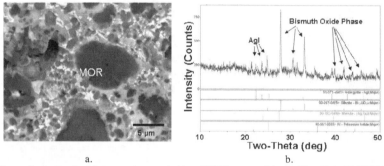

a.   b.

Figure 8. A scanning electron micrograph (a) and XRD pattern (b) of a 20 wt% AgI-MOR/glass sample sintered at 500 C.

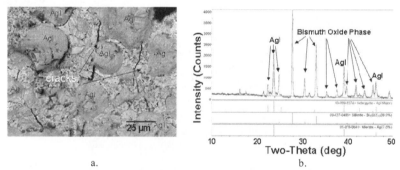

a.   b.

Figure 9. A scanning electron micrograph (a) and XRD pattern (b) of a 50 wt% AgI/glass sample sintered at 500 C.

The measured iodide concentration of samples of AgI-MOR (20 wt%)/glass and AgI (50 wt%)/glass after the PCT are shown in Fig. 10. For comparison, the results for an AgI/Ag phosphate glass are also shown. For the AgI-MOR/glass sample, the AgI-MOR was heated to 500 C before mixing with the glass powder to drive off any chemisorbed iodine. The results indicate that the sintered glass samples had lower levels of dissolved iodide than did the phosphate glass sample, in the several micro-molar range for the AgI-MOR/glass sample to below 1 micro-molar for the AgI/glass sample. However, ICP/MS analysis of the leachant from the low temperature sintering glass samples indicated relatively high concentrations of boron and sodium, although that of Bi and Zn were very low. Therefore, a bismuth oxide based glass without boron and sodium would be expected to give improved solubility results.

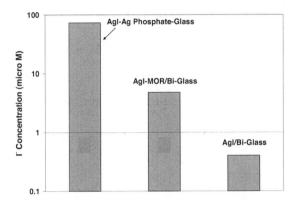

Figure 10. The measured I⁻ concentration of PCT samples of iodine waste forms.

CONCLUSIONS

Thermal analysis of iodine vapor loaded Ag-MOR with both ionic and metallic silver showed that when loaded to capacity in either dry or humid atmosphere, a significant fraction of the captured iodine is chemisorbed and is released as I_2 vapor upon heating between 200 and 550 C. Also, the AgI that does form is initially crystalline but becomes amorphous through interaction with the pores in the mordenite between 200 and 350 C. The use of bismuth oxide-based low temperature sintering glasses to encapsulate AgI-MOR and AgI has been investigated. Samples could be sintered to high density at 500 C, below the melting point of AgI and where its vapor pressure is low. Leaching test results showed low iodide solubility but relatively high solubility of B and Na.

ACKNOWLEDGEMENTS

The authors acknowledge the US DOE/NE-FCR&D program, Separations and Waste Forms Campaign for funding support. This work was performed at Sandia National Laboratories, Albuquerque, NM. Sandia is a multiprogram laboratory operated by Sandia Corporation, a Lockheed Martin Company, for the United States Department of Energy's National Nuclear Safety Administration under Contract DE-AC04-94AL85000.

REFERENCES

[1]CRC Handbook of Chemistry and Physics, 61st Ed., R.C. Weast ed., 1980.

[2]M.Y. Khalil, "Solidification of Loaded AC6120 in Borosilicate Glasses for Final Disposal of I-129," Unpublished work, 1991.

[3]T. Nishi, K. Noshita, T. Naitoh, T. Namekawa, K. Takahashi, M. Matsuda, "Applicability of V_2O_5-P_2O_5 Glass System for Low-Temperature Vitrification," Mat. Res. Soc. Symp. Proc. 465, 221-8 (1999).

[4]D. Perera, E. Vance, R. Trautman, B. Begg, "Current Research on I-129 Immobilization," Proc. of WM'04 Conference, WM-4089 (2004).

[5]H. Fujihara, T. Murase, T. Nishi, K. Noshita, T. Yoshida, and M. Matsuda, "Low Temperature Vitrification of Radioiodine Using AgI-Ag_2O-P_2O_5 Glass System," Mater. Res. Soc. Symp. Proc. Vol. 556, 375-382 (1999).

[6]K. Noshita, T. Nishi, T. Yoshida, H. Fujihara and T. Marase, "Vitrification Technique of Radioactive Waste Using AgI-Ag_2O-P_2O_5 Glass System", in International Conference Proc. Radioactive Waste Management and Environmental Remediation, pub., ASME, New York, 107-112 (1999).

[7]T. Sakuragi, T. Nishimura, Y. Nasu, H. Asano, K. Hoshino and K. Iino, "Immobilization of Radioactive Iodine Using AgI Vitrification Technique for the TRU Wastes Disposal: Evaluation of Leaching and Surface Properties," Mater. Res. Soc. Symp. Proc. Vol. 1107 (2008).

[8]R. D. Scheele, C. F. Wend, W. C. Buchmiller, A. E. Kozelisky, R. L. Sell, "Preliminary Evaluation of Spent Silver Mordenite Disposal Forms Resulting from Gaseous Radioiodine Control at Hanford's Waste Treatment Plant," Battelle - Pacific Northwest Division, Richland, Washington, December 2002.

[9]M. Fukumoto, "Method for Solidifying Waste Containing Radioactive Iodine," US Patent 5826203, 1998.

[10]G.P. Sheppard, J.A. Hriljac, E.R. Maddrell and N.C. Hyatt, "Silver Zeolites: Iodide Occlusion and Conversion to Sodalite – a Potential ^{129}I Waste Form?," Mat. Res. Soc. Symp. Proc. Vol. 932 (2006).

[11]E.R. Maddrell and P.K. Abraitis, "A Comparison of Wasteforms and Processes for the Immobilisation of Iodine-129," Mat. Res. Soc. Symp. Proc. Vol. 807 (2004).

[12]T.J. Garino, T.M. Nenoff, J.L. Krumhansl and D.X. Rademacher, "Development of Waste Forms for Radioactive Iodine," Proc. of the 8th Pacific Rim Conference on Ceramic and Glass Technology, Vancouver, WA, 2009.

[13]C.F. Baes and R.E. Mesmer, The Hydrolysis of Cations, John Wiley and Sons, Inc., 1976.

[14]Standard Test Methods for Determining Chemical Durability of Nuclear, Hazardous, and Mixed Waste Glasses and Multiphase Glass Ceramics: The Product Consistency Test (PCT), Designation: C 1285 – 02, ASTM Int., West Conshohocken, PA, (2008).

[15]E. Vance and D. Agrawal, X-Ray Studies of Iodine Sorption in Some Silver Zeolites, J. Mater. Sci. 17, 1889-1894 (1982).

[16]K.W.Chapman, P.J. Chupas, and T.M. Nenoff, Radioactive Iodine Capture in Silver-Loaded Zeolites Through Nanoscale Silver Iodide Formation. J. Amer. Chem. Soc. in press, (2010). DOI: 10.1021/ja103110y.

[17]J. P. Guha, S. Kunej, and D. Suvorov, "Phase Equilibrium Relations in the Binary System Bi_2O_3-ZnO," J. Mater. Sci. 39, 911– 918 (2004).

[18]G. Harvey and N. H. Fletcher, "Thermal Expansion of β-Silver Iodide at Low Temperatures," J. Phys. C: Solid St. Phys. 13, 2969-73 (1980).

Wind

NANOSTRENGTH® BLOCK COPOLYMERS FOR WIND ENERGY

Robert Barsotti, John Chen and Alexandre Alu
Arkema Corporation
King of Prussia, PA, USA

ABSTACT

One of the current challenges in the widespread adoption of wind energy is the ability to make larger, more reliable wind blades without significantly increasing the weight of the blades. Increase in service life is needed for both wind blade composites and adhesives. Thermoset composites and adhesives are valued for excellent strength, chemical resistance and high temperature properties but suffer from low toughness. For wind energy applications, it is necessary to improve the fracture toughness and fatigue performance of blades and adhesives without effecting mechanical properties such as strength or modulus or processing variable such as viscosity or curing kinetics. Although many additives exist for improving the toughness of thermosets, most are difficult to incorporate into formulations or result in a "trade-off" of properties. Arkema's controlled radical polymerization technology has been used to synthesize Nanostrength block copolymers additives, which provide excellent toughening to thermosets at low loading levels without sacrificing other properties. By controlling structuration of these polymers, a wide range of mechanical properties can be achieved while controlling the viscosity of the resin.

INTRODUCTION

Thermoset resins are valued for their high strength, excellent high temperature properties and outstanding chemical resistance. Due to these properties, the use of thermosets has greatly increased in wind energy applications in both the composite material used for the wind turbine blades and the structural adhesive used to attach the two halves of the blade together. One of the major challenges in wind energy is the creation of larger blades with increased reliability and service lifetime. An inherent weakness of the thermoset matrices is their brittle nature, often leading to rapid failure upon an impact event or continuous fatigue loading when a micro-crack or defect is formed. Solutions for toughening thermoset composites and thermoset matrices in general have typically followed one of two approaches: 1) modifications to the resin backbone by adding chain flexibilizers and reactive or non-reactive diluents; 2) use of rubber or thermoplastic additives to absorb and dissipate energy during impact events or fatigue loading. The 1st strategy greatly improves toughness but often does so at the expense of many of the favorable properties of the thermoset matrix: strength, modulus, glass transition temperature (Tg), and chemical resistance. The 2nd approach offers the possibility to improve toughening without adversely affecting other properties if the rubbery polymeric additive is well dispersed and phase separated in the host matrix at the nanoscale level. Solutions to thermoset toughening will allow for increased size and reliability of wind blades, lowering the overall cost of wind energy.

Reactive rubbers, such as carboxy terminated butadiene-co-acrylonitrile (CTBN) are one of the most widely used additives to toughen thermoset matrices, in particular epoxies. CTBN is initially soluble in the uncured resin. During cure, CTBN undergoes reaction induced phase separation to form micron-sized domains of rubber in the matrix. These rubber domains absorb energy during an impact event, undergoing cavitation. This cavitation causes plastic deformation of the surrounding thermoset matrix through the formation of shear bands, and this formation of shear bands absorbs additional energy to slow or stop crack propagation. This highly efficient toughening mechanism, commonly referred to as cavitation and shear band formation, is the primary manner in which rubber toughens a thermoset matrix.[1] While CTBN greatly improves the toughening of matrices it has several drawbacks. During cure, a percentage of the polymer remains miscible in the matrix. This results in a decrease in the strength, modulus and T_g of the material. A second drawback specific to composites deals with the relatively large particle size of the rubber domains. Due to its shorter cycle time and lower requirements for human capital, infusion processing is a rapidly becoming the preferred technique for manufacture of wind turbine blades. During an infusion process, fiberglass or carbon fiber is laid out in a mold. Resin is pulled via vacuum into the mold with positive or negative pressure. Critical to the process is the ability of the resin to flow freely between fiberglass bundles. Micron-sized rubber particles, such as those formed with CTBN, are often filtered out during the process, resulting in an uneven distribution of toughening agents in the finished part.

Thermoplastic spheres have also been utilized for the toughening of thermosets. These materials suffer from having a much higher modulus. Thus, they are not effective as stress concentrators and cannot easily cavitate. They can only toughen the thermoset resin by physical mechanism such as crack pinning or crack deflection, less efficient mechanisms than the previously described cavitation and shear band formation mechanism. Due to this poor efficiency, they are often used at high concentrations, where they can have a detrimental effect on properties critical to infusion processing such as viscosity.

In order to overcome the deficiencies associated with the aforementioned types of thermoset toughening agents, controlled radical polymerization has been used to design acrylic block copolymers for wind energy applications. These toughening agents typically consists of a) a rubbery low T_g block designed to toughen the matrix and b) a resin miscible block or blocks, allowing excellent compatibility with the host thermoset. The structure of these block copolymers can be optimized to allow for rubbery domains with sizes of 10-100 nm, giving excellent toughening to thermoset materials without sacrificing strength or thermal properties.[2,3] The polymers can be tuned to assemble into non-spherical shapes, such as worm-like vesicles, further increasing fracture toughness with only low loading levels. Wind energy adhesives employ high amounts of thixotropic material and filler to increase viscosity. The use of triblock copolymer toughening agents can increase toughening and viscosity simultaneously. For infusion, low viscosity resin is required. By careful design of diblock copolymers, excellent increase in toughening can be achieved in epoxy, vinyl ester (VER) or unsaturated polyester (UPR) resins.

EXPERIMENTATION

Materials and Processing

Block copolymers (BCP) were synthesized using nitroxide mediated controlled radical polymerization (M53, BCP1, BCP2, LV1, LV2, LV3, and LV4) and anionic polymerization (E21). All polymers utilized a polybutylacrylate (PBA) rubbery block, except for E21which utilized a polybutadiene rubbery block.

E21 is a commercial polystyrene-b-polybutadiene-b-polymethylmethacrylate (PMMA) triblock copolymer. M53 is a commercial PMMA-b-PBA-b-PMMA symmetric triblock polymer. BCP1 and BCP2 are symmetric triblock polymers with a central PBA block surrounded by two resin miscible blocks consisting of polymethylmethacrylate (PMMA)–co-dimethylacrylamide (DMA). LV1 and LV2 are asymmetric diblocks consisting of a PBA rubbery block and one PMMA-co-DMA resin miscible block. LV3 and LV4 are asymmetric diblocks consisting of a PBA rubbery block and a miscible block based on polymethylacrylate (PMA).

For wind energy adhesives, bisphenol A epoxy resin was used and cured with a tri-functional polyetheramine curative (Jeffamine T403). CTBN was used in adduct form, diluting down from a 40% masterbatch in a bisphenol A epoxy resin.

Low viscosity infusion grade epoxy, vinyl ester and unsaturated resins were used for wind energy composite studies. Curing of the epoxy was carried out with a low viscosity amine curative. Curing of UPR and VER was done with a methyl ethyl ketone peroxide (MEKP) and cobalt initiator.

Block copolymers are initially dissolved in the uncured thermoset resin. For epoxy, the resin and block copolymers are heated together to 150-160 C followed by shear mixing. For UPR and VER, shear mixing at room temperature is used to dissolve block copolymers. Viscosity measurements are carried out using a parallel plate strain rheometer at a frequency of 1 Hz and a heating rate of 3 C/minute.

Testing

Samples are cured into bars for fracture toughness testing (6 mm by 12 mm by 75 mm) and dynamic mechanical analysis or plaques (3 mm thick), which are then cut by a router into dog bone specimens for tensile testing. Single edge notched beam fracture toughness measurements were performed according to ASTM D 5045-99. Dynamic mechanical analysis was performed on a strain rheometer with torsion rectangular geometry at a frequency of 1 Hz and a heating rate of 2 C / minute. Tensile measurements were performed in accordance with ASTM D 638-02. Broken fragments from mechanical testing were used for atomic force microscopy (AFM) imaging of the morphology of the block copolymer in the resin. Samples were microtomed to allow for imaging of the morphology in the bulk material, significantly away from the fracture surface. Tapping mode AFM was carried out in phase mode to distinguish the rubber region against the stiff thermoset matrix.

RESULTS

Wind Energy Adhesives

The one inch thick adhesive used to bond the two halves of a wind turbine blade together has extremely high requirements for static and dynamic fracture toughness. It has been previously shown that block copolymers can assemble into nanometer sized spherical or non-spherical rubber domains in thermoset resins.[2,3,4,5,6] These nano-rubber domains can greatly increase the fracture toughness of the resin without adversely affecting other properties. A bisphenol A epoxy resin cured with a trifunctional polyetheramine (a curative often employed for wind adhesives) was used as a model system to test the effect of block copolymers on wind energy adhesive systems. E21, M53, BCP1 and CTBN were tested in this system. Mechanical and thermal data is presented in table I with the morphology of the rubber toughening agents is shown in figure 1.

Table I. Use of CTBN and block copolymers in an epoxy matrix at 10% loading.

Polyetheramine cured	NEAT	10% E21	10% M53	10% CTBN	10% BCP1
K_{IC} (MPa.m$^{1/2}$)	0.85	2.59	2.63	2.01	1.88
G_{IC} (J.m^{-2})	350	4184	4243	1940	TBD
Tg by DMA (°C)	87	90	91	85	91

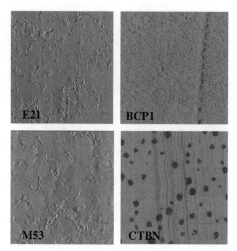

Figure 1. Tapping mode AFM phase image of rubber domains (dark regions) in a wind energy adhesive model system using CTBN or block copolymers additives at a 10% by weight loading.

All block copolymers modifiers show increased Tg as compared to the reactive rubber modifier. It is interesting to note that the spherical rubber particle, either micron sized with CTBN or nanometer-sized with BCP1, have similar effect on the fracture toughness of the resin. The "spider-web" morphology seen with E21 and M53 gives much better fracture toughness at equivalent loading. Previously, it has been observed with core shell modifiers, that slight aggregation of toughening agents can increase the fracture toughness of a resin.[7,8] The increased fracture toughness with increased Tg can potentially be of great interest to wind energy adhesive formulators for creating a tougher adhesive with increased fatigue life.

Wind Energy Composites

In order to allow the block copolymer toughening technology to be incorporated in wind turbine composites manufactured by infusion, it is necessary to greatly decrease the effect of block copolymer incorporation on resin viscosity. Two strategies were employed to decrease the effect on viscosity. First, the use of diblock polymers with only one miscible block is compared to the use of triblock polymers with two miscible blocks. Second, by changing the monomers used for the miscible block of the epoxy matrix, low MW, low Tg miscible blocks can limit the increase in resin viscosity.

Diblocks vs Triblocks

BCP1 and LV1 were designed as a triblock and its diblock equivalent. BCP1 is a triblock consisting of (PMMA-co-DMA)-b-PBA-b-(PMMA-co-DMA). LV1 is essentially BCP1 cut in half: same MW PMMA-co-DMA resin miscible block and ½ the MW of the PBA block in BCP1. In the epoxy resin, the immiscible PBA will self-assemble to form nano-micelles with dimension in the tens of nanometers. The PMMA-co-DMA blocks form the miscible shell of the micelle and will thus have the greatest effect on the viscosity of the resin.

Likewise, BCP2 and LV2 are triblock and diblock equivalent with higher molecular weight and higher content of dimethyl acrylamide than BCP1 and LV1.

Figure 2 shows the viscosity profile as a function of temperature for the triblock and diblock polymers when dissolved at 10% (by weight) in a bisphenol A epoxy.

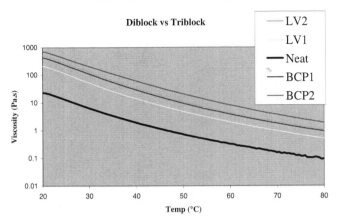

Figure 2. Comparison of effect of triblock (BCP1 and BCP2) and diblock polymers (LV1 and LV2) at 10% weight loading on the viscosity of an epoxy resin.

The data demonstrates that a significant difference exists between the viscosity effect of triblock and diblock polymers. One hypothesis for the increased viscosity of triblock polymers deals with the increased entanglement of chains in the PBA core triblocks as compared with diblocks whose free ends are in the core.

Effect of MW and Tg of the epoxy miscible block

In order to maintain nanostructured PBA domains during cure of the resin, the epoxy miscible block must be of a certain minimal length and polarity. The polarity and MW necessary to maintain structuration depends greatly on the type of resin and curative used.[3,6] By increasing the polarity of the epoxy miscible block, shorter MW blocks can be used and still maintain nanostructuration in the cured resin. This can be accomplished by replacing polymethylmethacrylate (PMMA) with polymethylacrylate (PMA) in the resin miscible block. (PMMA has a solubility parameter of 18.6 $(J/cm^3)^{1/2}$, while PMA is more polar with a solubility parameter of 20.7 $(J/cm^3)^{1/2}$. The MW of the thermoset miscible block has a significant effect on the viscosity of the resin as demonstrated in Figure 2. Simplistically, increasing the MW of the thermoset resin miscible block allows for increase chance of entanglement between two neighboring PBA nano-micelles.

It has also been seen, that, by decreasing the Tg of the thermoset resin miscible block, improvements in viscosity can be realized. For block copolymers with a Tg of the resin miscible block less than the measurement temperature, a decrease in viscosity is realized as compared to block copolymers using a resin miscible block with Tg above the measurement temperature, as shown in figure 3. The switch from PMMA to PMA again helps, as PMMA has a Tg of 105 C while PMA has a Tg of 10 C.

Viscosity of Diblock Copolymers in Epoxy

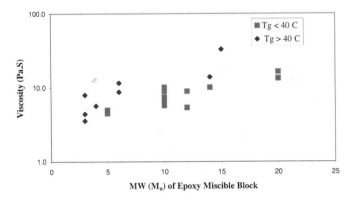

Figure 3. Comparison of the viscosity of epoxy resin with block copolymers containing resin miscible blocks of different molecular weight (M_n) and T_g above and below 40 C.

Using the above strategies block copolymers were synthesized using PMMA (LV1 and LV2) and PMA containing miscible blocks (LV3 and LV4), using the minimal miscible block MW necessary to achieve nanostructured PBA domains in cured thermoset resins. LV3 was designed with a co-monomer in the miscible block specifically for epoxies while LV1, LV2 and LV4 were designed with a universal co-monomer for epoxy, VER or UPR. Table II demonstrates the mechanical, thermal and rheological properties achieved in a bisphenol A epoxy resin cured with dicyandiamide (DICY) using triblock or diblock copolymer additives. It is seen that impressive reductions in viscosity with LV1 or LV3 can be realized while maintaining most of the mechanical advantages. (BCP 1 was selected for the control in this system as it gives an asymmetric structuration and maximum fracture toughness in the DICY cured system as compared to triblocks E21 and M53)

Table II. Use of triblock and diblock copolymers in an epoxy matrix at 10% loading.

Dicyandiamide cured	BCP1 10%	LV1 10%	LV3 10%
K_{1C} (MPa.m$^{1/2}$)	1.82	1.86	1.89
G_{1C} (J.m^{-2})	1867	1552	1778
Tg by DMA (°C)	135.4	128.4	127.7
Viscosity (Pa.s) at 40°C	28.6	14	7.45

Wind Blade Composite Infusion Resins

Testing was carried out in low viscosity infusion epoxy, vinyl ester and unsaturated polyester resins designed for wind blade composites. Table III shows the mechanical and rheological properties that are observed with a 5% loading of two low viscosity block copolymers in an epoxy infusion resin. Figure 4 shows the nano-scale morphology seen with block copolymers.

Table III. Use of diblock copolymers in an infusion epoxy resin at 5% loading.

	NEAT	LV1 5%	LV3 5%
K_{IC} (MPa.m$^{1/2}$)	0.98	2.50	2.14
G_{IC} (J.m^{-2})	481	3533	2276
Tensile Strain @ Break	8.68%	9.26%	10.03%
Tensile Stress @ Yield (MPA)	69	66	66
Tensile Modulus (MPA)	1289	1217	1161
Viscosity (Pa.s) at 40°C	0.28	0.82	0.57

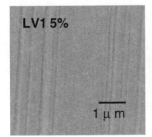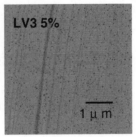

Figure 4. Short worm-like nano-micelles are seen with 5% loading of LV1, while perfect nanospherical rubber domains are seen with LV3 in an epoxy infusion resin.

Impressive improvements in fracture toughness are observed with low block copolymer loadings. The short asymmetric micelles with NM27 outperform the nano-spherical domains with XLV50. Work is ongoing to create more asymmetric block copolymers (short immiscible blocks with long immiscible blocks) to obtain the fracture toughness of LV1 with the viscosity increase of or below that of LV3.

Due to their excellent corrosion resistance, lower viscosity and lower price, vinyl ester resins and unsaturated polyester resins are competing against epoxy resins for use in wind blade composites. One of the primary drawbacks in VER and UPR is the brittle nature of the cured resin. Nanostructured block copolymers represent a promising approach to improve toughness without sacrificing other key properties.

Block copolymers were dissolved in a bisphenol A VER at room temperature with moderate shear stirring. Similar to epoxy, these block copolymers likely exist in the resin as nano-micelles of PBA (immiscible with the VER) surrounded by a resin miscible block (PMMA-co-DMA or PMA-co-DMA- both found to be miscible in cured VER depending on the DMA level). While the exact relationship between block copolymer composition and resin viscosity is still under investigation, it is believed that similar strategies as employed for epoxy resins will lead to lower viscosities. Due to this, the same block copolymers used for epoxy were investigated in VER.

In a low viscosity vinyl ester resin infusion system, diblocks LV1 and LV4 were trialed and compared to triblock BCP1 for fracture toughness and viscosity. Table IV shows the mechanical, thermal and rheological properties that are observed with a 5% loading of two low viscosity block copolymers. Figure 5 shows the nano-scale morphology seen with block copolymers.

Table IV. Use of block copolymer modified VER with 5% loading.

MEKP Cured	NEAT	5% BCP1	5% LV1	5%LV3	5% LV4
K_{1C} (MPa.m$^{1/2}$)	0.77	1.62	1.37	1.24	1.76
G_{1C} (J.m^{-2})	358	1506	1076	781	1952
Tg by DMA (°C)	120.2	120.1	119.1	TBD	117.5
Viscosity (Pa.s) at 25°C	0.15	3.35	1.2	TBD	1.76

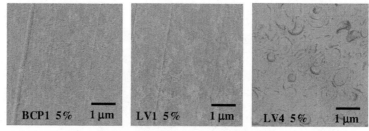

Figure 5. Fine filament structure is seen with 5% loading of BCP1 or LV1 in VER while vesicles structuration is observed with LV4.

Using BCP1 or its diblock partner, LV1, results in a resin with a fine rubber filament structure. To our best knowledge, this type of structure has not been observed previously using block copolymers in thermoset resins. Investigations are ongoing, but it is hypothesized that these block copolymers may be totally miscible (both PBA and PMMA/DMA blocks) in the uncured VER. This may be possible due to the high styrene monomer content present in the VER (Low MW PBA is miscible in styrene monomer). The fine filament structure may result from a reaction induced phase separation shortly before gelation of the resin.

With LV4, which contains higher MW PBA, it is likely that nanophase separation of the PBA domains occurs in the uncured resin or at a much earlier point in the curing process. LV4 shows a nano-vesicles structuration, consistent with what has been observed in the literature for highly asymmetric block copolymers with long immiscible blocks.[3,4]

The nano-vesicle structure present with LV4 provides the optimal in fracture toughness performance[3], closely followed by the fine filament structure with BCP1. However, the viscosity increase with the triblock polymer again is quite considerable, 22X as compared to the neat resin. Using LV diblock polymers, viscosity increases are limited to 8-12X. Tg of the VER drops very little (less than 3 C) with any of the block copolymers. Due to the very efficient toughening seen with 5% loading of the block copolymers, it was decided to investigate the fracture toughness and tensile properties with only 2.5% block copolymers after employing a heated post-cure on the sample. Results are presented in Table V.

Table V. Use of diblock copolymer modified VER (post-cured) with 2.5% loading.

MEKP Cured-Postcure	NEAT	2.5% LV1	2.5% LV4
K_{1C} (MPa.m$^{1/2}$)	0.87	1.22	1.64
G_{1C} (J.m^{-2})	375	679	1476
Tensile Strain @ Break	2.69%	9.11%	9.45%
Tensile Stress @ Yield (MPA)	34	78	69
Tensile Modulus (MPA)	1436	1397	1425
Viscosity (Pa.s) at 25°C	0.15	0.6	0.69

Impressively, using only 2.5% LV4, K_{1C} is increased by 90%, G_{1C} increased by 290% and elongation at break is increased by 250%. Tensile stress more than doubles and tensile modulus remains stable.

In a UPR infusion system, LV4 was again used to increase the fracture toughness of the resin with limited effect on viscosity. Table VI shows the improvement in mechanical properties as a function of loading level, while figure 6 shows the vesicle structure again attributable to highly asymmetric structure of LV4 with high MW immiscible block.

Table VI. Use of diblock copolymer modified UPR (post-cured) with 2.5% and 5% loading.

UPR Infusion Resin	Neat Resin	LV4 5%	LV4 2.5%
K_{1C} (MPa.m$^{1/2}$)	0.55	1.33	1.14
G_{1C} (J.m^{-2})	229	971	586
Tensile Strain @ Break (%)	3.5	/	6.9
Tensile Stress @ Yield (MPA)	46	/	58
Tensile Modulus (MPA)	1476	/	1323
Viscosity (Pa.s) at 20°C	0.2	1.11	0.57

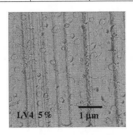

Figure 5. Vesicle morphology of LV4 in UPR infusion resin

CONCLUSION

Block copolymers provide a unique solution to increased reliability for wind turbine blades. Triblock polymers show outstanding increases in toughening as compared to competitive toughening technologies in a model wind blade adhesive formulation. By careful design of the architecture of the polymers, the increases of viscosity of the resin can be limited allowing for the adoption of the technology in infusion applications for wind blade composites. The technology is demonstrated to be effective in improving toughness in epoxy, VER and UPR systems at very low loading levels. Future work will focus on understanding the properties of filled composites with block copolymers, including quantifying improvements in interlaminar fracture toughness and fatigue lifetime.

REFERENCES

[1] R. Pearson, A. Yee, Influence of particle size and particle size distribution on toughening mechanisms in rubber-modified epoxies, *J. of Mat. Sci.,* **26**, 3828-3844 (1991).

[2] R. Hydro, R. Pearson, Epoxies Toughened with Triblock Polymers, *J. of Pol. Sci. B* **45**, 1470-1481 (2007_.

[3] S. Maiez-Tribut, J. Pascault, E. Soule, J. Borrajo, R. Williams, Nanostructured Epoxies Based on the Self Assembly of Block Copolymers: A New Miscible Block that can be Tailored to Different Epoxy Formulations, *Macromolecules* **40**,1268-1273 (2007).

[4] J. Dean, R.Grubbs, W. Saas, R. Cook, F. Bates, Mechanical Properties of Block Copolymer Vesicle and Micelle Modified Epoxies, *J. of Pol. Sci. B* **41**, 2444-2456 (2003).

[5] J. Wu, Y. Thio, F. Bates, Structure and Properties of PBO-PEO Diblock Copolymer Modified Epoxy, *J. of Pol. Sci. B*, **43**,1950-1965 (2005).

[6] V. Rebizant, A. Venet, F. Tournilhac, R. Girard-Reydet, C. Navarro, J. Pascault, L. Leibler, Chemistry and Mechanical Properties of Epoxy-Based Thermosets Reinforced by Reactive and Nonreactive SBMX Block Copolymers, *Macromolecules* **37**, 8017 (2004).

[7] R. Bagheri, R. Pearson, Role of blend morphology in rubber-toughened polymers, *J. of Mat. Sci.* **31**, 3945-3954 (1996).

[8] J. Qian, R. Pearson, V. Dimonie, O. Shaffer, M. El-Aasser, *Polymer*, **38**, 21-30 (1997).

Nanostrength is a registered trademark of Arkema, Inc.
Jeffamine is a registered trademark of Huntsman Petrochemical Corp.

DEVELOPMENT OF MULTIFUNCTIONAL NANOCOMPOSITE COATINGS FOR WIND TURBINE BLADES

Fei Liang[a], Yong Tang[a], Jihua Gou[a], Jay Kapat[b]

[a]Composite Materials and Structures Laboratory, Department of Mechanical, Materials and Aerospace Engineering, University of Central Florida, Orlando, FL 32816

[b]Center for Advanced Turbine and Energy Research, Department of Mechanical, Materials and Aerospace Engineering, University of Central Florida, Orlando, FL 32816

ABSTRACT

Wind energy has been growing at a rate of 25 to 30% annually, with installations in the U.S. now exceeding 10,000 MW in generation capacity, according to the American Wind Energy Association. The maintenances of wind turbines become a big issue and the optimization of the structural and aerodynamic properties of the blade is essential to the economic optimization of wind power production. The performance of the wind turbines can be significantly reduced by severely environmental conditions. Lightning strike seriously damages the blades, and results in accidents in which low voltage and control circuit breakdowns frequently occur in many wind farms. Vibrational damping is needed for the structural stability and dynamic response, position control, and durability of wind turbine blades. Surface erosion in desert, wind carrying large amounts of sands can erode the leading edge of a turbine blade and increase surface roughness, which deteriorate aerodynamic performance. In this paper, carbon nanofiber paper based nanocomposite coating was developed, which possessed excellent electrical conductivity, high damping ratio and good impact-friction resistance.

INTRODUCTION

Wind energy is a renewable energy source which produces no atmospheric pollution. As the fastest growing sustainable energy source, worldwide capacity of wind energy has reached 159,213 MW, out of which 38,312 MW were added in 2009, according to the World Wind Energy Association. Wind power showed a growth rate of 31.7 %, the highest rate since 2001. The maintenance of such large scale of wind turbine becomes a big issue which calls for the research into the environmental risks associated with the operation of large-scale commercial wind ventures. The structural and aerodynamic properties of the blades are fundamental to the efficient extraction of power from the wind. Ideal wind turbine blades are the combination of smooth surface which lead to aero-dynamical fluency and good mass distribution. For the weight consideration, glass fiber-reinforced plastics and carbon fiber-reinforced plastics are materials that mostly used for wind turbine blades. However, several problems are aroused by using these materials.

Firstly, it is easy for the tip of wind turbine blades to get stroked by lightning as its spine-like shape exposed on the open air. 85% of the downtime experienced by a second southwestern USA commercial wind farm was lightning-related during the startup period and into its first full year of operation. Direct equipment costs were $55,000, with total lightning-related costs totaling more than $250,000. In the area attacked directly on the blades, significant current concentration exists, which can heat and burn the area quickly. This is due to the poor conductivity of the fiber-reinforced plastics. In Japan, lightning damage to fiber-reinforced plastic blades in wind power generator has been increasing with the number of wind turbine generators installed in recent years [1]. Metallic coatings have been

325

used as the protection of non-conductive aircraft materials against lightning [2, 3]. Experimental results show that such a coating protects against lightning failure and damage [2]. However, a metallic coating suffers corrosion failure because of expose to chlorine in the marine environment. The surfaces suffer from pitting corrosion attack when an aluminum coating is exposed to an environment that riches chloride ion (Cl^-) [4–5].

Secondly, the smooth surface of blades can be greatly damaged by the erosion of wind carrying large amounts of sand and water droplets, which significantly deteriorates aerodynamic performance and reduces power output. The potential for erosion depends on the force at which the particulate matter impacts the airfoil which related to geometric shapes and relative velocities of both the airfoil and the impacting particles. Wind speed and rotational speed of the blade determine the impact velocity. According to the research of van Rooij and Timmer [6], the effect of roughness on a blade's aerodynamic performance depends on the geometric design of the blade. A blade may be adapted to induce minimal energy loss. However, surface roughness changes during operation due to erosion, and will typically always lead to unpredicted energy losses.

Damping properties of the wind turbine blades also should be taken into consideration. Resonance is greatly harmful for the wind turbine since it will reduce the efficiency of power output, especially at high speed rotating. Material with highly damping ratio is desirable.

In this work, carbon nanofiber paper based coating layer was developed, which possessed excellent electrical conductivity, high damping ratio and good impact-friction resistance. A unique paper made of carbon nanofibers and nickel nanostrands as surface layer on the composite panels increased the surface conductivity to provide a protection for lightning strike. To increase the impact-friction property, different types of ceramic nanoparticles, short carbon fibers, and nano-scaled graphite were incorporated into carbon nanofiber paper. In order to achieve good damping property, different fiber reinforcements were used.

EXPERIMENTAL

Materials

Pyrograf-III^{TM} carbon nanofibers (PR-PS-25) were supplied from Applied Sciences Inc., Cedarville, OH. The nanofibers have diameter of 50–100 nm and length of 30–100 lm. Nano-TiO_2 (Kronos 2310), alumina particles, short carbon fiber (Kureha M-2007s) and graphite flakes were incorporated into carbon nanofiber paper. The unidirectional carbon fiber mats, woven glass fiber, woven carbon fiber, and woven basalt fiber were used as fiber reinforcements. The unsaturated polyester resin supplied from Eastman Chemical Company was used as the matrix. The polyester resin was mixed with the MEK peroxide hardener at a weight ratio of 100:1.

Preparation of carbon nanofiber papers incorporated with other nano particles

For the study of lightning strike protection, both mono-layer and bi-layer papers were prepared with carbon nanofibers and nickel nanostrands. Three different types of papers labeled as CNFP-1, CNFP-2, and CNFP-3 were fabricated from the papermaking process, as shown in Table 1 and Table 2. The binders were used during the preparation of CNFP-1 and CNFP-2. The fabrication procedures of CNFP can be found in our previous research [7]. For the tribological study of carbon nanofiber paper, the coating layers with various compositions were made.

Table 1. Composition and structure of CNFPs with large size of 16"×16"

Sample	PR-PS-25 carbon nanofibers (g)	Nickel nanostrands(g)	CNFP structure
CNFP-1	9.75	9.75	Mono-layer paper with latex binder
CNFP-2	6.94	19.55	Bi-layer paper with latex binder
CNFP-3	6.94	19.55	Mono-layer without latex binder

Table 2. Composition of CNFPs containing different particles

Sample	PR-PS-25 carbon nanofibers (g)	Graphite flakes (g)	Short Carbon Fiber(g)	Nano-TiO_2 particles (g)	Nano-Al_2O_3 particles (g)
1	0	0	0	0	0
2	2.4	0	0	0	0
3	1.6	0.8	0	0	0
4	1.6	0	0.8	0	0
5	1.6	0.4	0.4	0	0
6	1.6	0	0	0.4	0.4
7	1.6	0.4	0.4	0.4	0.4

Manufacturing of carbon nanofiber paper based nanocomposites

Vacuum-assisted resin transfer molding (VARTM) process has been widely used to produce low-cost, high quality, and geometrically complicated composite parts. In this study, the VARTM process was used to fabricate the carbon nanofiber paper based nanocomposites, which were carried out in three steps. In the first step, glass fiber mats and carbon nanopaper sheet were placed on the bottom half of a mold. After the lay-up operation was completed, a peel ply, resin distribution media, and vacuum bag film were placed on the top of fiber mats. The vacuum film bag was then sealed around the perimeter of the mold and a vacuum pump was used to draw a vacuum within the mold cavity. The next step was the mold filling during which resin was sucked into the mold under atmospheric pressure. In the VARTM process, the distribution media provided a high permeability region in the mold cavity, which allowed the resin to quickly flow across the surface of the laminate and then wet the thickness of the laminate. Therefore, the dominant impregnation mechanism in the VARTM process was the through-thickness flow of resin. In the final step, the composite part was cured at the room temperature of 177°C for 24 hours and post-cured in the oven for another 2 hours at 100°C. In this study, composite laminates consisted of eight plies of fiberglass with a single layer of carbon nanopaper sheet on the surface.

Electrical resistivity measurement

In order to study their morphologies and network structures, the CNFPs were characterized with SEM (JEOL 6400F, at 5 kV). The samples were sputtered with about 10 nm of Au prior to SEM imaging. The electrical resistivities of the CNFP and composite panels were measured with the SIGNATONE QUADPRO system, incorporated with a four-point cylindrical probe. The four-point probe apparatus has four probes in a straight line with an equal inter- probe spacing of 1.56 mm. The probe needle radius is 100 lm. The electrical resistance of the paper, Rs, can be obtained from

$$Rs = 4.53 \times V \tag{1}$$

where I is a constant current that passed through the two outer probes and V is an output voltage measured across the inner probes with the voltmeter.

Lightning strike tests

Three 16"×16" panels were subjected to a Zone 2A (swept stroke) direct effects attachment using a peak current amplitude of 100 kA. For the Zone 2A test, all the areas of the aircraft surfaces where a subsequent return stroke is likely to be swept with a low expectation of flash hang on. During lightning strike tests, the composite panels were positioned with carbon nanofiber paper surface facing the lightning simulation hardware. The entire panel was heavily grounded along its perimeter to direct the current path. The experimental test rig was set up as shown in Figure 1.

Figure 1. Experimental rig for lightning strike tests.

Damping test of carbon nanofiber paper based nanocomposites

The regular composite beam and plate without carbon nanofiber paper and the nanocomposite beam and plate with carbon nanofiber paper were used as the specimens for damping test. For each beam, a PZT (lead zirconate titanate, a type of piezoceramic material) patch (20mm x 20mm) was attached on one side as an actuator to excite the beam and a smaller PZT patch (10mm x 8mm) was attached on the other side of the beam as a sensor to detect the beam's vibration, as shown in Figure 2. For each Plate, a 20mm × 20mm PZT (Lead Zirconate Titanate, a type of piezoceramic material) patch was attached at the center of the plate as an actuator to excite the plate and a smaller PZT patch (10mm× 8 mm) was attached at the corner of the plate as a sensor to detect the plate's vibration, as shown in Figure 3.

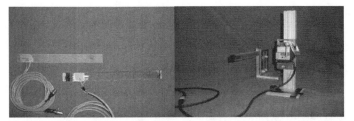

Figure 2. Regular composite beam and nanocomposite beam for damping test (left). Experimental setup for damping test (right).

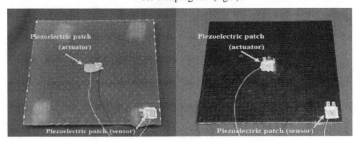

Figure 3. Regular composite plate for damping test (left), and nanocomposite plate with carbon nanopaper sheet for damping test (right)

Tribological test of fiber reinforced composites coated with carbon nanofiber paper

To evaluate many samples in a short amount of time, pin-on-plate testing was performed on a tribometer manufactured by Microphotonics (Irvine, CA). The schematic picture below (figure 4) shows pin-on-plate testing for tribology. Basically, in pin-on-plate mode, the pin (or ball) is loaded with dead weights and pressed against the plate (coated laminate) that is then spun at a given speed. During the test, the instrument gives real-time outputs of frictional force, coefficient of friction (COF), temperature of the pin/ball, wear depth, and wear factor for the sample.

Figure 4. The schematic picture of pin-on-plate testing for tribological application.

RESULTS AND DISCUSSION

Electrical resistivities of carbon nanofiber papers and composite panels

Table 3 shows the surface resistivities of the CNFP-1, CNFP-2, CNFP-3, and their composite panels measured from the four-point probe apparatus. Clearly, the CNFP-2 and the CNFP-3 had lower resistivities due to their high loading levels and uniform dispersion of nickel nanostrands within the paper. It was also found that the resistivities of the papers decreased slightly after they were integrated onto the surface of composite panels. The high pressure in the RTM process could make carbon nanofibers and nickel nanostrands more condensed within the paper, which led to the decrease in the resistivity of composite panels. These nanoparticles became more tightly packed and formed even better electrical pathway.

Table 3. Resistivities of top surface of CNFPs and composite panels.

CNFP identification	Conductivity (S/m)	Resistance (Ω/sq)	Panel identification	Conductivity (S/m)	Resistance (Ω/sq)
CNFP-1	3.33	116.40	CP-CNFP-1	222	17.34
CNFP-2	9090	0.42	CP-CNFP-2	31000	0.12
CNFP-3	8330	0.42	CP-CNFP-3	34100	0.10

Damage characterization of composite panels after lightning strike and ultrasonic testing of composite panels after lightning strike

The lightning strike tests were conducted on three large size composite panels with 16"×16"(CP-CNFP-1, CP-CNFP-2, and CP-CNFP-3). Figure 4 shows the surface damages on these composite panels after lightning strike. Clearly, the CP-CNFP-1 had the largest damaged area, where 5.9% in the area of the paper was damaged. The carbon fibers underneath the paper were obviously damaged. However, only 3.3% and 1% in the area of the paper were damaged for the CP-CNFP-2 and CP-CNFP-3, respectively. An ultrasonic testing was conducted on the composite panels in order to further confirm the damages caused by lightning strike. In the ultrasonic testing, the thickness readings were taken from the back of the panels to determine the thickness of the undamaged laminate. The thickness distribution of the undamaged laminate is shown in Figure 5. The 3D graphs clearly show that the CP-CNFP 1 had the most seriously damaged area and depth. It had a damaged area of 17 grids and each gird is 2:54 cm×2:54 cm. However, the CP-CNFP-2 and the CP-CNFP-3 had a damaged area of 10 and 7 grids, respectively. However, the CP-CNFP-2 had its damaged depth in the range of 0.15–0.18 cm. The CP-CNFP-3 had their damaged depth in the range of 0.08–0.10 cm. Based on the original thickness of the panels of 0.25 cm, the CP-CNFP-1 had its depth of 60–70 % damaged on the area of 110 cm^2. The CP-CNFP-3 was damaged in 30–40% of its thickness on the area of 45 cm^2.

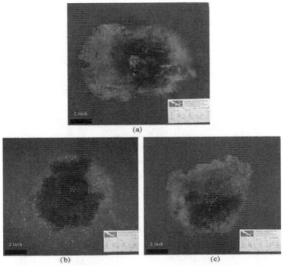

Figure 4. Surface damages of composite panels: (a) CP-CNFP-1; (b) CP-CNFP-2; (c) CP-CNFP-3.

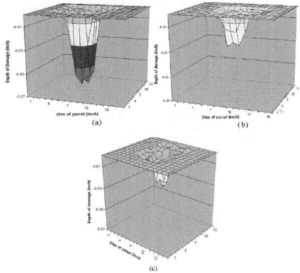

Figure 5. Damage area and thickness of composite panels after lightning strike test: (a) CP-CNFP-1;(b) CP-CNFP-2; (c) CP-CNFP-3

Damping properties of carbon nanofiber-enabled composite laminates

The damping test was conducted on the composite laminates with carbon nanofiber paper as mid-layer and surface layer. During the damping test, the sweep sinusoidal signals were used as excitation source for the PZT actuator to get the frequency response of the system. For the nanocomposite beam with carbon nanofiber paper as mid-layer, the time responses of both sweep sine excitations are shown in Figures 6 and 7, respectively. The peak value in the sweep sine response represents resonance at a certain natural frequency. From the sweep sine responses, it can be clearly seen that the peaks of first mode, second mode, and third mode are significantly reduced for the nanocomposite beam, which indicates that the nanocomposite beam has improved damping property.

To estimate the damping ratio for each mode, the half-power bandwidth method was used. Corresponding to each natural frequency, there is a peak in the magnitude frequency plot of the system. 3 dB down from the peak, there are two points corresponding to half-power point. A larger frequency range between these two points means a larger damping ratio value. The damping ratio is calculated by using the following equation:

$$2\zeta = \frac{\omega_2 - \omega_1}{\omega_n} \qquad (2)$$

where ω_1, ω_1 are the frequencies corresponding to the half-power point, ω_n is the natural frequency corresponding to the peak value, and ζ is the damping ratio. Table 4 shows the first three modal frequencies and associated damping ratio of the two beams. From the damping ratio comparison, it is clear that the damping ratio of the nanocomposite beam has increased up to 200–700% at the 2nd mode and 3rd mode frequencies. However, there is little change in mode frequencies, which means that there is slight change in the stiffness of the composites. This demonstrates an advantage of nanocomposite over regular composite with viscoelastic layers. The regular composites with viscoelastic layers will sacrifice in reduced stiffness, though damping is improved. For the nanocomposite beam with carbon nanofiber paper as surface layer, the analysis shows good agreement with the test data for the nanocomposite beam with carbon nanofiber paper as mid-layer, as shown in Figure 8. Therefore, it is concluded that the incorporation of carbon nanofiber paper could result in a significant increase in structural damping of conventional fiber reinforced composites. For the nanocomposite plate with carbon nanopaper sheet as the surface layer, the time responses of both the sweep sine excitation are shown in Figure 9. The peak value in the sweep sine response represents resonance at a certain natural frequency. From the sweep sine responses, it can be clearly seen that the peak of first mode, second mode and third mode are significantly reduced for the nanocomposite beam, which indicates that the nanocomposite plate has improved damping property.

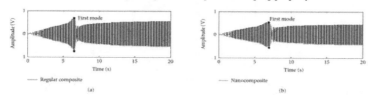

Figure 6. Sweep sine response (0.1 Hz to 100 Hz) of the composite beam without carbon nanofiber paper and the nanocomposite beam with carbon nanofiber paper as mid-layer.

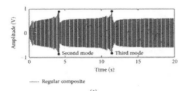
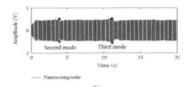

Figure 7. Sweep sine response (10 to 1000 Hz) of the composite beam without carbon nanofiber paper and the nanocomposite beam with carbon nanofiber paper as mid-layer.

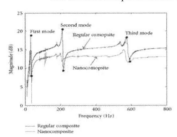
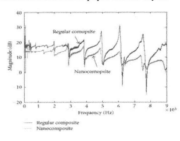

Figure 8. Frequency response of the first three modes for the composite beam without carbon nanofiber paper and the nanocomposite beam with carbon nanofiber paper as mid-layer (left). Frequency response of the first three modes for the composite beam without carbon nanofiber paper and the nanocomposite beam with carbon nanofiber paper as surface layer (right).

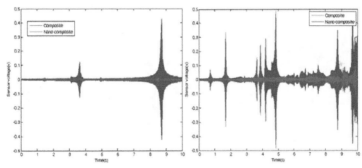

Figure 9. Sweep sine response (100-2000Hz) of the composite plate without carbon nanopaper sheet and the nanocomposite plate with carbon nanopaper sheet as surface layer (left), Sweep sine response (100-10,000Hz) of the composite plate without carbon nanopaper sheet and the nanocomposite plate with carbon nanopaper sheet as surface layer (right).

Table 4. Damping ratio calculated by half-power bandwidth method.

	1st mode frequency (Hz)	1st mode damping ratio	2nd mode frequency (Hz)	2nd mode damping ratio	3rd mode frequency (Hz)	3rd mode damping ratio
Regular composite beam	31.93	0.0278	210.8	0.0251	572.5	0.0349
Nanocomposite beam	34.60	0.0373	198.5	0.1985	558	0.1104

Morphology of coating layer for tribological application

SEM images of carbon nanopaper with different kinds of nanopaticles are showed in Figures 10-14. In Figure 10 (left), CNFs were well dispersed without any clusters which indicate that the coating layers are homogenous. Because of its strong and tough, carbon nanofiber can act as micro-crack reducer. In Figure 10 (right), the layered structure graphite flakes were evenly distributed into CNFs. Because of layered structure, graphite flakes propose self-lubricant ability which can act as friction reducer. In Figure 11 (left), micron sized short carbon fiber were incorporated with CNFs. Short carbon fiber has good compressive strength and creep resistance. Besides, it acts as a frame of the whole matrix system which greatly improves the stress transfer of the composites. In Figure 12, nano-Al_2O_3 and nano-TiO_2 were dispersed on the surface of CNFs. The surface hardness can be significantly increased by adding those ceramic nanoparticles. They also act as a spacer to protect matrix and create a rolling effect, then reduce friction coefficient. To study the synergistic effect of those particles, all particles mentioned above were mixed together with certain composition ratio. Figures 12 and 13 are the SEM images of coating materials after resin infusion. Good adhesion between resin and nanoparticle are obviously shown.

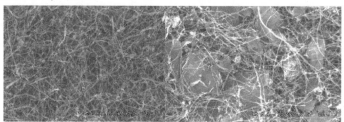

Figure 10. SEM images of pure CNF (left) and CNF with graphite flakes (right)

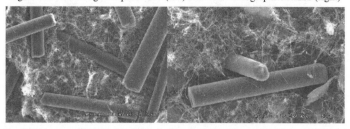

Figure 11. SEM images of CNF with SCFs (left) and CNF with graphite flakes, SCF (right)

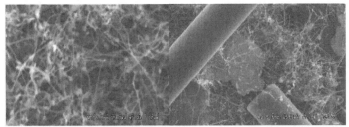

Figure 12 SEM images of CNF with Al$_2$O$_3$ and TiO$_2$ (left), and CNF with all above nanoparticles (right)

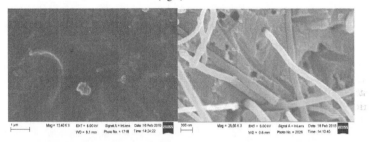

Figure 13. SEM images of pure resin (left) and CNF with resin (right)

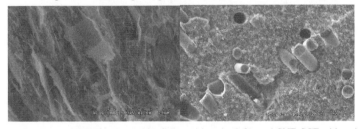

Figure 14. SEM images of CNF and graphite flakes with resin (left), and CNF, SCF with resin (right)

CONCLUSIONS

In this study, multifunctional carbon nanofiber paper was developed and optimized as platform material. Carbon nanofiber paper was integrated into polymer composites through liquid composites molding process. Carbon nanofiber paper can be used for lightning strike protection of wind turbines by incorporating Nickel nanostrands because the highly conductive surface could provide safe conductive paths. Our ongoing work is to align nickel nanostrands and carbon nanofibers to increase the surface conductivity of the paper. Vibrational test results have shown significant increase in damping ratios of the nanocomposite laminates, which is critical for the structural stability and dynamic response, position control, and durability of wind turbine blades. Carbon nanofiber paper can be used for erosion protection of wind turbines by incorporating other nanoparticles, which will provide a significant protection especially for wind turbines installed in desert.

REFERENCES

[1] M. Hanai, H. Ikeda, M. Nakadate, H. Sakamoto, IEEJ Trans. Power Energy 127- 3 (2007).

[2] M. Hanai, IEEE (2005) 289 1E-2-2F.

[3] H. Sakamoto, H. Kubo, Y. Hashimoto, I. Suzuki, Y. Ueda, M. Hanai, Proc. 2004 World Renewable Energy Conference, Denver, Colorado, (2004).

[4] B. Shaw, G. Davis,W. Moshier, G. Long, R. Black, J. Electrochem. Soc.138-3288 (1991).

[5] G. Davis, B. Shaw, B. Rees, M. Ferry, J. Electrochem. Soc. 138-3194 (1993).

[6] van Rooij RPJOM, TimmerWA. Roughness sensitivity considerations for thick rotor blade airfoils. Trans ASME;125:468–78(2003).

[7] Zhao ZF, Gou J, Khan A. Processing and structure of carbon nanofiber paper. J Nanomater 2009,7(2009).

Biomass

VOLATILITY OF INORGANICS DURING THE GASIFICATION OF DRIED SLUDGE

C.Bourgel*, J.Poirier*, F. Defoort**, J-M Seiler**, C. Peregrina***

* CEMHTI (conditions Extrêmes et Matériaux : Haute Température et Irradiation), 1D Avenue de la Recherche Scientifique, Orléans, 45000, France

**CEA, DTN, 17 rue des Martyrs, 38054 Grenoble, France

***CIRSEE, Suez Environnement, 38 rue du président Wilson, 78230 Le Pecq, France

ABSTRACT

The context of the present study relates to the gasification of dried sludge under high temperature.

The aim of this work is to shed new light on the impacts of sludge ashes in gasification process. The purification sludge can contain up to 50% of inorganic matter. The objective of this study is to understand the role of these inorganics during the heating.

Several techniques are used to solve this problem. First, using thermodynamic calculations (Factsage®) the evolution of the volatility of the inorganics is observed and the condensed phases formed during the heat treatment are determined. The simulations are done under atmospheric pressure condition, from 500 to 1500°C. Second, to compare with the calculus, an XRD and XRF in situ measurement experiment is developed in the CEMHTI Laboratory in order to determine which species volatilize.

The major inorganic elements of the sludge are Si, Al, Fe, P and Ca. The calculations allow us to determine the formed gases, and the condensed phases during heat treatment. The major formed gases are CO, H_2, H_2O, CO_2, CH_4 and N_2. The minor formed gases are H_2S, COS, HF, HCl and NH_3. On the contrary, Na, K, Fe and P elements don't volatilize entirely during the heating. Indeed, they are found in the condensed phases formed of sodium and potassium alumino-silicate, iron silicate and calcium phosphate. Finally, these results will be confronted to the experimental measurements.

INTRODUCTION

This study deals with the gasification of dried sludge under high temperature (i.e. at temperatures higher than 1000°C).

The aim of this work is to shed new light on the impact of municipal sewage sludge ashes in the gasification process. In fact, 30 to 50% of the dry solids of sewage sludge is composed of mineral matter. When it is heated at temperatures may cause sintering of the reacting sludge or produce inorganic vapors that must be removed away from the syngas, or corrode the refractory lining. The objective of this study is to understand the role of these inorganics during gasification.

Using a thermodynamic modeling code (Factsage®), the evolution of the volatility of the inorganics is studied and the condensed phases formed during the heat treatment are determined. The simulations are done under atmospheric pressure conditions, from 500 to 1500°C using oxygen enriched air gasification agent.

RESULTS AND DISCUSSION

In the municipal sewage sludge inorganic fraction, Si, Al, Fe, P and Ca are the major inorganic elements [1]. The calculations make it possible to determine which gases are formed, and which phases are condensed during the thermal treatment.

Two different sludge were chosen for the calculations. Their inorganic contents are presented in the table 1. The major difference comes from the amount of phosphorus which is much higher in the second than in the first sludge.

Element (mol)	Si	Ca	Fe	Al	P
Sludge 1	0.22	0.15	0.039	0.065	0.036
Sludge 2	0.101	0.085	0.045	0.052	0.087

Table 1 : composition of the major elements in two sludges chosen for calculations

The rest of the composition is consisted of C and O, but also : Mg, K, Na, Ti, Mn, B, S, Cu, Cd, Zn, Ag, Ba, Cr, Cl, F and Ni elements in very small proportions (less than 0.01 mol). All of them were taken into account for the next calculations.

In order to simulate an entrained flow reactor, the calculations are made for temperatures varying between 500 and 1500°C, under atmospheric pressure. The air is considered with partial oxygen enrichment. The conditions taken for the calculation are thus :

> 1 mol of biodry ($C_6H_9O_4$)
>
> Ashes (the equivalent of 100g of sludge)
>
> 2.5 mol of O_2 and 0.5 mol of N (which represent the air)

In this study, the aim is to understand the phenomena of volatility occurring during the gasification. The calculations will allow us to determine which elements evaporate and which ones remain in solid phase during the heat treatment. Moreover, the software used calculates the phases present at the thermodynamic equilibrium for each temperature of the heating. The results will thus distinguish the major gases (when the molar fraction of gas exceeds 0.05), the minor gas (it concerns all the other gases) and the condensed phases.

In order to know in which phase is each element (gas or condensed phase), it is interesting to calculate the release fraction F_i of each element (quantity of element i in the gas on its initial quantity).

The Figure 1-a and 1-b present the evolution of the release fraction of each element in gas in function of the temperature, respectively, on the sludge 1 and 2.

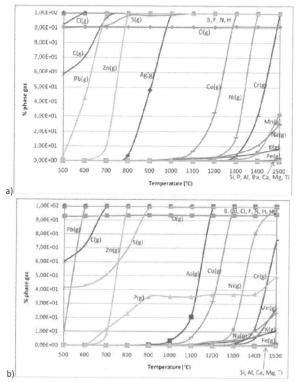

Figure 1 : Evolution of fraction of elements present in gas phase with the temperature. a) Calculation on the sludge 1. b) Calculation on the sludge 2.

On the figure 1-a, the elements B, F, N and H are completely in gas phase ($F_i=100\%$). On the contrary, the elements Si, P, Al, Ba, Ca, Mg and Ti are completely in condensed phase ($F_i=0\%$).
In opposition to the Cl, C, Pb, Zn, S, Ag, O and Cu elements which are in gas phase at 1300°C (range of temperature in an entrained flow reactor), the Fe and Mn elements are in form of condensed phase. The Ni, Cr, Mn, K and Na elements are partially in gas and condensed phases.
In the figure 1-b, the elements B, Cd, Cl, F, N, Hg and H are completely in gas phase ($F_i=100\%$) and the elements Si, Al, Ca, Mg and Ti are completely in condensed phase ($F_i=0\%$).
At operating temperature (i.e. 1 300°C) Cu, As, S, Zn, C, Pb and O elements are gas while Fe is condensed. Moreover, P, Ni, Cr, Mn, K and Na elements are partially in gas and condensed phases.

Note that, only in the second sludge, the P element is partially in gas phase and begins to evaporate at 700°C.

The Figure 2-a and 2-b present the evolution of the condensed phases with the temperature, respectively, on the sludge 1 and 2. The carbon C(s) is not indicated in the plot because of a scale problem (30g at 500°C and 21g at 600°C for the first sludge and 27g at 500°C and 18g at 600°C for the second sludge).

The quantity of solid phases is equivalent in both cases. Indeed, in figure 2-a, 24.8% of the mass is in solid phase at 500°C and decreases until 12.5% at 1500°C. In figure 2-b, 21.7% of the mass is in solid phase at 500°C and decreases until 8.3% at 1500°C.

This study shows also that the calcium is present in all major compounds for both cases. Moreover, in the figure 2-a, these calcium compounds contain Fe, Mg, P and Al elements. In the figure 2-b, the calcium compounds contain Al and P. The elements Mg, Fe and K are in alumino-silicate compounds.

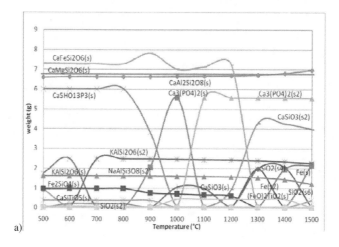

a)

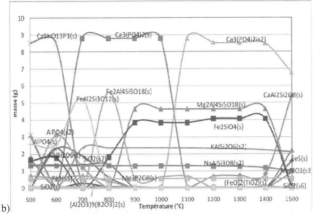

b)

Figure 2 : Evolution of condensed phases with the temperature. a) Calculation on the sludge 1. b) Calculation on the sludge 2.

The Figure 3-a and 3-b show the composition of the major gases in function of the temperature, respectively, on the first and second sludge.

In both cases, the CO, H_2, H_2O, CO_2 and CH_4 gases are formed in majority. This result is in accordance with the release of most of the organic matter of sewage sludge[2][3]. Among these gases, the formation of H_2 and CO exceeds the others from 650°C. Besides, the temperature range 700-900°C shows both a major release of H_2 and CO and a maximum LHV (lower heating value) value of the syngas. However, no equilibrium is reached. Finally, it can be expected that around 0.7-1.5% of the initial organic fraction of the dried sewage sludge is converted into tars and CH4 within this range of temperature [4].

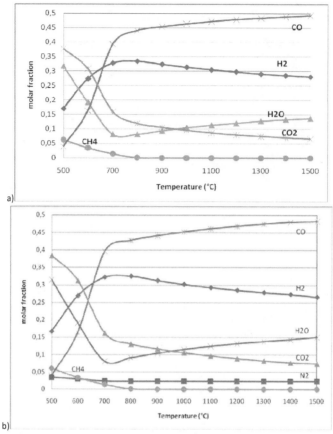

Figure 3 : Evolution of the molar fraction of the major gases with the temperature. a) Calculation on the sludge 1. b) Calculation on the sludge 2.

The Figure 4-a and 4-b present the evolution of the minor gas compounds in function of the temperature, respectively, on the first and second sludge. N_2 has the highest concentration and varies just a little with temperature (concentration ranging from 2.3 to 3.4 w.% in the Figure 4-a from 2.3 to 3.4 w.% in the Figure 4-b).

H_2S, COS, HF, HS, HCl and NH_3 are produced in the volatilization of both sludge 1 and 2. Their evolutions are similar in the two cases.

If one does not consider N_2, H_2S is the most concentrated gas. Moreover, its value, in Figure 4-a, is around 580 ppm and almost constant from 500°C. In Figure 4-b, the behavior of H_2S is the same than before, at a value around 1160 ppm. Then, the HCl and HF are also constant with, respectively, 220 and 120 ppm, in Figure 4-a and 440 and 170 ppm in Figure 4-b, between 500 and 1500°C.

During this simulation, a part of the initial chlorine is found in form of KCl(g) and NaCl(g). Indeed, these gases concentrations reach 50 and 60 ppm respectively in the first and second sludge, at 1500°C. Moreover, there is formation of NaOH and KOH. The contents of these are relatively high : at 1500°C, nearly 10 ppm of KOH and 11 ppm of NaOH, in Figure 4-a and nearly 5 ppm of KOH and 6 ppm of NaOH in Figure 4-b.

Only in the second sludge, the gasification of phosphorus is observed in form of $(P_2O_3)_2$, PO_2 and PO. Note that $(P_2O_3)_2$ attains 316 ppm from 700°C and almost reaches the H_2S amount (990 ppm) at 1500°C. The other P rich gases just attain some ppm from 1400°C.

In both cases, all the other gases are found with compositions under 100 ppm. NH_3 concentration decreases until less than 1 ppm at 1500°C.

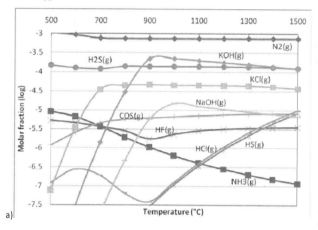

a)

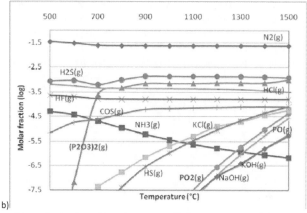

b)

Figure 4 : Evolution of the molar fraction of the minor gases with the temperature. a)Calculation on the sludge 1. b)Calculation on the sludge 2.

CONCLUSION

The major difference in the evolution of the release fraction of each element in the gas concerns the phosphorus. In fact, this element partially evaporates only for the sludge 2, having a gas with a composition of 20.6% of P, which is much higher than for sludge 1.

In both sludge, the major gases formed are the same and show similar behaviors.

As for the minor gases, the formation of H_2S, COS, HF, HS, HCl and NH_3 is observed in both types of sludge. Moreover, Na and K vaporize in form of KCl, NaCl, KOH and NaOH in both calculations. Nevertheless, the evaporation of the phosphorus takes place in the second sludge in form of PO_2, PO and $(P_2O_3)_2$. This last one appears from 700°C and overtakes all the others.

Focusing the condensed phases, the calcium is present in all major compounds. Indeed, Mg rich calcium compounds are found in both sludges. The results of the calculations show the presence of Al and P elements in the calcium compounds. The elements Mg, Fe and K, are, as for them, either in calcium rich compounds, or in alumino silicate ones.

In order to bring to light the phosphorus volatility, which may cause corrosion of the refractory lining, a comparison between the calculus and an XRD and XRF in situ measurement experiment will be done. First, by focusing the ternary system $CaO-SiO_2-P_2O_5$. Indeed, the calcium and silicone are the major elements of the inorganics in the sludge. Several samples of this system will thus been synthesized to observe the phosphorus behavior during the heat treatment. Then, when the ternary system will be understand, each element of sludge will be add one by one to observe their influence on the P volatility until to reach the complex compositions of the sludges.

REFERENCES

[1] Large, M. Peregrina, C. 2009 (Characterization of the mineral part of sewage sludge).
[2] Petersen, I. (2003) Shaker Verlag. 136 pages.
[3] Rezaiyan, J. and Cheremisinoff, N.P. (2005) Gasification technologies. 336 pages.
[4] A. Adegoroye, N. Paterson*, X. Li1, T. Morgan, A.A. Herod, D.R. Dugwell, R. Kandiyoti. 2004 Fuel 83 (2004) 1949–1960

CATALYSTS AND SORBENTS FOR THERMOCHEMICAL CONVERSION OF BIOMASS TO RENEWABLE BIOFUELS—MATERIAL DEVELOPMENT NEEDS

Singfoong Cheah[1], Stefan Czernik, Robert M. Baldwin, Kimberly A. Magrini-Bair, and Jesse E. Hensley
National Bioenergy Center, National Renewable Energy Laboratory, 1617 Cole Blvd., MS 3322, Golden, CO 80401, USA

ABSTRACT

Rising world demands for oil and a finite petroleum reserve have renewed interest in alternate liquid fuel sources that are diverse, secure, and affordable. Biomass is a renewable source of carbon that is abundant in many regions of the world. A focal point in alternative fuel research is the production of biofuels derived from lignocellulosic material and especially biofuels made with waste biomass or feedstocks grown on marginal land. Several different routes of thermochemical biomass conversion, including production of bio-oil through pyrolysis with subsequent bio-oil upgrading and production of syngas through gasification with subsequent gas to liquid fuel synthesis will be discussed. The material development needs in each stage will be identified.

INTRODUCTION

The development of alternative and/or renewable fuels is of high priority because of rising oil prices, national security concerns, and the need to reduce greenhouse gas emissions. A promising path to near-term renewable fuel production is through conversion of lignocellulosic feedstocks such as corn stover, switch grass, and wood chips to mixed alcohols or hydrocarbon fuels.

Several pathways are available to convert lignocellulosic biomass into renewable biofuel. These include biochemical,[1,2] aqueous phase processing,[3] and thermochemical conversions[1,4,5]. Biochemical conversion of biomass to ethanol proceeds through a number of steps, including steam, acid, alkali, and/or enzymatic pretreatment, saccharification, and fermentation.[1,2] Aqueous phase processing and reforming convert the sugars produced from biomass deconstruction to hydrocarbon fuels.[3] Thermochemical conversion can occur through the gasification and pyrolysis pathways. They have the potential to be cost competitive with gasoline and are the focus of this review.[1,4]

In gasification, biomass is pulverized at high temperature in steam or oxygen, producing a product gas that consists of syngas (CO and H_2), CH_4, CO_2, tar, and inorganic impurities such as H_2S, HCl, and NH_3.[1,4] The inorganic impurities must be removed from the product gas as they can corrode equipment and poison downstream catalysts (Figure 1). Tar must also be removed to protect equipment, to minimize hazardous waste effluents, and to maximize yields of fuels produced downstream.[4] Once cleaned, the product gas can be used to produce liquid fuels like mixed alcohols or gasoline/diesel/jet fuels. The fundamentals of biomass gasification are similar to those of coal gasification, however, significant differences in coal and biomass compositions (e.g., oxygen and sulfur content)[6] warrant the development of different materials and processes.

When biomass is heated to temperatures in the range of 400–600 °C at rates approaching 1000 °C/s in the absence of oxygen and added catalysts, a material known as pyrolysis oil or bio-oil is produced in high yield (up to 70% of the energy of the biomass feed).[5] The crude bio-oil contains high levels of oxygen, rendering it low in energy density, unstable, and incompatible with standard petroleum fuels.[7] The bio-oil must be upgraded to be used directly or to be blended with gasoline or diesel. There are several methods and pathways to upgrade bio-oil, including hydrotreating, hydrocracking, hydrothermal conversion, and vapor-phase upgrading (Figure 2). These pathways can

---

[1] Corresponding author, e-mail: Singfoong.cheah@nrel.gov

be followed sequentially or in parallel, with parallel processing conducted after the bio-oil is fractionated into pyrolytic lignin and water-soluble organic phases. Finally, bio-oil with improved properties can be produced by using a catalyst in the pyrolysis step, a process known as catalytic pyrolysis. A recent study identified potential processing routes and conducted cost and life cycle analyses for the conversion of bio-oil into transportation fuel.[8]

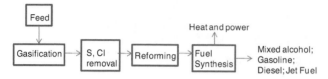

Figure 1. The gasification pathway for converting biomass to liquid fuel.

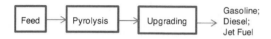

Figure 2. A simplified pyrolysis pathway for converting biomass to liquid fuel.

All of the major thermochemical conversion steps involve catalysis, and a critical requirement for process and economic viability is the development of highly efficient catalysts that are optimized for the chemical feeds and intermediates present in biomass conversion. In the following sections, material development needs for catalysts used in bio-oil upgrading, sorbents used in syngas cleanup, and catalysts used in reforming and mixed alcohol synthesis are discussed. Due to space limitations, some processes like Fischer-Tropsch synthesis and aqueous phase reforming are not included in this review.

CATALYSTS FOR CATALYTIC PYROLYSIS AND BIO-OIL UPGRADING

Although bio-oil produced from pyrolysis of biomass superficially resembles heavy fuel oil, it contains large amounts of oxygen (about 50 wt%). The major groups of oxygen-containing compounds are water (15–30 wt%), carboxylic acids, sugars, hydroxyaldehydes, ketones, and oligomers of phenolic compounds.[7] Oxygenates impart several undesirable characteristics to the bio-oil including:

- Acidity (as measured by total acid number or TAN): petroleum crude oils generally have TAN<1 while bio-oils can have a TAN of 100–200;
- Instability in storage and transport: condensation reactions that create larger molecules in the oil can increase the oil viscosity with time;
- Low volatility in distillation: the carbon residue upon fractional distillation is typically 50 wt% or greater and the residue content increases with time.

Therefore, oxygen elimination is necessary to render bio-oil useful as an intermediate or finished liquid fuel that is compatible with existing refining and distribution infrastructure. Bridgwater described two types of processes that have been used to remove oxygen from pyrolysis oil: catalytic hydrodeoxygenation (HDO) and catalytic vapor-phase upgrading.[9] Hydrodeoxygenation uses hydrogen to remove oxygen as water while catalytic vapor-phase upgrading removes oxygen as carbon oxides. Both reactions produce hydrocarbons. Several reviews on hydrotreating[10] and catalytic upgrading[11,12] provide details on the principles, conditions, and product distributions of these processes.

Catalytic Hydrodeoxygenation

The HDO reaction can be represented by Reaction (1). The exact composition of the bio-oil depends on pyrolysis conditions and feedstock. The stoichiometries in Reactions (1) are representative of real bio-oil upgrading reactions but should be regarded as conceptual, for illustration only. Water is included in the conceptual bio-oil formula but it is written in a form where the water is "separated" ($\cdot 2H_2O$) to show that the water molecules in bio-oil do not take part in hydrotreating and do not consume hydrogen.

$$C_6H_8O_3 \cdot 2H_2O + 6H_2 \rightarrow C_6H_{14} + 5H_2O \qquad (1)$$

The catalysts and process conditions currently used for HDO are similar to those used in the petroleum hydroprocessing operations of hydrodesulfurization (HDS), hydrodenitrogenation (HDN), hydrotreating, and hydrocracking. Typical petroleum hydroprocessing catalysts include materials such as nickel-molybdenum (NiMo) or cobalt-molybdenum (CoMo) on a high-surface-area alumina supports. When used as sulfides, these catalysts are very effective in HDS, HDN, and hydroprocessing of petroleum-derived intermediates.[13] Extensive research on catalytic HDS has yielded catalysts that selectively abstract sulfur from heterocyclic organo-sulfur compounds without substantial saturation of aromatic moieties, resulting in favorable product streams and decreased hydrogen consumption. However, crude petroleum contains very low quantities of organic oxygen (often less than 0.1 wt%), so oxygen removal has not been thoroughly researched.

With its high oxygen content, crude bio-oil is chemically very different from crude petroleum. When catalytic HDO is applied to fast pyrolysis oil using the catalysts developed for petroleum, the oil cokes severely in a single stage process.[14] One solution to the coking problem was the development of a two-stage process, though at the cost of high hydrogen use.[15,16] In the two-stage process, the oil was stabilized via saturation of C=C and C=O bonds in a lower temperature reactor (150–280 °C) before it was fed to a high temperature reactor (350–400°C) where the majority of the oxygen removal took place. This process yielded 37 – 44% liquids (v/v) and consumed 85 m³ (3,000 standard cubic feet) of hydrogen per barrel of feed at pressures of 2000 psig. The product had a reduced oxygen content of 0.5 – 2.3% (w/w). Up to 87 vol% of the product distilled in the gasoline range. Similar studies have also reported substantial reduction in oil acidity.[17-19]

It has become clear that the catalytic functions for sulfur and nitrogen abstraction will not be as efficient for organic oxygen removal; hence, this is an area where research is needed. Development of a catalyst that can minimize both hydrogen usage and coke formation with oxygenates feedstocks is necessary. In addition, the nature of the support materials used for HDO catalysts is an area of significant interest. Materials such as γ-alumina (which are often used to support petroleum hydrotreating catalysts) are unstable in the high water environment typical of bio-oil hydrotreating. Recent research at the National Renewable Energy Laboratory has found traces of alumina and nickel in hydroprocessed bio-oil when conventional alumina-supported NiMo catalysts were used.[20] Other support materials that are stable in high steam conditions have been studied, including carbon, ceria, and zirconia and all have been found to be effective. In a recent study, Pt on carbon and Ru on TiO_2 catalysts were found to be superior to $NiMo/Al_2O_3$ in both mild and deep hydrotreating.[21] Additional research on catalyst formulations and support materials is needed to develop catalysts that are active, selective, and cost effective in HDO of bio-oil.

Vapor Phase Upgrading

Vapors produced from pyrolysis can be deoxygenated and upgraded using catalysts and reaction conditions that are dramatically different from those used in catalytic HDO. The exact formula of the pyrolysis vapor depends on experimental conditions. Reaction (2) conceptually represents the best case scenario for deoxygenation of pyrolysis vapors through selective decarboxylation, where two oxygen atoms are removed per carbon atom lost to gaseous products:

$$C_6H_8O_3 \rightarrow C_5H_6 + CO_2 + H_2O \tag{2}$$

The yield of hydrocarbons can theoretically reach 50 wt% of the pyrolysis vapor (with yield defined as mass of hydrocarbon per mass of pyrolysis vapor) or 35 wt% of the biomass feedstock. This is less than the yields from hydrotreating, but the vapor upgrading process has the advantage of not requiring hydrogen and occurring at close to atmospheric pressure conditions, which are both economically favorable.

Small-pore zeolites such as ZSM-5 have been the most effective in transforming biomass pyrolysis vapors to hydrocarbons. In the temperature range of 350 – 450 °C, and with catalyst, oxygenates in the bio-oil can undergo cracking, dehydration, decarboxylation, aromatization, alkylation, condensation, and polymerization reactions. The product is a mixture of aromatic hydrocarbons and low molecular weight olefins. Deoxygenation of pyrolysis oils using zeolites gave hydrocarbon yields of 12–15%.[20] High coke production and rapid catalyst deactivation were reported, consistent with hydrogen deficiency in the biomass pyrolysis oils[22-24] (the effective hydrogen index of pyrolysis oil is 0.3, with effective hydrogen index being defined as $(H/C)_{eff} = (H - 2O - 3N - 2S)/C$ where H, C, O, N, and S are the number of moles of hydrogen, carbon, oxygen, nitrogen, and sulfur respectively[11]).

The National Renewable Energy Laboratory pioneered an integrated process in which vapors from a biomass pyrolyzer were fed to a catalytic reactor containing ZSM-5 and converted to aromatic and olefinic hydrocarbons.[25,26] Using a pilot-scale vortex reactor in series with a fixed-bed of commercial Mobil MCSG-2 catalyst at 450 °C a total hydrocarbon yield of 12.7% (based on wood feedstock) was achieved, or about one third of the theoretical yield. Excessive coking was observed, which was attributed to the high acidity of the HZSM-5 catalyst, and dehydration reactions that are catalyzed by acidic groups.

A number of studies have been conducted on individual classes of compounds to elucidate which oxygenates are most challenging to upgrade. Using a fixed bed of ZSM-5 catalyst, high conversions (>90%) were obtained for alcohols, aldehydes, ketones, acids, and esters while phenols and ethers remained mostly unconverted. Alcohols and ketones produced high yields of aromatic hydrocarbons.[27] Acids and esters produced permanent gases, water, and coke with low yields of hydrocarbons. Coke formation was the highest and hydrocarbon formation the lowest with oxygenates that have low effective hydrogen indexes.

Recent promising development includes co-processing of biomass pyrolysis vapors with compounds of high effective hydrogen index like methanol and alkanes.[11] Catalytic vapor phase upgrading is still less developed than hydrotreating and has potential for greater improvement if more effective catalysts are invented.

Catalytic Pyrolysis

Catalytic pyrolysis is a process where a catalyst is used during pyrolysis, with the objective of producing a bio-oil that has been substantially upgraded and deoxygenated. Catalytic fast pyrolysis has been studied in batch reactor systems, circulating fluid bed reactors, and bench-scale bubbling bed systems. Lappas et al. reported on catalytic fast pyrolysis of biomass with ZSM-5 and found improved yields of liquids compared to non-catalytic pyrolysis.[28] Carlson et al. carried out investigations of the effect of heterogeneous catalysts on the pyrolysis of cellulose and glucose in a pyroprobe batch reactor, and found that at very high catalyst loading (c.a. 20/1 catalyst to biomass by weight) a hydrocarbon product containing a significantly high content of aromatic hydrocarbons could be produced.[29] The same research group also reported that catalytic fast pyrolysis of glucose involves two steps.[30] The first is the decomposition of glucose and the second is the formation of aromatics within the pores of the zeolite catalyst.[30]

Other research reported reduced coking when zeolites were replaced by transition metal oxides, though at the expense of less deoxygenation during pyrolysis.[31] In a recent paper, French and Czernik found that the highest yield of hydrocarbons (approximately 15 wt%, including 3.5 wt% of toluene) was achieved using nickel, cobalt, iron, and gallium-substituted ZSM-5.[32]

KiOR is attempting to commercialize catalytic pyrolysis using a proprietary catalyst,[33] indicating the promise of this process for commercial applications. However, the experimental yield of catalytic pyrolysis is still low, and there is a significant need for more effective catalysts to enhance the production of hydrocarbons and to reduce coke formation.

HIGH TEMPERATURE SORBENTS FOR SYNGAS CLEAN UP

During biomass gasification, inorganic content is partially transformed into gaseous forms including H_2S, HCl, NH_3, HCN, and alkali metals.[34-38] Organic sulfur and nitrogen heteroatom species like thiophene and pyridine are also produced. The levels of sulfur and chlorine in the gasification product gas are 20–600 ppmv[34,36,38] and 1–200 ppmv,[36,38] respectively. Hydrogen sulfide and HCl are harmful to downstream catalysts and need to be removed. Hydrogen sulfide can be removed via amine scrubbing or ZnO sorbents. Amine scrubbing occurs below 100 °C and ZnO-based sorbents typically operate below 300 °C because higher temperatures cause zinc metal vaporization from ZnO.[39,40] Typical temperatures for tar and methane reforming, however, are in the range of 700–900 °C. Consequently, thermal efficiency and economics will improve with H_2S and HCl sorbents that function at high temperatures, so that gas need not be cooled out of the gasifier and reheated for reforming.[41]

The development of high temperature sorbents has historically focused on syngas derived from coal gasification, with the development of sorbents for H_2S being most active.[42-48] The materials being developed as high temperature sulfur sorbents remove H_2S via surface adsorption or reaction of a metal oxide with gas phase H_2S according to the general reaction:

$$M_xO_y\,(s) + yH_2S\,(g) \Leftrightarrow M_xS_y\,(s) + yH_2O\,(g) \qquad (3)$$

where M denotes a metal. While syngases derived from coal and biomass share similarities, there are also significant differences. Because of differences in gasification temperature (biomass gasification generally occurs at lower temperature than coal gasification), biomass syngas contains hydrocarbon (> 10 %)[41] and tar, both of which can have negative effects on sulfur and chlorine sorbents. In addition, both direct oxygen-blown and indirect steam-driven gasifiers are being developed and operated with biomass. Syngas produced from indirect steam-driven gasifiers are rich in steam (30–65 vol%). Because high level of steam is thermodynamically less favorable for the forward reaction in Reaction (3), the effectiveness of H_2S removal in high steam-containing syngas at high temperature can be impacted.

Westmoreland et al. conducted thermodynamic modeling and kinetic studies of a series of metal oxides and carbonates of barium, calcium, cobalt, copper, iron, manganese, molybdenum, strontium, tungsten, vanadium, and zinc. These studies indicate that a number of oxides are good candidates.[49,50] However, some of these materials are unstable and lose activity in the high temperature, highly reducing gasification product.

Zinc oxide, a material that has been widely used for sulfur removal from natural gas, is known to undergo reduction, then vaporization of zinc metal in high temperature syngas, resulting in loss of both material and activity.[39] Iron-, cobalt-, and copper-oxide based materials readily reduce in syngas at high temperature.[51] Metallic copper is thermodynamically less active in its reaction with H_2S than copper oxides. In the development of sorbents targeted for coal syngas applications in the last two to three decades, the focus has been on producing high capacity, regenerable materials that are stable in high temperature, reducing environments. The research efforts found promising sorbents based on zinc titanate, copper- and iron-containing mixed oxides, manganese-, and rare earth-based materials.

Zinc titanate, whose active phase is not easily vaporized, is one of the materials developed to address the zinc vaporization problem.[45-47] Copper chromite ($CuCr_2O_4$), a material that has the lowest reducibility of all the copper-oxide containing compounds, was specifically developed to address the copper oxide reduction issue.[52,53] Other developments include investigations into manganese and rare earth based materials that are especially suitable for high temperatures.[41] The investigation into the mechanisms of desulfurization of substituted and unsubstituted rare earth oxides indicates that the reduced form of cerium oxide is more effective in sulfur sorption. Consequently copper, which promotes the reduction of ceria, was incorporated into the formulation to improve the effectiveness of ceria as sulfur sorbent.[48,52]

All of these newly developed materials are good candidates for further research and testing at the pilot-scale. Other research priorities for high temperature sorbent materials include compatibility with other impurities in the syngas, regenerability, and an ability to remove multiple impurities.[41] While it is generally accepted that there are predominantly one or two mechanisms for H_2S sorption (based on Reaction 3 or surface adsorption reactions), there are multiple regeneration schemes and mechanisms depending on the gases used for regeneration and the process design.[41] There is a significant economic advantage with a sorbent that is efficient at removing multiple contaminants, but developing a good regeneration procedure for multiple contaminants can be complicated and will require considerable additional research.

CONDITIONING BIOMASS DERIVED SYNGAS

Purification of biomass-derived syngas is one of the major technical challenges facing thermochemical conversion of biomass to fuels. Tars and other unwanted hydrocarbons need to be removed or catalytically converted into useable species in order to protect downstream fuel synthesis catalysts and to efficiently utilize the carbon and hydrogen in biomass. Catalytic tar reforming provides both of these functions by converting unwanted hydrocarbon and tar species to additional product gas (CO, H_2, and CO_2).

Tar cracking consists of several reactions (steam reforming, dry reforming, and partial oxidation) which occur to different extents depending on the catalyst and reaction conditions. Most steam reforming catalysts are metal based with the most commonly used catalysts comprising nickel supported on metal oxides. Commercial nickel/alumina (Ni/Al_2O_3) catalysts have been used extensively in the petroleum industry to reform methane and naphtha and have been evaluated with both model and raw syngas[54] though coking, sulfur deactivation and attrition present significant operational problems.[55] For biomass derived syngas, catalytic tar cracking can be performed during gasification by placing catalysts directly in the gasification unit[56,57] or after gasification using a separate, downstream reactor. Either configuration requires robust catalysts that can withstand process temperatures to > 800°C, steam, and resist coking.

To improve reforming activity, a variety of metals have been evaluated in raw or model syngas to reform its significant hydrocarbon components (benzene, naphthalene, and methane).[55,58,59] Recent work evaluated Rh, Ru, Pt, Pd, and Ni supported on CeO_2-SiO_2 for steam reforming of biomass derived tars and found the order of activity to be Rh > Pt > Pd ~ Ni > Ru at 650 °C.[60] Unfortunately rhodium and platinum based catalysts are hindered by their high cost. As a result, most studies have focused on Ni-based catalysts, and for these, deactivation due to coking, attrition from fluidization, sintering, and sulfur species in the syngas must be minimized before commercialization.

Recent work has focused on enhancing nickel's reforming properties by addition of a second metal to form bimetallic supported catalysts. Several different approaches--synthesis of surface gold-nickel alloy,[61] or tin-nickel alloy,[62] or addition of small amounts of ruthenium,[63] or boron[64] to nickel--have been found to decrease coke deposition on nickel catalysts. Copper also increases reducibility and slows sintering of nickel alumina catalyst used for auto-thermal reforming of ethanol.[65] Overall, the addition of a second metal to nickel can improve its performance through enhanced reducibility, decreased coking, and decreased rate of sintering.

Catalyst support materials provide mechanical strength/integrity and surfaces on which to disperse active catalytic materials, and may also play a chemical role in catalysis. The acidity/basicity, surface area, pore structure, and electronic structure of the support can affect the reducibility, metal dispersion, mechanical strength, and the overall nature of the active sites on a catalyst. These parameters influence the activity of the catalyst and can be manipulated for catalyst design. For catalytic gasification (tar cracking during gasification) applications, promising support materials include natural materials such as dolomite and olivine.[56,57] Another study investigated the effect of support materials such as Al_2O_3, ZrO_2, TiO_2, CeO_2, and MgO on tar cracking in cedar-derived syngas.[66] It was found that support materials have large impacts on catalyst dispersion, conversion rate, and coke formation. To address the issue of coke formation, fluidized tar reforming operation has been experimented and a durable nickel based catalysts with a spherical α-alumina support that withstands the harsh environment of fluidized tar cracking has been developed.[67]

To advance catalytic conditioning technology towards commercialization, future work needs to address extended reforming and regeneration operations in closely modeled or actual biomass derived syngas to provide the data for industrial applications. The effects of poisons and trace impurities also need further investigation. With raw syngas, the primary limitation of hot gas catalytic tar cracking is catalyst deactivation via sulfur poisoning from H_2S and organosulfur species.[67-70] Among the other inorganic contaminants present in biomass-derived syngas (HCl, alkali metals, NH_3, etc.), several are typical catalyst poisons and their impact on catalyst performance should be evaluated.

CATALYSTS FOR SYNTHESIS OF ETHANOL AND HIGHER ALCOHOLS FROM SYNGAS

After the syngas is cleaned and conditioned, it can be used for fuel production, and mixed alcohol is a fuel that has shown promise. Known mixed-alcohol synthesis catalysts show potential for high productivity and/or selectivity to ethanol, but suffer from various complications that degrade overall process economics. Many require extreme pressures (e.g. homogenous catalysts and sulfide catalysts), difficult catalyst recovery (homogenous catalysts), low activity (some Rh-based catalysts, modified Fischer-Tropsch (FT) catalysts), or addition of hazardous materials that increase separation duty downstream (sulfide catalysts). None of the currently available catalysts produce the highest yield of alcohols at highest selectivity at lowest temperature and pressure. Typically, productivity increases with temperature and pressure, and selectivity increases with decreased temperature and lower conversion. Heterogeneous catalysts that convert syngas to ethanol at lower temperatures and pressures with high productivity are still needed. Two recent reviews highlight the history and current status of mixed-alcohol synthesis catalysts.[71,72]

Mechanisms for Higher Alcohol Synthesis

To understand the challenges of developing an improved mixed-alcohol synthesis catalyst, a study of the mechanisms for higher alcohol synthesis is useful. Many mechanisms for higher alcohol synthesis have been proposed. Several well-accepted studies in this area have come from Klier and group,[73-75] with other research supporting their proposed mechanisms.[76] It is generally thought that modified methanol synthesis catalysts produce alcohols via coupling reactions,[73,77,78] which yield predominantly methanol, 2-methyl-1-propanol, and propanols. Methanol is the overwhelming reaction product and significant quantities of esters are produced as well. Modified FT catalysts, molybdenum carbides, and the sulfide catalysts are thought to produce alcohols via a classical CO insertion process,[74,79] which yields C_1-C_4 linear alcohols, C_1-C_3 linear paraffins, and small amounts of esters. Methanol and methane are the dominant reaction products at low conversion but ethanol starts to become a significant product as methanol reaches its equilibrium concentration in the reactor. Both alcohols and hydrocarbons tend to follow an Anderson Shultz Flory-like distribution[80] with low alpha, meaning the lower carbon number species are the primary products, and higher carbon number species have a probability of formation that decreases logarithmically with carbon chain length. Rhodium catalysts are also thought to produce higher alcohols through a classical CO insertion process and have shown some evidence of production of aldehydes which are subsequently hydrogenated to form alcohols.[72]

With all of the above mechanisms, methanol is the most preferred alcohol while ethanol is not a favored product (an exception is rhodium catalysts, which produce ethanol as the main oxygenate product and methane as the main product). Tradeoffs exist for each system. Modified methanol catalysts do not produce significant quantities of hydrocarbons, but they produce little ethanol and only some higher, branched alcohols. Sulfide catalysts force a tradeoff between low conversions (little hydrocarbon formation but mostly methanol formed) and high conversions (less methanol relative to other alcohols but more C_1-C_4 hydrocarbons). Promoters have been shown to increase the ratio of ethanol to methanol in the product, but not to alter the mechanism of ethanol formation.[81] Experimental and modeling studies have shown that promoters are needed to suppress the high methanation activity of rhodium catalysts.[82,83] In general, breakthroughs are needed to reduce the methane selectivity on rhodium-type catalysts.

An improved higher alcohol synthesis catalyst, and in particular an ethanol synthesis catalyst, will promote carbon chain growth and hydrogenation (to form $-CH_x$ groups), CO dissociation (to provide removal of O from CO for C chain growth), CO insertion (to retain C–O bonds for alcohol chain termination), hydrogenation of unimolecular O and $-C-O$ to form H_2O and alcohols, and will prevent final hydrogenation of adsorbed $-CH_x$ species to free hydrocarbons. This may require catalysts which produce alcohols via a different mechanism than those listed above, promoters which further suppress hydrocarbon liberation, or both.

Thermodynamic considerations in mixed-alcohol synthesis

Mixed alcohol synthesis must also be considered in light of the thermodynamics of the process. An energy minimization of syngas to all potential products will yield methane and CO_2 with small amounts of larger hydrocarbons to balance imperfect ratios of H_2 and CO. Therefore, thermodynamics does not afford us a straightforward route from syngas to mixed alcohols and a highly specific catalyst is necessary. Figure 3 shows the logarithm of the thermodynamic equilibrium constant as a function of temperature for a number of potential alcohol-forming reactions. Positive values indicate that products are favored over reactants and negative values indicate that reactants are favored over products. All negative values and positive values that are small (< 2 or 3) are thermodynamically limited. That is, the products cannot be made by any catalyst once a modest conversion of CO is achieved. If byproducts are made (decreasing the amount of syngas in the reactor), this limiting conversion decreases.

Figure 3 shows that two products of interest, methanol and ethanol from syngas (lines a and b), are thermodynamically unfavorable across a range of reactor temperatures. Reactions that produce ethanol from methanol or dimethyl ether (DME) (lines c, e, and f) are more thermodynamically favorable, and catalysts which promote ethanol formation through these intermediates are more desirable (methanol and/or DME produced upstream using mature commercial processes, for instance). San et al. recently considered a route to ethanol via carbonylation of DME to methyl acetate and subsequent hydrogenation to ethanol and methanol, as an example.[84] Reactions which convert alcohols to hydrocarbons are very favorable (lines g and h), so catalysts which promote these reactions should be avoided, if possible. Since the equilibrium constant is a function of temperature only, yields might be improved if catalysts are developed that are active at lower temperatures.

It should be noted that if catalysts are developed to convert methanol or DME to ethanol, it will be advantageous for those catalysts to have limited or no activity for methanol synthesis from syngas. If the catalyst is active in this reaction, most of the feed methanol will revert to syngas as the syngas to methanol reaction has a very small equilibrium constant.

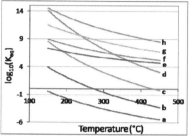

Figure 3. Thermodynamic equilibrium of various thermochemical reactions: (a) $CO + 2H_2 \leftrightarrow CH_3OH$, (b) $2CO + 4H_2 \leftrightarrow CH_3CH_2OH + H_2O$, (c) $CH_3OH + CO + 2H_2 \leftrightarrow CH_3CH_2OH + H_2O$, (d) $CO + 3H_2 \leftrightarrow CH_4 + H_2O$, (e) $CH_3OCH_3 \leftrightarrow CH_3CH_2OH$, (f) $2CH_3OH \leftrightarrow CH_3CH_2OH + H_2O$, (g) $CH_3CH_2OH + H_2 \leftrightarrow C_2H_6 + H_2O$, (h) $CH_3OH + H_2 \leftrightarrow CH_4 + H_2O$.

Economics

It is important to also consider the economics of the mixed-alcohol synthesis process. Phillips et al. prepared a baseline process and economic model of a biomass to ethanol plant using a sulfide catalyst with certain improvable characteristics.[85] A sensitivity analyses was performed and showed that the largest economic drivers were catalyst cost, activity, and selectivity to alcohols. Therefore, emphasis should be placed on lower cost materials followed by increased activity and improved selectivity. Of the known mixed alcohols catalysts, performance tends to decline with less expensive materials and as such, a significant catalyst breakthrough may be needed to achieve cost-effective gas to mixed alcohols synthesis.

SUMMARY

Catalysts and sorbents are essential in the thermochemical conversion of biomass. Experimental results have shown that catalysts developed for petroleum feedstock, e.g., HDS and cracking catalysts are not optimal for biomass conversion. Sorbents that were developed for coal syngas will also need to be further developed to operate effectively in biomass syngas. Catalysts and sorbents that target the process conditions and the molecules unique to biomass conversion must be developed to improve yield, operation efficiency (e.g., less extreme temperature and pressure conditions), and catalyst life span.

Some catalysts are in the later stages of development and require validation at the pilot scale (e.g. reforming catalysts) while others still require breakthroughs at the bench scale (e.g. HDO, catalytic cracking, mixed alcohol synthesis). To maximize the yield of liquid fuel from biomass and to make renewable biofuel economical and sustainable, the development of next generation, highly selective and active catalysts is of utmost importance.

ACKNOWLEDGMENTS

Funding for this research was provided by the Office of the Biomass Program, U.S. Department of Energy, under contract number DE-AC36-99GO10337 with the National Renewable Energy Laboratory.

REFERENCES

1. T.D. Foust, A. Aden, A. Dutta, and S. Phillips, An Economic and Environmental Comparison of a Biochemical and a Thermochemical Lignocellulosic Ethanol Conversion Processes. *Cellulose*, **16**, 547–65, (2009).
2. M.E. Himmel, S.Y. Ding, D.K. Johnson, W.S. Adney, M.R. Nimlos, J.W. Brady, and T.D. Foust, Biomass Recalcitrance: Engineering Plants and Enzymes for Biofuels Production. *Science*, **315**, 804–7, (2007).
3. J.N. Chheda, G.W. Huber, and J.A. Dumesic, Liquid-Phase Catalytic Processing of Biomass-Derived Oxygenated Hydrocarbons to Fuels and Chemicals. *Angew. Chem.-Int. Edit.*, **46**, 7164–83, (2007).
4. S.D. Phillips, Technoeconomic Analysis of a Lignocellulosic Biomass Indirect Gasification Process to Make Ethanol Via Mixed Alcohols Synthesis. *Ind. Eng. Chem. Res.*, **46**, 8887–97, (2007).
5. A.V. Bridgwater and G.V.C. Peacocke, Fast Pyrolysis Processes for Biomass. *Renew. Sust. Energ. Rev.*, **4**, 1–73, (2000).
6. C. Higman and M. van der Burgt, *Gasification*. 2nd ed., Burlington, MA: Gulf Professional Publishing, an imprint of Elsevier. 435 pp., (2008).
7. S. Czernik and A.V. Bridgwater, Overview of Applications of Biomass Fast Pyrolysis Oil. *Energy Fuels*, **18**, 590–8, (2004).
8. J. Holmgren, R. Marinangeli, P. Nair, D. Elliott, and R. Bain, Consider Upgrading Pyrolysis Oils into Renewable Fuels. *Hydrocarb. Process.*, **87**, 95–103, (2008).
9. A.V. Bridgwater, Production of High Grade Fuels and Chemicals from Catalytic Pyrolysis of Biomass. *Catal. Today*, **29**, 285–95, (1996).
10. D.C. Elliott, Historical Developments in Hydroprocessing Bio-Oils. *Energy Fuels*, **21**, 1792–815, (2007).
11. A. Corma, G.W. Huber, L. Sauvanaud, and P. O'Connor, Processing Biomass-Derived Oxygenates in the Oil Refinery: Catalytic Cracking (FCC) Reaction Pathways and Role of Catalyst. *J. Catal.*, **247**, 307–27, (2007).
12. G.W. Huber and A. Corma, Synergies between Bio- and Oil Refineries for the Production of Fuels from Biomass. *Angew. Chem.-Int. Edit.*, **46**, 7184–201, (2007).
13. B.C. Gates, *Catalytic Chemistry*, The Wiley Series in Chemical Engineering, New York, NY: John Wiley Sons. 480 pp., (1991).
14. D.C. Elliott and E.G. Baker, Catalytic Hydrotreating of Biomass Liquefaction Products to Produce Hydrocarbon Fuels: Interim Report, Pacific Northwest National Laboratory, Richland, WA, USA, p. 76, (1986).
15. E.G. Baker and D.C. Elliott, Catalytic Upgrading of Biomass Pyrolysis Oils. in *Research in Thermochemical Biomass Conversion*, A.V. Bridgwater and J.L. Kuester, Editors, Elsevier Science Publishers: Barking, England. p. 883–95, (1988).

16.   E.G.R. Baker, WA), Elliott, Douglas C. (Richland, WA). Method of Upgrading Oils Containing Hydroxyaromatic Hydrocarbon Compounds to Highly Aromatic Gasoline. US5,180,868, 07/577781, Battelle, Memorial Institute (Richland, WA), (1993).

17.   L. Conti, G. Scano, J. Boufala, and S. Mascia. Experiments of Bio-Oil Hydrotreating in a Continuous Bench-Scale Plant. in *EU-Canada Workshop on Thermal Biomass Processing*: CPL Press, (1996).

18.   X. Fu, Z. Dai, S. Tian, J. Long, S. Hou, and X. Wang, Catalytic Decarboxylation of Petroleum Acids from High Acid Crude Oils over Solid Acid Catalysts. *Energy Fuels*, **22**, 1923–9, (2008).

19.   T.L. Marker and J.A. Petri. Gasoline and Diesel Production from Pyrolytic Lignin Produced from Pyrolysis of Cellulosic Waste. U.S. 7,578,927 B2, 11/468813 UOP LLC (Des Plaines, IL, US), (2009).

20.   R. French and R.M. Baldwin, *Environ. Prog.*, (in press).

21.   J. Wildschut, F.H. Mahfud, R.H. Venderbosch, and H.J. Heeres, Hydrotreatment of Fast Pyrolysis Oil Using Heterogeneous Noble-Metal Catalysts. *Ind. Eng. Chem. Res.*, **48**, 10324–34, (2009).

22.   P. Chantal, S. Kaliaguine, J.L. Grandmaison, and A. Mahay, Production of Hydrocarbons from Aspen Poplar Pyrolytic Oils over H-Zsm5. *Appl. Catal.*, **10**, 317–32, (1984).

23.   N.Y. Chen, D.E. Walsh, and L.R. Koenig. Fluidized-Bed Upgrading of Wood Pyrolysis Liquids and Related-Compounds. in *Pyrolysis Liquids from Biomass, ACS Symposium Series*: American Chemical Society, (1988).

24.   M. Renaud, J.L. Grandmaison, C. Roy, and S. Kaliaguine. Low-Pressure Upgrading of Vacuum-Pyrolysis Oils from Wood. in *Pyrolysis Liquids from Biomass, ACS Symposium Series*: American Chemical Society, (1988).

25.   J. Diebold and J. Scahill. Biomass to Gasoline - Upgrading Pyrolysis Vapors to Aromatic Gasoline with Zeolite Catalysts at Atmospheric-Pressure. in *Pyrolysis Liquids from Biomass, ACS Symposium Series*: American Chemical Society, (1988).

26.   J.P. Diebold, H.L. Chum, R.J. Evans, T.A. Milne, T.B. Reed, and J.W. Scahill. in *Energy from Biomass and Wastes X*: IGT Chicago and Elsevier Applied Sciences Publishers, (1987).

27.   J.D. Adjaye and N.N. Bakhshi, Catalytic Conversion of a Biomass-Derived Oil to Fuels and Chemicals. 1. Model-Compound Studies and Reaction Pathways. *Biomass Bioenerg.*, **8**, 131–49, (1995).

28.   A.A. Lappas, M.C. Samolada, D.K. Iatridis, S.S. Voutetakis, and I.A. Vasalos, Biomass Pyrolysis in a Circulating Fluid Bed Reactor for the Production of Fuels and Chemicals. *Fuel*, **81**, 2087–95, (2002).

29.   T.R. Carlson, T.R. Vispute, and G.W. Huber, Green Gasoline by Catalytic Fast Pyrolysis of Solid Biomass Derived Compounds. *ChemSusChem*, **1**, 397–400, (2008).

30.   T.R. Carlson, J. Jae, Y.C. Lin, G.A. Tompsett, and G.W. Huber, Catalytic Fast Pyrolysis of Glucose with HZSM-5: The Combined Homogeneous and Heterogeneous Reactions. *J. Catal.*, **270**, 110–24, (2010).

31.   M.C. Samolada, A. Papafotica, and I.A. Vasalos, Catalyst Evaluation for Catalytic Biomass Pyrolysis. *Energy Fuels*, **14**, 1161-7, (2000).

32.   R. French and S. Czernik, Catalytic Pyrolysis of Biomass for Biofuels Production. *Fuel Process. Technol.*, **91**, 25–32, (2010).

33.   P. O'connor, D. Stamires, and J.A. Moulijn. Process for Converting Carbon-Based Energy Carrier Material. US20090308787, BIOeCon INTERNATIONAL HOLDING B.V. (CURACAO, AN), (2009).

34.   D.L. Carpenter, R.L. Bain, R.E. Davis, A. Dutta, C.J. Feik, K.R. Gaston, W. Jablonski, S.D. Phillips, and M.R. Nimlos, Pilot-Scale Gasification of Corn Stover, Switchgrass, Wheat Straw,

and Wood: 1. Parametric Study and Comparison with Literature. *Ind. Eng. Chem. Res.*, **49**, 1859–71, (2010).

35. J. Leppalahti, Formation and Behavior of Nitrogen-Compounds in an IGCC Process. *Bioresource Technology*, **46**, 65–70, (1993).

36. K. Sato, T. Shinoda, and K. Fujimoto, New Nickel-Based Catalyst for Tar Reforming, with Superior Resistance to Coking and Sulfur Poisoning in Biomass Gasification Processes. *J. Chem. Eng. Jpn.*, **40**, 860–8, (2007).

37. W. Torres, S.S. Pansare, and J.G. Goodwin Jr., Hot Gas Removal of Tars, Ammonia, and Hydrogen Sulfide from Biomass Gasification Gas. *Catal. Rev.*, **49**, 407–56, (2007).

38. A. van der Drift, J. van Doorn, and J.W. Vermeulen, Ten Residual Biomass Fuels for Circulating Fluidized-Bed Gasification. *Biomass Bioenerg.*, **20**, 45–56, (2001).

39. J.B. Gibson, III and D.P. Harrison, Reaction between Hydrogen-Sulfide and Spherical Pellets of Zinc-Oxide. *Ind. Eng. Chem. Process Des. Dev.*, **19**, 231–7, (1980).

40. D.P. Harrison, Performance Analysis of ZnO-Based Sorbents in Removal of H_2S from Fuel Gas, in *Desulfurization of Hot Coal Gas*, A.T. Atimtay and D.P. Harrison, Editors, Springer: Berlin. p. 213–42, (1998).

41. S. Cheah, D.L. Carpenter, and K.A. Magrini-Bair, Review of Mid- to High-Temperature Sulfur Sorbents for Desulfurization of Biomass- and Coal-Derived Syngas. *Energy Fuels*, **23**, 5291–307, (2009).

42. R. Ben-Slimane and M.T. Hepworth, Desulfurization of Hot Coal-Derived Fuel Gases with Manganese-Based Regenerable Sorbents. 1. Loading (Sulfidation) Tests. *Energy Fuels*, **8**, 1175–83, (1994).

43. M. Flytzani-Stephanopoulous and Z. Li, Kinetics of Sulfidation Reactions between H_2S and Bulk Oxide Sorbents, in *Desulfurization of Hot Coal Gas*, A.T. Atimtay and D.P. Harrison, Editors, Springer: Berlin. p. 179–211, (1998).

44. M. Flytzani-Stephanopoulous, M. Sakbodin, and W. Zheng, Regenerative Adsorption and Removal of H_2S from Hot Fuel Gas Streams by Rare Earth Oxides. *Science*, **312**, 1508–10, (2006).

45. S.K. Gangwal, S.M. Harkins, M.C. Woods, S.C. Jain, and S.J. Bossart, Bench-Scale Testing of High-Temperature Desulfurization Sorbents. *Environ. Prog.*, **8**, 265–9, (1989).

46. S. Lew, A.F. Sarofim, and M. Flytzani-Stephanopoulos, Sulfidation of Zinc Titanate and Zinc-Oxide Solids. *Ind. Eng. Chem. Res.*, **31**, 1890–9, (1992).

47. S. Lew, A.F. Sarofim, and M. Flytzani-Stephanopoulos, The Reduction of Zinc Titanate and Zinc-Oxide Solids. *Chem. Eng. Sci.*, **47**, 1421–31, (1992).

48. Z. Wang and M. Flytzani-Stephanopoulos, Cerium Oxide-Based Sorbents for Regenerative Hot Reformate Gas Desulfurization. *Energy Fuels*, **19**, 2089–97, (2005).

49. P. Westmoreland and D.P. Harrison, Evaluation of Candidate Solids for High-Temperature Desulfurization of Low-Btu Gases. *Environ. Sci. Technol.*, **10**, 659–61, (1976).

50. P.R. Westmoreland, J.B. Gibson, and D.P. Harrison, Comparative Kinetics of High-Temperature Reaction between H_2S and Selected Metal Oxides. *Environ. Sci. Technol.*, **11**, 488–91, (1977).

51. J.H. Swisher and K. Schwerdtfeger, Review of Metals and Binary Oxides as Sorbents for Removing Sulfur from Coal-Derived Gases. *J. Mater. Eng. Perform.*, **1**, 399–408, (1992).

52. Z.J. Li and M. Flytzani-Stephanopoulos, Cu-Cr-O and Cu-Ce-O Regenerable Oxide Sorbents for Hot Gas Desulfurization. *Ind. Eng. Chem. Res.*, **36**, 187–96, (1997).

53. J. Abbasian and R.B. Slimane, A Regenerable Copper-Based Sorbent for H_2S Removal from Coal Gases. *Ind. Eng. Chem. Res.*, **37**, 2775–82, (1998).

54. M.P. Aznar, M.A. Caballero, J. Gil, J.A. Martin, and J. Corella, Commercial Steam Reforming Catalysts to Improve Biomass Gasification with Steam-Oxygen Mixtures. 2. Catalytic Tar Removal. *Ind. Eng. Chem. Res.*, **37**, 2668–80, (1998).

55. D. Dayton, A Review of the Literature of Catalytic Biomass Tar Destruction, National Renewable Energy Laboratory, Golden, CO, USA, p. 33, (2002).

56. J. Corella, J.M. Toledo, and R. Padilla, Olivine or Dolomite as in-Bed Additive in Biomass Gasification with Air in a Fluidized Bed: Which Is Better? *Energy Fuels*, **18**, 713–20, (2004).

57. J.N. Kuhn, Z. Zhao, A. Senefeld-Naber, L.G. Felix, R.B. Slimane, C.W. Choi, and U.S. Ozkan, Ni-Olivine Catalysts Prepared by Thermal Impregnation: Structure, Steam Reforming Activity, and Stability. *Appl. Catal. A-Gen.*, **341**, 43–9, (2008).

58. T.A. Milne, R.J. Evans, and N. Abatzoglou, Biomass Gasifier Tars: Their Nature, Formation, and Conversion, National Renewable Energy Laboratory, Golden, CO, p. 204, (1998).

59. L. Devi, K.J. Ptasinski, and F.J.J.G. Janssen, A Review of the Primary Measures for Tar Elimination in Biomass Gasification Processes. *Biomass Bioenerg.*, **24**, 125–40, (2003).

60. K. Tomishige, T. Miyazawa, M. Asadullah, S. Ito, and K. Kunimori, Catalyst Performance in Reforming of Tar Derived from Biomass over Noble Metal Catalysts. *Green Chem.*, **5**, 399–403, (2003).

61. F. Besenbacher, I. Chorkendorff, B.S. Clausen, B. Hammer, A.M. Molenbroek, J.K. Norskov, and I. Stensgaard, Design of a Surface Alloy Catalyst for Steam Reforming. *Science*, **279**, 1913–5, (1998).

62. E. Nikolla, J. Schwank, and S. Linic, Promotion of the Long-Term Stability of Reforming Ni Catalysts by Surface Alloying. *J. Catal.*, **250**, 85-93, (2007).

63. J.H. Jeong, J.W. Lee, D.J. Seo, Y. Seo, W.L. Yoon, D.K. Lee, and D.H. Kim, Ru-Doped Ni Catalysts Effective for the Steam Reforming of Methane without the Pre-Reduction Treatment with H_2. *Appl. Catal. A-Gen.*, **302**, 151–6, (2006).

64. J. Xu and M. Saeys, Improving the Coking Resistance of Ni-Based Catalysts by Promotion with Subsurface Boron. *J. Catal.*, **242**, 217-26, (2006).

65. M.H. Youn, J.G. Seo, P. Kim, J.J. Kim, H.I. Lee, and I.K. Song, Hydrogen Production by Auto-Thermal Reforming of Ethanol over Ni/γ-Al_2O_3 Catalysts: Effect of Second Metal Addition. *J. Power Sources*, **162**, 1270–4, (2006).

66. T. Miyazawa, T. Kimura, J. Nishikawa, S. Kado, K. Kunimori, and K. Tomishige, Catalytic Performance of Supported Ni Catalysts in Partial Oxidation and Steam Reforming of Tar Derived from the Pyrolysis of Wood Biomass. *Catal. Today*, **115**, 254–62, (2006).

67. K.A. Magrini-Bair, S. Czernik, R. French, Y.O. Parent, E. Chornet, D.C. Dayton, C. Feik, and R. Bain, Fluidizable Reforming Catalyst Development for Conditioning Biomass-Derived Syngas. *Appl. Catal. A-Gen.*, **318**, 199-206, (2007).

68. N. Koizumi, K. Murai, T. Ozaki, and M. Yamada, Development of Sulfur Tolerant Catalysts for the Synthesis of High Quality Transportation Fuels. *Catal. Today*, **89**, 465–78, (2004).

69. R.L. Bain, D.C. Dayton, D.L. Carpenter, S.R. Czernik, C.J. Feik, R.J. French, K.A. Magrini-Bair, and S.D. Phillips, Evaluation of Catalyst Deactivation During Catalytic Steam Reforming of Biomass-Derived Syngas. *Ind. Eng. Chem. Res.*, **44**, 7945–56, (2005).

70. P.K. Cheekatamarla and A.M. Lane, Catalytic Autothermal Reforming of Diesel Fuel for Hydrogen Generation in Fuel Cells Ii. Catalyst Poisoning and Characterization Studies. *J. Power Sources*, **154**, 223–31, (2006).

71. J.J. Spivey and A. Egbebi, Heterogeneous Catalytic Synthesis of Ethanol from Biomass-Derived Syngas. *Chem. Soc. Rev.*, **36**, 1514-28, (2007).

72. V. Subramani and S.K. Gangwal, A Review of Recent Literature to Search for an Efficient Catalytic Process for the Conversion of Syngas to Ethanol. *Energy Fuels*, **22**, 814-39, (2008).

73. J.G. Nunan, C.E. Bogdan, K. Klier, K.J. Smith, C.W. Young, and R.G. Herman, Methanol and C_2 Oxygenate Synthesis over Cesium Doped Cu/ZnO and Cu/ZnO/Al$_2$O$_3$ Catalysts - a Study of Selectivity and C^{13} Incorporation Patterns. *J. Catal.*, **113**, 410-33, (1988).

74. J.G. Santiesteban, C.E. Bogdan, R.G. Herman, and K. Klier. Mechanism of C_1-C_4 Alcohol Synthesis over Alkali/MoS$_2$ and Alkali/Co/MoS$_2$ Catalysts. in *Proceedings of the 9th International Congress on Catalysis*, (1988).

75. K.J. Smith, R.G. Herman, and K. Klier, Kinetic Modeling of Higher Alcohol Synthesis over Alkali-Promoted Cu/ZnO and MoS$_2$ Catalysts. *Chem. Eng. Sci.*, **45**, 2639-46, (1990).

76. G.C. Chinchen, P.J. Denny, D.G. Parker, M.S. Spencer, and D.A. Whan, Mechanism of Methanol Synthesis from CO$_2$/CO/H$_2$ Mixtures over Copper/Zinc Oxide/Alumina Catalysts - Use of C$_{14}$-Labeled Reactants. *Appl. Catal.*, **30**, 333-8, (1987).

77. J.G. Nunan, C.E. Bogdan, K. Klier, K.J. Smith, C.W. Young, and R.G. Herman, Higher Alcohol and Oxygenate Synthesis over Cesium-Doped Cu/ZnO Catalysts. *J. Catal.*, **116**, 195-221, (1989).

78. J.G. Nunan, R.G. Herman, and K. Klier, Higher Alcohol and Oxygenate Synthesis over Cs/Cu/ZnO/Al$_2$O$_3$, Cs/Cu/ZnO/Cr$_2$O$_3$ Catalysts. *J. Catal.*, **116**, 222-9, (1989).

79. J.G. Santiesteban, Alcohol Synthesis from Carbon Monoxide and Hydrogen over MoS$_2$-Based Catalysts, in *Chemistry*, Lehigh University, Bethlehem, PA, p. 253, (1989).

80. D.L. King, J.A. Cusumano, and R.L. Garten, A Technological Perspective for Catalytic Processes Based on Synthesis Gas. *Catal. Rev.-Sci. Eng.*, **23**, 233-63, (1981).

81. J.M. Christensen, P.M. Mortensen, R. Trane, P.A. Jensen, and A.D. Jensen, Effects of H$_2$S and Process Conditions in the Synthesis of Mixed Alcohols from Syngas over Alkali Promoted Cobalt-Molybdenum Sulfide. *Appl. Catal. A-Gen.*, **366**, 29-43, (2009).

82. Y. Choi and P. Liu, Mechanism of Ethanol Synthesis from Syngas on Rh(111). *J. Am. Chem. Soc.*, **131**, 13054-61, (2009).

83. M.A. Gerber, J.F. White, M. Gray, and D.J. Stevens, Evaluation of Promoters for Rhodium-Based Catalysts for Mixed Alcohol Synthesis, Pacific Northwest National Laboratory, Richland, WA, p. 116, (2008).

84. X.G. San, Y. Zhang, W.J. Shen, and N. Tsubaki, New Synthesis Method of Ethanol from Dimethyl Ether with a Synergic Effect between the Zeolite Catalyst and Metallic Catalyst. *Energy Fuels*, **23**, 2843-4, (2009).

85. S. Phillips, A. Aden, J. Jechura, D. Dayton, and T. Eggeman, Thermochemical Ethanol Via Indirect Gasification and Mixed Alcohol Synthesis of Lignocellulosic Biomass, National Renewable Energy Laboratory, Golden, CO., p. 126, (2007).

MATERIAL CHARACTERIZATION AND ANALYSIS FOR SELECTION OF REFRACTORIES
USED IN BLACK LIQUOR GASIFICATION

James G. Hemrick and James R. Keiser
Oak Ridge National Laboratory
Oak Ridge, TN, USA

Roberta A. Peascoe-Meisner
University of Tennessee
Knoxville, TN, USA

ABSTRACT
        Black liquor gasification provides the pulp and paper industry with a technology which could
potentially replace recovery boilers with equipment that could reduce emissions and, if used in a
combined cycle system, increase the power production of the mill allowing it to be a net exporter of
electrical power. In addition, rather than burning the syngas produced in a gasifier, this syngas could
be used to produce higher value chemicals or fuels. However, problems with structural materials such
as the refractory lining of the reactor vessel have caused unplanned shutdowns and resulted in
component replacement much sooner than originally planned. Through examination of exposed
materials, laboratory corrosion tests and cooperative efforts with refractory manufacturers, many
refractory materials issues in high-temperature black liquor gasification have been addressed and
optimized materials have been selected for this application. In this paper, an updated summary of the
characterization and analysis techniques used for refractory screening and selection will be discussed
along with characteristic results from these methods which have led to the selection of optimized
materials for both the hot-face and back-up linings used in this application.

INTRODUCTION
        The pulp and paper industry is among the nation's most energy intensive manufacturing
industries using an estimated 3,248 TBtu/year[a] (only behind the chemical and petroleum industries).
Thompson recovery boilers have traditionally been used for the recovery of pulping chemicals (sodium
sulfide and sodium hydroxide) needed for separation of wood fibers for papermaking by the Kraft
process from the waste product (black liquor) of the process. As has been pointed out in the past, these
boilers suffer from many shortcomings including inefficiency with respect to production of steam and
power, high pollutant emission levels, and inherent danger due to boiler explosions. Black liquor
gasification has been suggested as an alternative technology to the Thompson recovery boiler with
possible energy, environmental, and financial benefits. The potential economic and environmental
benefits of black liquor gasification incorporated with a combined cycle system have been described in
previous reports[1,2], where these gasifiers are used to replace chemical recovery boilers in kraft mills
and the additional energy recovered through use of the combined cycle mode results in sufficient
power production that a mill could become a net exporter of electric power. It is estimated that
gasification, in a combined cycle configuration, could lead to a 50-60% increase in energy efficiency.
Additionally, the concept of the "Integrated Forest Biorefinery" has been proposed where the
traditional pulp and paper mill becomes a facility which, in addition to the traditional line of wood pulp
and paper products, is also capable of producing streams of higher value chemicals, fuels and/or
electric power including the possible production of a syngas from black liquor.[3]

---

[a] According to U.S. Department of Energy, Energy Information Administration

Two gasification processes, as described elsewhere[4], have been used commercially in North America. These are characterized as "low-temperature" and "high-temperature". The focus of this paper is on the refractory issues associated with a high-temperature gasification process designed by Chemrec AB of Sweden and employed at a Weyerhaeuser Company mill in New Bern, North Carolina. This unit, which was used to supplement the black liquor capacity of the existing recovery boiler, was operated at slightly above atmospheric pressure and temperatures on the order of 900-1000°C. Black liquor, steam, and air were injected into the top of the gasifier, while the products (including syngas and the molten, inorganic sodium-rich salts that are contained in the black liquor) were removed from the bottom of the refractory-lined vessel shown schematically in Figure 1. Additional information on the process can be found in the literature.[5-7]

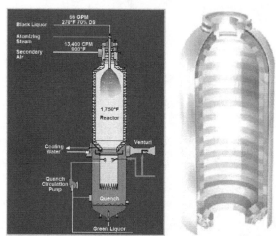

Figure 1. Drawings of the high-temperature black liquor gasifier with original design shown on left indicating where materials are injected or removed and a cut-away view of the gasifier vessel on the right after the vessel was redesigned and rebuilt in 2003 (drawings provided by Weyerhaeuser, Inc.).

Several issues exist regarding the refractory lined combustion vessel, as will be further discussed below. The first is containment of the molten smelt reaction product by the primary lining. Additionally issues exist regarding penetration of the mortar joints between the bricks of the primary lining by the molten smelt (not discussed here). Finally issues were discovered regarding the insulative back-up lining found behind the primary lining.

REFRACTORY SELECTION AND APPLICATION AT THE NEW BERN GASIFIER

A history of the refractory materials used in the New Bern gasifier is given in Table I. The original materials used for the construction of this unit were selected by Chemrec, AB based on testing conducted at a research institute in Sweden. Within months of the initial start up, the mullite-based ($3Al_2O_3*2SiO_2$) refractory brick used for the hot-face lining of the vessel was found to be experiencing significant thinning. Material from this original hot-face lining was subsequently removed and Oak

Ridge National Laboratory (ORNL) personnel were consulted to perform failure analysis of these samples.

Table I. History of Refractory Selection and Installation at New Bern, NC Gasifier

Time period	Reactor shell	Back-up refractory	Hot-face refractory
Dec 1996 – Early 1999	316L stainless steel	Super duty refractory brick	60% Al_2O_3, SiO_2 brick
Early 1999 – Dec 1999	316L stainless steel	Fusion cast β-Al_2O_3	Fusion cast α/β-Al_2O_3 – manufacturer A
June 2003 – Sept 2004	Carbon steel	High-alumina bonded brick	Fusion cast α/β-Al_2O_3 – manufacturer B
Oct 2004 – October 2006	Carbon steel	Bonded mullite – based refractory	MgO-Al_2O_3 fusion cast spinel
Oct 2006 – present	Carbon steel	Calcium aluminate – based refractory	MgO-Al_2O_3 fusion cast spinel

Through analysis by X-ray diffraction (PANalytical X'Pert Pro diffractometer with Cu Kα radiation), this material was found to show evidence of reaction between the refractory brick and sodium salts from the smelt resulting in the formation of the compounds nosean ($Na_8Al_6Si_6O_{24}SO_4$) and nepheline ($NaAlSiO_4$) as described in a previous report[8] and shown with respect to distance from the refractory hot-face in Figure 2. Patterns were generated for samples representative of material at various distances away from the fireside surface, providing information on the phases that were present or absent as you moved through the refractory brick. From these patterns, estimates could be made of the quantity of each phase present at individual sampling locations.

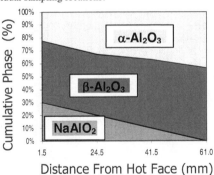

Figure 2. Estimated concentration of phases at and below the fireside (hot-face) surface of the original New Bern hot-face refractory.

Repair/replacement campaigns were carried out over the next several years using the same mullite-based bricks which continued to rapidly degrade. Thus, an alternate hot-face refractory material was sought. To this end, a testing program was undertaken to identify and screen candidate materials for this application. A laboratory immersion test system was constructed to simulate actual gasifier exposure as shown in Figure 3. This system had been previously demonstrated to successfully reproduce the corrosion products observed to form on refractories exposed in operating gasifiers using commercially generated smelt (nominal composition: 60-75% Na2CO3, 20-38% Na2SO4, 1-4% Na2S, and 1-4% Na2S2O3).[8] Samples (102 mm x 25 mm x 12.5 mm, 4" x 1" x 0.5") were immersed in undiluted smelt obtained from the Weyerhaeuser mill at temperatures of 900-1000°C for 50-250 hours.

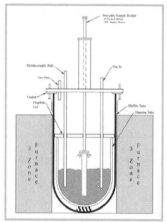

Figure 3. ORNL Immersion test system utilized for rapid screening of candidate black liquor gasification containment materials.

The above analyses led to the selection of α/β and β-alumina fusion cast refractories for use as hot-face and back-up refractory lining materials, respectively. Following the gasifier rebuild, spalling of the α/β-alumina refractory used for the hot-face was again noted shortly after restarting of the unit. Therefore, core-drilled samples (shown in Figure 4) from both the hot-face and back-up lining were collected about a half year after the refractory was installed.

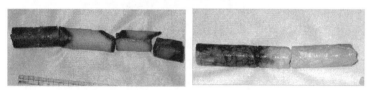

Figure 4. Core drilled samples of fusion cast α/β-alumina hot-face lining refractory (left) and β-alumina back-up lining refractory (right).

Discoloration was found on the hotter end of both samples and considerable cracking was present throughout the samples. The discoloration along the cracks in the hot-face sample suggests that these cracks existed while the refractory was in service and were not just a result of damage during sample collection. Further analysis by X-ray diffraction confirmed that $NaAlO_2$ was present at the surface and in decreasing amounts with distance below the surface. In addition, the amount of α-alumina present at the surface was decreased significantly below that found to be present in un-reacted material. The phase diagram in Figure 5 of the Na_2O-Al_2O_3 system, shows the stable phases as the concentration of NaO is increased.

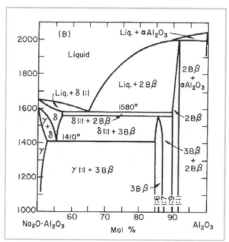

Figure 5. Phase diagram showing the $Na_2O*Al_2O_3$ - Al_2O_3 system[9].

When Na_2O is added to Al_2O_3 a volume increase occurs with each phase change as shown in Equation 1. Through such reactions, it is expected that compressive stresses develop in the hot-face lining as Na_2O from the environment reacts with the alumina refractory leading to the observed spalling.

$$\alpha - Al_2O_3 \rightarrow \beta - Al_2O_3 \text{ resulting in an expansion of } 6.5\,\overset{o}{A}{}^3 \text{/Al atom}$$

$$\beta - Al_2O_3 \rightarrow \beta^{'} - Al_2O_3 \text{ resulting in an expansion of } 0.3\,\overset{o}{A}{}^3 \text{/Al atom} \qquad (1)$$

$$\beta^{'} - Al_2O_3 \rightarrow NaAl_2O_3 \text{ resulting in an expansion of } 21.4\,\overset{o}{A}{}^3 \text{/Al atom}$$

Based on the above identified reactions occurring between the alumina refractory and the smelt, a new fusion cast α/β-alumina refractory provided by a different manufacturer was selected for use as the hot-face refractory material during the next gasifier rebuild. This material was screened using the laboratory smelt immersion test system described above and found to show superior performance to the currently used fusion cast alumina brick. Additionally, through additional laboratory smelt immersion studies other alternative materials were identified that appeared to be more corrosion resistant than the current fusion cast α/β-alumina material being used. Consequently, samples of several of these alternate refractories were also included in the hot-face lining that was installed.

During a subsequent gasifier shut down for repairs of other systems, the opportunity was taken to collect core-drilled samples of the alternate refractories, as well as samples of the primary hot-face lining. Cross-sections of four of the eight samples collected are shown in Figure 6.

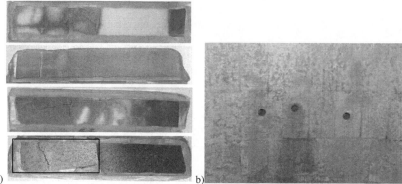

a)   b)

Figure 6. Longitudinal cross-sections of four core-drilled samples (a) removed from the gasifier hot-face lining (b).
(fusion cast α/β-alumina – top, fusion cast magnesia-alumina spinel - second, alternate fusion cast α/β-alumina – third, and bonded magnesia-alumina spinel - outlined portion (left half) on bottom)

Extensive cracking and discoloration was found in both the standard fusion cast α/β-alumina refractory and the alternate fusion-cast alumina sample suggesting that sodium compounds penetrated into and reacted with the alumina refractory. The other two alternate materials contained magnesia-alumina spinel (MgO-Al_2O_3); one being dense fusion cast brick like the current α/β-alumina material and one being a bonded brick. Based on appearance it appeared the fusion cast spinel was less affected by the service in the molten sodium-rich salts.

Analysis by optical microscopy and electron microprobe as shown in Figure 7 through Figure 10 confirmed that the fusion cast spinel had reacted less with the sodium compounds than any of the other refractories examined to date. Additionally, Scanning Electron Microscopy with Energy Dispersive Spectroscopy analysis was performed on the surfaces of the hot-face refractories using a Hitachi S-4100 as shown in Figure 11. For the fusion cast alumina sample (Figure 7, Figure 8, and Figure 11(a)) it can be seen that smelt collects at the sample surface and penetrates along the cracks that form from the hot-face of the material leading to significant levels of sodium and sulfur being present in the bulk region of the sample. The spinel sample (Figure 9, Figure 10, and Figure 11(b)) shows a small amount of solidified smelt on the sample surface, but only limited penetration of smelt (along mortar joint) and collection in a tangential crack approximately 25 mm (1") into the sample. This is supported by the limited amounts of sodium and sulfur found in the bulk region of this sample.

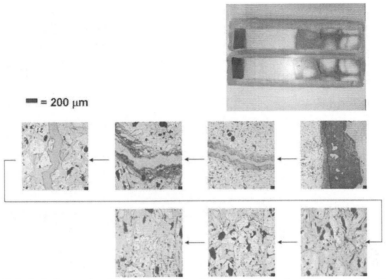

Figure 7. Optical micrographs of fusion cast α/β-alumina core-drilled specimen taken from New Bern gasifier.

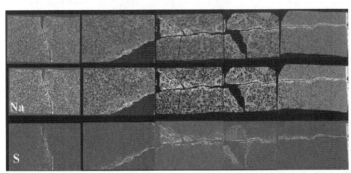

Figure 8. Electron microprobe image for fusion cast α/β-alumina core-drilled specimen taken from New Bern gasifier.

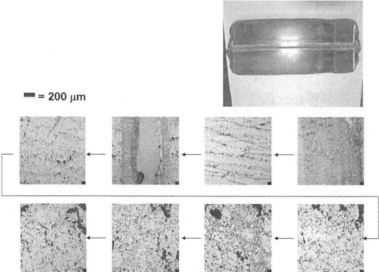

= 200 μm

Figure 9. Optical micrographs of and fusion cast spinel (MgO-Al$_2$O$_3$) core-drilled specimen taken from New Bern gasifier.

Figure 10. Electron microprobe image for fusion cast spinel (MgO-Al$_2$O$_3$) core-drilled specimen taken from New Bern gasifier.

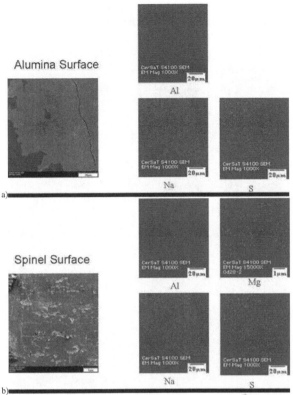

Figure 11. SEM/EDS analysis of fusion cast α/β-alumina (a) and spinel (MgO-Al$_2$O$_3$) (b) hot-face surfaces from core-drilled specimens taken from New Bern gasifier.

More importantly, it was found that the fusion cast spinel reacted significantly less than the standard fusion cast α/β-alumina refractory. As a result of these observations and measurements, the fusion cast magnesia-alumina spinel refractory was selected for use in the hot-face lining of the subsequent rebuild of the gasifier. Along with the spinel hot-face refractory, a back-up lining of a low thermal conductivity bonded mullite brick was installed. Following installation of the new lining, the gasifier operated with no evidence of degradation of the spinel refractory. Strain measurements were carried out on the shell of the gasifier vessel, as reaction of the previous α/β-alumina refractories resulting in expansion of the refractory lining had been shown to create loading on the shell. Strain measurements on the shell with the current lining showed that the strain was increasing at a slower rate than was seen when the previous α/β-alumina lining was in place. However, even though the hot-face lining did not show evidence of significant reaction, the strain was increasing at a measurable rate.

To analyze the cause of the increasing strain, after nine months of operation a core-drilled sample was taken that included a portion of the back-up lining as well as the hot-face lining as shown in Figure 12.

Figure 12. Core-drilled sample of the back-up mullite-based lining.

From this sample, it was determined that the portion of the back-up lining nearest the hot-face lining (the region on the right of the photograph) had been extensively physically (crumbled) and chemically (discolored) degraded. Further analysis by X-ray diffraction (Figure 13) showed that sodium-containing reaction products were present in the degraded portion of the back-up lining leading to the observed increasing strain on the gasifier vessel shell. As a result of this observation, a series of tests were conducted to evaluate other candidate refractories for the back-up lining.

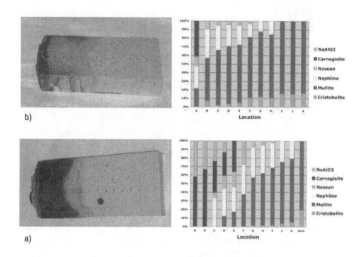

Figure 13. Refinement of X-ray diffraction patterns showing sodium-containing reaction products present in the degraded portion of the back-up lining from a lower (a) and upper (b) location.

Smelt immersion testing (100 hours in molten salt at 900 or 1000°C) was again used to screen a number of candidate refractories for use in the gasifier back-up lining. Even though mullite materials had been previously tested and found to react rapidly with the molten smelt[10,11], it was thought that recently identified higher purity materials may be considerably less reactive. Immersion testing of the newly identified mullite materials found them still to be rapidly degraded by the molten smelt as shown in Figure 14. Calcium-aluminate materials were investigated based on previous studies at ORNL conducted on an experimental pressed and sintered refractory that had very good resistance to molten smelt. At the time of this initial work, no refractory manufacturer was found that had a commercial product, but such a manufacturer was identified several years later and this manufacturer agreed to work with the authors to find an appropriate composition and firing temperature for a refractory that would have the desired corrosion resistance. Initial immersion testing results of the prepared calcium-aluminate refractory materials showed warpage due to smelt immersion as shown in Figure 14.[12]

Figure 14. Degraded mullite (left) and calcium-aluminate (right) smelt immersion samples.

Further work was performed to refine the structure of the calcium-aluminate material. X-ray diffraction of samples produced through several iterations of changes in particle size distribution and the firing temperature (shown in Figure 15 (b)) was used to optimize the phase formation and structure within the refractory, thus improving the corrosion resistance of these materials significantly resulting in a product that had very good resistance to molten smelt as shown in Figure 15 (a). At the conclusion of this last round of testing, calcium-aluminate refractories were selected for use as the back-up refractory lining. These materials were installed along with a fusion-cast spinel hot-face lining and were found to operate successfully for over three years.

a)

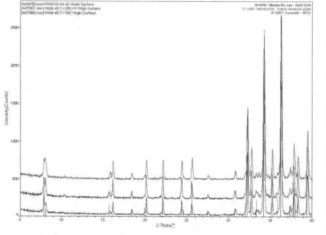

b)

Figure 15. Calcia-alumina refractory sample with optimized composition and fabrication after smelt immersion testing (a). X-ray diffraction patterns for calcia-alumina refractory samples produced through iteration of particle size distribution and firing temperature (b).

SUMMARY

Refractory issues associated with the high-temperature black liquor gasification process employed by Weyerhaeuser Company in New Bern, North Carolina were discussed along with efforts undertaken at Oak Ridge National Laboratory (ORNL) to evaluate numerous refractory ceramic materials for use in black liquor gasification. Additionally, work to identify new materials which may show improved lifetimes in the smelt environment was discussed. Screening and analysis of materials was performed using X-ray diffraction, optical microscopy, scanning electron microscopy with energy dispersive spectroscopy (SEM/EDS), electron microprobe and smelt immersion testing. All these techniques together have allowed for screening of candidate refractory systems for use as containment lining materials and the final selection of optimized materials.

Analysis of samples removed from the original hot-face lining of the New Bern gasifier was accomplished by X-ray diffraction through analysis of samples representative of material at various distances away from the fireside surface of the refractory hot-face lining. From these patterns, estimates could be made of the quantity of each phase present at individual sampling locations, with the results of these estimates forming a map which showed the reaction product $NaAlO_2$ which leads to a volume increase in the refractory structure and the subsequent spalling of the bricks seen in the original lining. These findings led to refractory screening of candidate alternate materials to be used in place of the current mullite-based refractories using a laboratory immersion test system. Results of this testing led to the finding that improved corrosion resistance could be achieved using fusion-cast α/β and α-alumina materials.

Yet, the new α/β–alumina lining was still seen to experience spalling due to the same chemical reactions between the alumina refractory and the smelt previously identified. This led to selection of an alternative fusion cast α/β-alumina refractory provided by a different manufacturer for use in the hot-face lining. Additional refractory screening through immersion testing also identified materials that appeared to be more corrosion resistant than the fusion cast α/β-alumina being used. This allowed for samples of several of these alternate refractories to be included in the hot-face lining along with the installed fusion cast α/β-alumina refractory. During a subsequent gasifier shut down for repairs of other systems, core-drilled samples of the alternate refractories, as well as samples of the primary hot-face lining, were collected and analyzed at ORNL by optical microscopy, SEM/EDS, and electron microprobe. This analysis confirmed that the fusion cast spinel had reacted less with the sodium compounds than any of the other refractories previously examined. These findings led to the selection of the fusion cast magnesia-alumina spinel refractory for use as the hot-face lining for the subsequent rebuild of the New Bern gasifier.

Following the rebuild, strain was still present on the gasifier shell despite the fact that the spinel refractory lining showed limited degradation. Therefore, a core-drilled sample was taken of the back-up lining to determine if the back-up lining was influencing the strain. X-ray diffraction of the core-drilled specimen showed that sodium-containing reaction products were present in the degraded portion of the back-up lining leading to the explanation that reaction and expansion of the back-up refractory was responsible for the increasing strain on the gasifier vessel shell. Smelt immersion testing was again used to screen a number of alternate refractories for use in the back-up lining. Immersion testing of newly identified mullite materials found them still to be rapidly degraded by the molten smelt and initial immersion testing results of the prepared calcium-aluminate refractory materials showed rapid wastage and warpage due to smelt immersion. Yet, subsequent refinement of the calcium-aluminate materials showed promising results leading to optimization of this material through X-ray diffraction analysis of samples produced through several iterations of changes in particle size distribution and the firing temperature. Based on favorable laboratory results, calcium-aluminate refractories were selected for use as the back-up refractory lining. These materials were installed along with a fusion-cast spinel hot-face lining and operated successfully for over three years.

ACKNOWLEDGMENTS
The authors would like to acknowledge immersion smelt testing performed by Adam Willoughby of ORNL and sample preparation/light microscopy examinations of refractory samples performed by Hu Longmire of ORNL. Contribution to this work was also provided by Ingvar Landälv of Chemrec AB. The contribution of test materials by refractory manufacturers including Vesuvius Monofrax, Westmoreland Advanced Materials, Harbison-Walker, SEPR and MinTeQ International is also greatly appreciated. The efforts of reviewers Andrew Wereszczak and Jy-An Wang are also gratefully acknowledged. Research sponsored by the U.S. Department of Energy, Assistant Secretary for Energy Efficiency and Renewable Energy, Industrial Technologies Program, Industrial Materials for the Future. Oak Ridge National Laboratory is managed by UT-Battelle, LLC, for the U.S. Department of Energy under contract DE-AC05-00OR22725, and Chemrec AB of Sweden.

REFERENCES
1. E.D. Larson, S. Consonni and T.G. Kreutz, "Preliminary Economics of Black Liquor Gasifier/Gas Turbine Cogeneration at Pulp and Paper Mills," Paper 98-GT-346, Paper presented at the International Gas Turbine & Aeroengine Congress & Exhibition, Stockholm, Sweden, June 2-5, 1998.

2. E.D. Larson, S. Consonni and R.E. Katofsky, "A Cost-Benefit Assessment of Biomass Gasification Power Generation in the Pulp and Paper Industry", Final Report, October 8, 2003.
3. "The Integrated Forest Biorefinery - A Preliminary Business Case - Updated September 2005", Prepared by The Forest Biorefinery Pathways Working Team as part of the Agenda 2020 Program.
4. J.G. Hemrick, J.R. Keiser, R.A. Peascoe, C.R. Hubbard, and E. Lara-Curzio, "Gasification Containment Materials Related Research at Oak Ridge National Laboratory" Refractories Applications and News, Vol. 9, No. 6, (2004).
5. K. Whitty, and C. L. Verrill, "A Historical Look at the Development of Alternative Black Liquor Recovery Technologies and the Evolution of Black Liquor Gasifier Designs," 2004 TAPPI Int'l Chemical Recovery Conf., June 6-10, 2004, Charleston, S.C., 2004.
6. C.A. Brown and W.D. Hunter, "Operating Experience At North America's First Commercial Black Liquor Gasification Plant", Proceedings of the International Chemical Recovery Conference, Tampa, Florida, pg 655-662, June 1-4, 1998.
7. L. Stigsson, "Chemrec™ Black Liquor Gasification", Proceedings of the International Chemical Recovery Conference, Tampa, Florida, pg 663-692, June 1-4, 1998.
8. C.M. Hoffmann, J.R. Keiser and C.R. Hubbard, "Characterization of the Degradation Behavior of High Alumina Silica-Bonded Refractory Used as Gasifier Lining for Black Liquor Gasification", CRADA Final Report C/ORNL/97-0481, January, 1999.
9. E.M. Levin and H.F. McMurdie, Phase Diagrams for Ceramists Volume III, The American Ceramic Society, Inc., 1975.
10. R.A. Peascoe, J.R. Keiser, J.G. Hemrick, M.P. Brady, P. Sachenko, C.R. Hubbard, R.D. Ott, C.A. Blue and J.P. Gorog, "Materials Issues in High Temperature Black Liquor Gasification", 2003 TAPPI Fall Technical Conference, Chicago, IL, October 26-30, 2003.
11. James R. Keiser, James G. Hemrick, Roberta A. Peascoe-Meisner, Camden R. Hubbard and J. Peter Gorog, "Studies and Selection of Containment Materials for High-Temperature Black Liquor Gasification", Proceedings of the 2006 TAPPI Engineering, Pulping and Environmental Conference, November 5-8, 2006, Atlanta, GA, 2006.
12. James R. Keiser, Roberta A. Meisner, James G. Hemrick, J. Peter Gorog, W. Ray Leary, Craig A. Brown, Ingvar Landälv, Amul Gupta and Kenneth A. McGowan, "Recent Experience With Structural Materials In Commercial-Scale Black Liquor Gasifiers", Proceedings of the International Chemical Recovery Conference, May 29-June 1, 2007, Quebec City, QC, Canada, 2007.

ADDRESSING THE MATERIALS CHALLENGES IN CONVERTING BIOMASS TO ENERGY

Cynthia Powell
National Energy Technology Laboratory, U.S. Department of Energy
Pittsburgh, PA, U.S.A.

James Bennett
National Energy Technology Laboratory, U.S. Department of Energy
Albany, OR, U.S.A.

Bryan Morreale
National Energy Technology Laboratory, U.S. Department of Energy
Pittsburgh, PA, U.S.A.

Todd Gardner
National Energy Technology Laboratory, U.S. Department of Energy
Morgantown, WV, U.S.A.

ABSTRACT

Converting biomass to energy offers an intriguing opportunity to reduce CO_2 levels in the atmosphere, while simultaneously producing a value-added product in the form of electricity and liquid fuels. Utilized in combination with coal as the feedstock for gasification technology, biomass provides a carbon-neutral source of renewable energy, without concerns for seasonal variations in feedstock availability. Utilized alone, biomass crops such as algae can be converted to fuels while simultaneously capturing CO_2 emitted from point sources such as power plants. However, the utilization of biomass-based feedstocks frequently results in an increase in the severity of the process environment, requiring the discovery and development of improved performance materials in order to make such processes economically viable. This paper provides an overview of on-going research at the National Energy Technology Laboratory that addresses the materials challenges in converting biomass and coal-biomass mixtures to electricity and transportation fuels. Reviewed are materials issues in gasification refractory development, biomass contaminant effects on Fischer-Tropsch fuel production technology, the efficient treatment of Fischer-Tropsch off-gases and transportation fuels derived from algae.

INTRODUCTION

Biomass is the only renewable energy resource that is available in sufficient quantities for the sustainable, large scale production of both energy and transportation fuels. The application of biomass to gasification based processes provides a carbon neutral source of renewable energy that reduces CO_2 levels in the atmosphere. However, biomass feedstock utilization in processes that are principally designed for coal results in an increase in the severity of the process environment and new materials challenges exist that must be addressed to efficiently utilize this feedstock.

Critical to the development of reliable, long-life, next generation gasification technology is an understanding of the influence of mixed coal and biomass feedstock mineral impurities on slag formation, slag physico-chemical properties and the impact that mixed slags have on refractory liner materials performance. For refractory materials, the interaction that coal and biomass mixed slags have on refractory liner performance is not well understood and provides the driver for continued research and development in the area of refractory wear reduction. Biomass introduction into coal gasification processes, not only affects the materials used in the gasification process, but also influences the composition of the syngas that is produced including increased concentrations of vapor

377

phase alkali impurities. In the Fischer-Tropsch (F-T) synthesis process, the influence that contaminants have on catalyst performance is not well understood which drives the technical evaluation of risk associated with the introduction of contaminates into the F-T process and how they influence the F-T product yields. In addition to biomass contaminant effects, the efficient integration of the F-T synthesis process into integrated gasification combined cycle power production increases the energy conversion efficiency and reduces CO_2 emissions while using waste products, i.e., the light F-T off-gases. When utilized alone; biomass crops, such as algae, that have facile growth kinetics can be used to photosynthetically capture CO_2 from integrated gasification combined cycle power plant effluent. The lipid in the algae can be converted thermocatalytically into transportation fuels and the solid algae residue can be re-used as carbon feedstock in the gasifier.

This paper provides an overview of on-going research at the National Energy Technology Laboratory (NETL) that addresses the unique materials challenges associated with converting biomass and coal-biomass mixtures into electricity and transportation fuels. Reviewed are materials issues in gasification refractory development, biomass contaminant effects on F-T fuel production technology, the efficient treatment of F-T off-gases and transportation fuels derived from algae.

GASIFICATION REFRACTORY DEVELOPMENT

Gasification technology is used to convert low cost carbon feedstocks into electricity, chemical products and transportation fuels. Gasification is a critical technology in the DOE's integrated gasification combined cycle (IGCC) power plants and could play a key role in defining the nation's long-term energy security in both electricity and fuels. In addition, gasification is considered a leading candidate for H_2 production in a hydrogen based economy[1,2]. The gasification process occurs at high temperature and high pressure in a chemical reactor that converts carbon feedstocks such as coal, petroleum coke, and/or biomass, steam, and oxygen into a syngas rich in CO and H_2. Slagging gasifiers operate at temperatures between 1,325 to 1,575 °C in a reducing environment, and at pressures between 300 to 1,000 psig which necessitates the use of refractory liners materials which are composed of durable, thermally stable ceramic compounds comprised mainly of Cr_2O_3, Al_2O_3 and/or Zr_2O_3 with Cr_2O_3 contents typically between 60 and 95 wt%.

Mineral impurities in the carbon feedstock can range from up to 10 wt%, or higher for coal, up to 1 wt% for petcoke, and up to 20 wt% for biomass. The impurities of concern in these feedstocks include oxides of Si, Al, Ca, Mg, Fe and V and impurities found at higher concentrations in biomass, such as Na and K, that may form aggressive molten slags at high temperatures that chemically dissolve or penetrate the refractory liner and result in wear and failure. Refractory liner technology development has been identified as a leading research need in gasification technology[3] and is considered a roadblock to increased gasifier service life. The two most pervasive wear mechanisms lead to refractory failure through structural spalling, which is caused by slag penetration into the refractory liner, and chemical dissolution of the refractory liner which is induced by the corrosive nature of the impurities present in the slag. These two causes of refractory failure are shown in Figure 1.

As biomass utilization in coal-based gasification processes is increased, the impact that coal and biomass mixed feedstock slags have on refractory liners is unknown and needs to be evaluated. In addition, a fundamental understanding of the interactions between the mixed slag mineral impurities, the slag properties and the impact that mixed slags have on the wear mechanisms of refractory liner materials is critical for the continued development of reliable gasifiers and for ensuring long service life. To develop this fundamental understanding, NETL's refractory materials research program is focused in two broad areas which includes the study of slag-refractory interactions and the predictive modeling of slag physico-chemical properties. The goal of this research is to improve refractory service life, gasifier on-line availability and to understand slag flow, slag-refractory corrosion, and how slag can be controlled relative to gasifier operation. The most recent direction for NETL's refractory program is the study of carbon feedstock flexibility, improved long term performance, environmentally friendly materials (low/no-chrome ceramic materials and the minimization of Cr^{+6} formation) and the development of low cost refractory materials. The modeling of mixed gasifier slags includes predictive chemistry, viscosity and phase formation. These models are used to study the slag-refractory interactions and slag viscosity that may lead to increased refractory service life. An example of the coal and biomass mixed feedstock modeling effort is shown in Figure 2, where the effect of temperature on potassium partitioning in the gas, slag and solid phases is shown for a coal-biomass mixed feedstock comprised of 70 wt% western coal and 30 wt% switchgrass[4].

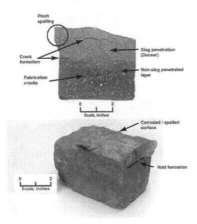

Figure 1. Wear mechanisms in high chrome ceramic oxide refractory leading to failure.

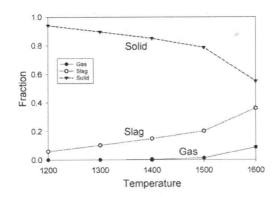

Figure 2. Predicted high temperature potassium phases produced from a coal-biomass feedstock mixture comprised of 70 wt% western coal (with 9 wt% ash) and a 30 wt% switchgrass biomass (with 9 wt% ash).

CONTAMINANT EFFECTS ON FISCHER-TROPSCH FUELS PRODUCTION

The co-production of electricity and fuel from co-gasified coal and biomass is a promising method to reduce CO_2 emissions in F-T fuel[5,6,7] and integrated gasification combined cycle power co-production power plants[5]. The advantage of biomass addition to existing coal gasification systems is that it utilizes domestic coal and biomass feedstocks without the concerns for seasonal variations in biomass availability and when utilized with carbon capture sequestration, has a lower greenhouse gas footprint than conventional petroleum fuels[7]. However, the introduction of biomass into an existing coal gasification plant affects not only the performance of gasification reactor materials, but also the composition of the syngas is strongly influenced by the characteristics of the mixed coal and biomass feedstock. In addition, the impurities that are present in the product syngas are a function of the coal and biomass composition. In biomass feedstock, alkali compounds containing Na and K, are high in concentration relative to coal and, correspondingly, the concentration of impurities containing these elements in the co-gasified syngas are also potentially higher. Low temperature, wet cleanup systems, presently in use in coal to liquids plants are effective at removing alkali species; however, the alkali removal efficiency of newer, more energetically favorable syngas cleanup systems, that utilize warm or hot gas cleanup technology, has not been assessed.

At the gasifier exit, the concentration of contaminants in the raw syngas is always greater than the contaminant specifications for F-T synthesis catalyst materials and syngas cleaning is always necessary before utilization in chemical co-production processes. The extent to which the syngas cleaning is required to remove contaminant species ultimately influences fuel production economics and potentially catalytic performance and longevity. In F-T fuel production catalysts; in particular cobalt-containing catalysts, are very sensitive to the trace impurities that are present in the syngas[8,9,10]. Therefore, the specification for cleaning the syngas is based on the economics of investing in gas cleanup systems versus the risk associated with increased alkali contaminant levels that can be adsorbed, not only onto the catalyst, but also onto the reactor walls which can result in catalyst re-exposure even after change out.

The specifications for syngas cleaning sulfur, nitrogen, halide, alkali and soot, ash and tar compounds are given in Table I for F-T catalyst fuel production. The removal specification does vary slightly between the different investigations which suggest that syngas cleaning below the specifications listed in this table is critical for fuels processing technologies that use sensitive and

Table I. Syngas contaminant removal specifications for F-T synthesis.

Syngas Contaminant	Cleanup Specification	Ref.
Total Sulfur: $H_2S + COS + CS_2$	0.05 - 0.1 ppmv	9
	1 ppmv	11
	0.2 ppmv	12
	60 ppbv	13
Total Nitrogen: $NH_3 + CN^-$	1 ppmv	11
NH_3	10 ppmv	13
NO_x	0.2 ppmv	13
CN^-	10 ppbv	13
Total Halide: $Cl^- + Br^- + F^-$	10 ppbv	11
Alkaline metals	10 ppbv	11
Soot and ash	Complete removal	11
Tars	Below dew point	11

expensive catalytic materials for biomass to energy and fuels conversion. Alternatively, the development of new catalyst formulations that have a reduced sensitivity to contaminants without significant increases in cost also remain a developmental pathway to improving the biomass materials compatibility issues.

Ongoing research at the NETL is addressing biomass specific syngas contaminant materials compatibility issues by assessing the risk associated biomass specific contaminants on F-T catalysts from poisons that may result from the co-gasification of coal and biomass mixtures. These studies include the assessment of F-T catalyst robustness to variances in syngas composition caused by the seasonal nature of biomass availability and the study of catalytic behavior in the presence of unique coal and biomass derived syngas impurities. These studies are necessary to process syngas optimally into F-T liquids. The effect that individual contaminants, their concomitant interactions and the effect of contaminant concentration are being quantified through studies in the F-T catalytic activity and selectivity behavior. The contaminants that are presently under investigation include H_2S, NH_3, Na, K and Cl.

TREATMENT OF FISCHER-TROPSCH LIGHT OFF-GASES

Mixed hydrocarbon streams that contain CO_2 are plentiful in fuel production processes. These mixed streams can either be burned, to form additional CO_2 and the power recovered; or alternatively, they can be converted via catalytic reforming into useful syngas or hydrogen and reutilized in the process to improve efficiency and to minimize the net CO_2 emitted per energy content of coal and biomass fed into the gasifier. For F-T synthesis, the chemical conversion of CO_2 and light off-gases derived from F-T into synthesis gas can be efficiently accomplished via the dry reforming:

$$CO_2 - CH_4 - 2H_2 - 2CO \qquad (1)$$

$$\Delta H_{298K}^{\circ} = +245.9 \text{ kJ/mol}$$

If the syngas derived from reforming is recycled back into the F-T process, it is possible to increase the amount of carbon that is incorporated in the fuel product; thereby, minimizing the carbon that is emitted from the co-production plant as CO_2. In fuel co-production power plants, CH_4 and light hydrocarbons are generated during F-T synthesis. The light product gases represent approximately 16 to 37 % C atom in the form of methane, olefins and low molecular weight paraffins[14].

NETL is presently assessing the utilization and integration of catalytic dry reforming technology, where CO_2 is used as an oxidant, to treat F-T light off-gases present in coal and biomass mixed feed fuel co-production processes. The barrier issues that are being addressed are the demonstration of efficient thermal integration of this technology into the overall fuel co-production process and CO_2 emissions, relative to the energy content of mixed carbon feedstock fed into the gasifier, are lower with this CO_2 re-use concept. The key technical characteristics which are desired for the dry reforming process include low cost, high efficiency, maximum thermal integration, low maintenance intervals, and acceptable response features.

The catalyst material which is employed in the dry reformer is a critical component of this system and must be able to provide a clean, tailored synthesis gas for re-integration back into the power and fuel production process. The development of a catalyst for this process is application specific and the unique conditions associated with F-T tailgases which includes high-pressure operation as well as reforming coking-prone olefins and higher molecular weight paraffinic compounds[14]. When reforming a feedstock with many different types of hydrocarbon compounds, controlling the residence time of hydrocarbons that stabilize on the surface of the catalyst is a key property to control and optimize. This property is related to the metal ensemble structure on the surface of the catalyst[15]. Olefinic compounds found in F-T light off-gases tend to form pyrolytic carbon that encapsulates the catalyst[15] occluding active catalytic sites. The risk of carbon formation is, therefore, minimized since the relative rate of hydrocarbon adsorption then becomes slower than the rate of surface reaction. The materials issues presently being addressed by NETL are related to the design of a coking tolerant dry reforming catalytic material.

The approach presently under investigation by NETL is to incorporate active transition metals into the framework lattice of refractory solid oxide compounds. Catalysts of this type are synthesized by doping the framework lattice of refractory compounds, such as Ba-β-alumina with catalytically active transition metals. This approach disperses the active metals in low coordination which destabilizes coke precursors[16] and the metal solid oxide support interaction may serve to minimize sintering[17,18] of small metallic clusters. Figure 3 gives the reaction stability of Ba-β-alumina that has been substituted with Ni. The catalyst that is produced by doping is $BaNi_{0.4}Al_{11.6}O_{19}$ and its stability performance over 18 hours of operation is given in Figure 3. Details of the catalyst synthesis and physico-chemical properties are reported previously[17,18]. The dry reforming reaction was performed isothermally at a temperature of 900 °C, a pressure of 2 atm, and a CO_2/CH_4 of 1:1. From this figure, for the first 5 hours of operation the catalyst undergoes changes before the performance stabilizes. From 5 hours to 18 hours of operation the catalyst exhibits exceptional stability.

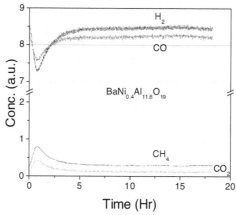

Figure 3. $BaNi_{0.4}Al_{11.6}O_{19}$ catalyst stability at 900 °C, 2 atm and CO_2/CH_4 = 1:1.

FUELS FROM ALGAE

Fast growing biomass crops such as algae can be used to photosynthetically capture CO_2 from integrated gasification combined cycle power plant effluent[19]. Once captured by the algae, the lipid component in the algae can be converted thermocatalytically into transportation fuels and the solid algae residue, left after lipid extraction, can be used as a carbon re-use feedstock in the gasifier. Several critical issues exist in the development of algae-to-fuels technology including the accurate assessment of the rate of CO_2 capture in algae, which is directly related to algae growth kinetics, and an accurate assessment of the lipid content present in the different strains of algae. Overestimation of algae growth rates, approaching the theoretical limits of the photosynthetic rate, provide an overly optimistic estimate of the quantity of algal-biomass that can be produced in a given time period. The accurate laboratory measurement of the different strains of algae growth kinetics is critical to determining the footprint of an algal reactor.

Similarly, the overstatement of lipid production rates by algae can lead to the unrealistic prediction of fuel yields. The source of these issues is related to the analytical methodology used to determine lipid production rates and the lack of standard analytical techniques that can be used to accurately and consistently determine algal lipid production rates. These issues are being addressed by NETL and are central to the advancement of this technology. The accurate measurement of growth kinetics and lipid and oil production is critical to properly sizing raceway ponds or photo-bioreactors and to suggest the potential engineering of new strains of algae.

Table II. Comparison between gravimetric and direct extraction-esterification lipid content analytical methods.

Species and Extraction Method	Raw Gravimetric (wt% lipid)	Direct Extraction-Esterification (wt% lipid)
Scenedesmus sp.		30.4 ± 0.5
Bligh Dyer	32.5 ± 0.4	
Methanol + HCl	55.3 ± 0.9	
Hexane + HCl	7.6 ± 0.4	
NS-1		2.3 ± 0.2
Bligh Dyer	5.8 ± 1.7	
Methanol + HCl	52.0 ± 1.3	
Hexane + HCl	2.0 ± 1.0	
Chlorella vulgaris		12.3 ± 4.7
Bligh Dyer	12.5 ± 1.1	
Methanol + HCl	20.4 ± 2.9	
Hexane + HCl	1.7 ± 0.2	
Chlorella pyrenoidosa		14.1 ± 3.0
Bligh Dyer	17.2 ± 4.0	
Methanol + HCl	24.9 ± 0.7	
Hexane + HCl	4.7 ± 0.3	

Conventional methods for determining algae lipid content are based on solvent extraction and gravimetric results and are highly variable and in some cases the results are observed to vary significantly for identical algae species. As shown in Table II, NETL's comparative study between four different microalgae species shows that depending on the solvent system used during the extraction, the gravimetric lipid content is highly variable and can lead to significant overestimates of algal lipid contents.

Present research at NETL is addressing this challenge through the development of analytical methods that provide more lipid-specific extraction information. The method presently under development is a derivatization analytical methodology; where lipids undergo an esterification reaction to form fatty acid methyl esters, and gas chromatography (GC-FID) is used to identify the species present, providing more accurate concentration information. Results using this analytical methodology are also shown in Table II. By comparing the data obtained by direct extraction-esterification to raw gravimetric analysis, a more lipid-specific quantity is obtained as illustrated by comparison of the two data sets. From this table, in most cases, the gravimetric determination overestimates the lipid content, in particular, when polar reagent mixtures are used. Additionally, non-polar reagents such as hexane are unable to completely extract the lipid from the algae species which affects the lipid yields. This finding is relevant considering that non-polar hexane solvent is typically applied to the extraction of fuel oils from other biomass crops.

CONCLUSION

Biomass and coal-biomass mixtures are critical to development of next generation CO_2 reduction technology for the production of energy and fuels and biomass compatibility issues drive NETL's innovative efforts for materials improvements. However, the use of biomass-containing feedstock results in an increase in the severity of the process environment, requiring the discovery and development of improved performance materials that make energy and fuel production processes economically viable.

Four enabling biomass to energy and fuels production technologies are reviewed and the unique challenges associated with each technology and materials issues were discussed. Biomass materials compatibility issues in gasification refractory development are related to increased wear through refractory chemical dissolution and spalling that result from the increased corrosivity of inorganic biomass constituents in the mixed slag. These issues are being addressed through the development and testing of novel, robust gasifier refractory materials. The effect that biomass specific contaminants, such as Na and K present in the syngas, that potentially slip through syngas cleanup infrastructure and impact catalyst performance in the F-T fuel co-production technology. The issue of biomass specific syngas contaminants influencing catalytic performance is being addressed through a risk assessment of contaminant interactions on F-T catalysis. In addition, the optimization of F-T fuel co-production is being investigated through novel integration strategies that re-use CO_2 via dry reforming. The materials issue associated with this technology is the development of coking tolerant reforming catalyst materials. Preliminary assessment of transition metal doped structural oxides, such as Ni-substituted Ba-β-alumina, has given promising early results. The issues related to the utilization of algae are the accurate measurement of growth kinetics and lipid content. To address these issues, improved analytical methodologies are under development that provides more lipid specific information and accuracy.

REFERENCES

[1] F. Mueller-Langer, E. Tzimas, M. Kaltschmitt, and S. Peteves, Techno-Economic Assessment of Hydrogen Production Processes for the Hydrogen Economy for the Short and Medium Term, *Int. J. Hyd. Engy.*, **32**, 3797-3810 (2007).

[2] Hydrogen from Coal Multi-Year Research Development & Demonstration Plan: For the Period 2008 to 2016, *U.S. Department of Energy, National Energy Technology Laboratory*, September (2008).

[3] S. Clayton, G. Stiegel, and J. Wimer, Gasification Markets and Technologies - Present and Future: An Industry Perspective, *U.S. Department of Energy Report No. DOE/FE-0447*, July (2002).

[4] S. Bale, A. Pelton, W. Thompson, G. Eriksson, K. Hack, P. Chartrand, S. Decterov, I-H. Jung, J. Melancon, and S. Peterson, FactSage 6.1, *Thermfact and GTT-Technologies* (2009).

[5] R. Williams, E. Larson, and H. Jin, Synthetic Fuels in a World with High Oil and Carbon Prices, *8th Int. Conf. on Greenhouse Gas Control Tech.*, Trondheim, Norway, June (2006).

[6] E. Larson, R. Williams, H. Jin, Fuels and Electricity from Biomass with CO_2 capture and storage, *8th Int. Conf. on Greenhouse Gas Control Tech.*, Trondheim, Norway, June (2006).

[7] D. Gray, C. White, G. Thomlinson, M. Ackiewicz, E. Schmetz, and J. Winslow, Increasing Security and Reducing Carbon Emissions of the U.S. Transportation Sector: A Transformational Role for Coal and Biomass, *U.S. Department of Energy Report No. DOE/NETL-2007/1298*, August (2007).

[8] C. Visconti, L. Lietti, P. Forzatti, and R. Zennaro, Fischer-Tropsch Synthesis of Sulfur Poisoned Co/Al$_2$O$_3$ Catalyst, *Appl. Catal. A*, **330** 49-56 (2007).

[9] C. Bartholomew and R. Bowman, Sulfur Poisoning of Cobalt and Iron Fischer-Tropsch Catalysts, *Appl. Catal.*, **15** 59-67 (1985).

[10] M. Dry, The Fischer–Tropsch process: 1950–2000, *Catal. Today*, **71** 227-41 (2002).

[11]H. Boerrigter, H. den Uil, and H.-P. Calis, H.-P., Green Diesel from Biomass via Fischer-Tropsch Synthesis: New Insights in Gas Cleaning and Process Design, *Proceedings of Pyrolysis and Gasification of Biomass and waste, Expert Meeting*, Strasbourg, FR, September (2002).

[12]M. Dry, and J. Hoogendoorn, Technology of the Fischer-Tropsch Process, *Catal. Rev. Sci. Eng.*, **23** 65-78 (1981).

[13]B. Turk, T. Merkel, A. Lopez-Ortiz, R. Gupta, J. Portzer, G. Kishnam, B. Freeman, and G. Fleming, Novel Technologies for Gaseous Contaminants Control, *Final Report for U.S. Department of Energy Contract No. DE-AC26-99FT40675*, September (2001).

[14]B. Jager, M. Dry, T. Shingles, and A. Steynberg, Experience with a New Type of Reactor for Fischer-Tropsch Synthesis, *Catal. Lett.*, **7** 293-302 (1990).

[15]D. Shekhawat, D. Berry, T. Gardner, and J. Spivey, Catalytic Reforming of Liquid Hydrocarbon Fuels for Fuel Cell Applications, *Catalysis Vol. 19*, Eds. J. Spivey and K. Dooley, RSC Publishing, Cambridge, UK 184-244 (2006).

[16]T. Gardner, D. Shekhawat, D. Berry, M. Smith, and E. Kugler, Effect of Nickel Hexaaluminate Mirror Cation on Structure Sensitive Reactions During n-Tetradecane Partial Oxidation, *Appl. Catal. A*, **323** 1-8 (2007).

[17]T. Gardner, D. Shekhawat, D. Berry, M. Salazar, M. Smith, E. Kugler, D. Haynes, and J. Spivey, Catalytic Partial Oxidation of CH4 over Ni-Substituted Hexaaluminate Catalysts, *Proceedings of the North American Catalysis Society, 20th North American Meeting*, Houston, TX, June (2007).

[18]T. Gardner, E. Kugler, J. Hissam, A. Campos, J. Spivey, and A. Roy, Catalytic Partial Oxidation of CH_4 over Ni-Substituted Barium Hexaaluminate Catalysts, *Proceedings of the 6th World Congress on Oxidation Catalysis*, Lille, France, July (2009).

[19]Y. Chisti, Biodiesel from Microalgae, *Biotech. Adv.*, **25** 294-306 (2007).

Geothermal

EXPERIENCE WITH THE DEVELOPMENT OF ADVANCED MATERIALS FOR GEOTHERMAL SYSTEMS

Toshifumi Sugama, Thomas Butcher, and Lynne Ecker
Brookhaven National Laboratory
Upton, NY 11973

ABSTRACT

For nearly the past 2 decades, the emphasis of the geothermal material programs at DOE's Brookhaven National Laboratory (BNL) has been directed toward resolving the material-related problems confronting the geothermal-drilling and -power plant industries in conventional natural hydrothermal systems.

In the field of drilling technology, BNL developed high-temperature, highly chemical-resistant well cementing materials that withstood hydrothermal environments containing CO_2 and H_2S (pH ~2.0) at temperatures up to 300°C. As a result, this material offered a significant lifetime extension of well-casing cement and substantially reduced the cost of well maintenance. Recently, work was devoted to designing cost-effective, inorganic polymer-based, cementitious materials by using recycled industrial byproducts, such as fly ash and slag. A new type of material called "Geopolymer," possessing advanced properties such as outstanding resistance to acid and readily-controllable setting behavior at high temperature was developed. This information motivated us to evaluate and validate its potential as a sealing material in Enhanced Geothermal Systems (EGS).

In the field of energy conversion, our focus centered on developing advanced coating materials, which provide upgraded corrosion-, erosion-, and fouling-prevention performance for carbon steel- and aluminum power plant components. These components include wellhead, heat exchangers, and condensers in very harsh geothermal environments. As a result, several coating systems were developed. Among these were highly thermally conductive, self-healing, multifunctional coatings for wellheads and heat exchangers at hydrothermal temperatures up to 250°C and self-assembling nanocomposite coatings for air-cooled condensers.

INTRODUCTION

Conventional well cements, consisting of calcium silicate hydrates (CaO-SiO_2-H_2O system) and calcium aluminum silicate hydrates (CaO-Al_2O_3-SiO_2-H_2O system), have been historically used in hydrothermal geothermal wells. The cementing industries were troubled by their poor performance in mechanically supporting the metallic well casing pipes and in mitigating the pipe's corrosion in very harsh geothermal reservoirs. These difficulties are particularly acute in two geological regions: One is the deep hot downhole area (~ 10 km depth at temperatures of ~ 300 C) that contains hyper saline water with high concentrations of CO_2 (> 40,000 ppm); the other is the upper well region between the well's surface and ~ 5 km depth at temperatures up to 200 C. The specific environment of the upper well region is characterized by highly concentrated H_2S (pH < 2.0) brine containing at least 5000 ppm CO_2. When conventional cements are subjected to these harsh environments, their major shortcoming is the susceptibility to reactions with hot CO_2 and acid which cause their deterioration through carbonation- and acid-initiated erosion. Such degradation not only rapidly reduces the strength of cements, decreasing the mechanical support of casing pipes, but also increases the permeability of the brine through the cement layer, increasing the pipe's corrosion rate. Severely carbonated and acid eroded cements often impaired the integrity of a wellbore structure in less than one year. In the worst cases, casings have collapsed within three months, leading to the need for costly and time-consuming repairs or re-drilling operations. These are the reasons why the geothermal well drilling and cementing industries were concerned about using conventional well cements. Furthermore, their deterioration was a major impediment in the development of geothermal energy resources.

R&D work at Brookhaven National Laboratory (BNL) in the U.S. Department of Energy's (DOE's) Geothermal Drilling program has been focused on developing and characterizing new types of cementitioius materials that confer outstanding resistance to CO_2 and acid at brine temperatures up to 300 C. The aim of the program is to reduce the drilling and reservoir management costs by 25%. To achieve this goal, three major geothermal industries, Halliburton, Unocal Corporation, and CalEnergy Operating Corporation, supported BNL's work by cost-sharing collaborative efforts. Halliburton played a pivotal role in evaluating the technical and economic feasibility of BNL-developed cements and, consequently, in formulating field-applicable ones. Unocal provided us with information on whether the developed cements are compatible with conventional well cementing operations and processes in the field. CalEnergy conducted a field exposure test for validating the integrity and reliability of newly-developed cements. As a result, two advanced cementitious materials were developed for completing wellbore structures: one was the CO_2-resistant calcium aluminate phosphate cement; the other, referred to as alkali-activated cement, is an acid-resistant cementitious material.

In the new generation of geothermal technology, known as "Enhanced Geothermal Systems (EGS)," two material-related projects are underway: one is to develop temporary sealing materials; the other is related to the study of carbonation mechanisms of reservoir rock and clay minerals by supercritical CO_2.

Corrosion, erosion, oxidation, and fouling by scale deposits are critical issues for the metal components used at geothermal power plants operating at brine temperatures up to 250°C. Replacing these components is very costly and time consuming. Currently, titanium alloy- and stainless steel plant components are commonly used for dealing with these problems. However, these metals are considerably more expensive than carbon steel and the corrosion-preventing oxide layer on the outermost surface serves as a reaction site for brine-induced scales, such as silicate, silica, and calcite. Such reactions lead to the formation of a strong interfacial bond between the scales and oxide layer and can lead to the accumulation of multiple scale layers and the impairment of the plant components' function and efficacy. Cleaning operations to remove the scales are essential, but time consuming and increase the plant maintenance costs. If inexpensive carbon steel components could be coated and lined with cost-effective, high-hydrothermal, temperature stable, anti-corrosion, -oxidation, and –fouling materials, this would improve the power plants profitability. Carbon steel parts would result in a considerable reduction in the capital investment and a decrease in the costs of operations and maintenance.

Emphasis was directed towards developing advanced coating material systems with upgraded corrosion-, erosion-, and fouling-prevention performances that extend the lifecycle of carbon steel-based plant components including heat exchangers, wellheads, and condensers. Since these plant components operate in chemically, physically, and thermally different environments, the material performance criteria of the coating systems to be developed depend on the components. The work in this area provides information on new material synthesis, processing technologies, material properties and the scale up of coating technologies for plant components.

ADVANCED CEMENTS

BNL synthesized, hydrothermally, two new cements in response to the program goal: one was calcium aluminate phosphate (CaP) cement ($CaO-Al_2O_3-P_2O_5-H_2O$ system); the other was sodium silicate-activated slag/fly ash blend (SSASF) cement ($Na_2O-CaO-Al_2O_3-SiO_2-MgO-H_2O$ system). The CaP cements has three basic components: calcium aluminate cement, sodium polyphosphate, and water. It was designed as CO_2-resistant cement for use in mildly acidic (pH ~ 5.0), CO_2-rich downhole environments. The SSASF cement has five starting materials: slag, Class F fly ash, metakaoline, sodium silicate, and water. They were designed to resist a hot, strong acid environment containing low levels of CO_2. They are also cost-effective cements because of the use of inexpensive cement-forming by-products from coal combustion and steel-manufacturing processes.

CO_2-resistant cement

Calcium aluminate phosphate (CaP) cementitious material (CaO-Al_2O_3-P_2O_5-H_2O system, US patent[1]) was synthesized from calcium aluminate cement (CAC) (the basic solid reactant which is a proton-acceptor cation-leachable powder) and the sodium- or ammonium-polyphosphate solution (the acidic reactant which serves as a proton-donator anion-yielding liquid)[2-4]. The process is a three-step reaction consisting of hydrolysis, acid-base interaction, and hydrothermal hydration.The major chemical ingredients of CAC used in this work consisted of monocalcium aluminate ($CaO.Al_2O_3$, CA) and calcium bialuminate ($CaO.2Al_2O_3$). Using the sodium polyphosphate, $-[-(-NaPO_3)_n-$, NaP] as the acidic reactant, the three-step reaction between the CAC and NaP, shown in Figure 1, led to the formation of hydroxyapatite [$Ca_5(PO_4)_3(OH)$, HOAp] and boehmite ($-AlOOH$). Concurrently, hydrogarnet, $3CaO.Al_2O_3.6H_2O$, was formed through the hydrothermal reaction between the hydrolysate species; $3Ca^{2+} + 2Al(OH)_4^- + 4OH^- \quad 3CaO.Al_2O_3.6H_2O$.

Step 1. Hydrolysis of reactant:
$CaO.Al_2O_3 + 4H_2O \rightarrow Ca^{2+} + 2Al(OH)_4^-$,
$CaO.2Al_2O_3 + 7H_2O + 2OH^- \rightarrow Ca^{2+} + 4Al(OH)_4^-$,
$-[-NaPO_3-]_n^- + nH_2O \rightarrow nNaHPO_4^- + nH^+$,

Step 2. Acid-base reaction:
$Ca^{2+} + NaHPO_4^- \rightarrow Ca(HPO_4)$,
$Al(OH)_4^- + H^+ \rightarrow AlH(OH)_4$,

Step 3. Hydrothermal hydration:
$Ca(HPO_4) + 4Ca^{2+} + 2NaHPO_4^- + xH_2O \rightarrow$
$Ca_5(PO_4)_3(OH) + 2Na^+ + xH^+ + yOH^-$,
$AlH(OH)_4 \rightarrow \gamma\text{-}AlOOH + 2H_2O$.

Figure 1. Three-step reaction for synthesizing the calcium aluminate cement

When the HOAp phase came in contact with the CO_2-derived CO_3^{2-} anions in an aqueous media, the OH groups within crystalline HOAp phase were replaced by CO_3^{2-}; thereby forming the CO_3-intercalated HOAp ($2Ca_5(PO_4)_3(OH) + CO_3^{2-} \quad 2Ca_5(PO_4)_3.CO_3$). The CO_3^{2-} was intercalated and sequestered in the HOAp molecular structure without deteriorating its structure. This is why the HOAp was responsible for minimizing the rate of cement's carbonation[5,6]. Conversely, the boehmite phase was unsusceptible to reactions with CO_2. These reaction products not only alleviate disintegration of the cements due to carbonation, but also provide the high compressive strength (> 90 MPa hydrothermal temperature between 150° to 300°C) and a low water permeability (< 1 x 10^{-4} Darcy). In addition, the CaP cement met the following material criteria:1) maintenance of pumpability for at least 3 hours[7] ; 2) compressive strength > 3.5 MPa after 24 hours of curing time; 3) water permeability < 1 x 10^{-4} Darcy; 4) bond strength to steel casing > 0.3 MPa; 5) carbonation rate <

5wt% after 1 year in 40,000 ppm CO_2-laden brine at 300°C; 6) fracture toughness > 0.008 MN/m$^{3/2}$ after 24 hours of curing time[8]; 7) resistance to mild acid (pH ~ 5.0) with < 5 wt% loss after 30 days exposure at 300°C; and 8) slurry density of foamed cement < 1.3 g/cc[9].

Acid-resistant cement

Industrial by-products, such as the finely granulated blast-furnace slag produced during steel-manufacturing and Class F fly ash generated from coal-combustion power plants, are very attractive for use as cementitious materials because of the ability to convert into a cementitious structure by adding alkaline activators and low price.

Sodium silicate, $-(Na_2O.SiO_2-)_n-$, is used as an alkali activator. It dissolves in water into polysilicic acid, $-[-O-Si(OH)_2-]_n-$, and sodium hydroxide, n NaOH. The sodium hydroxide hydrolysate then acts as an alkali activator that promotes rupture of the bonds in the network structure of slag and fly ash. This is followed by the dissociation of ionic calcium, silica, magnesium, and aluminum species from the slag and fly ash in the activated cement slurry. This dissociation leads to the evolution of the pozzalanic property of slag and fly ash. The dissociated calcium ion then hydrothermally reacts with the ionic silica derivative to form calcium silicate hydrates in the $CaO-SiO_2-H_2O$ system. Simultaneously, the free Na ions react with the ionic silica and aluminum derivatives to form the Na-zeolite phase in the $Na_2O-Al_2O_3-SiO_2-H_2O$ system. Thus, the phase composition of the hydrated cement consists of the $CaO-SiO_2-H_2O$ and $Na_2O-Al_2O_3-SiO_2-H_2O$ systems.

The sodium silicate-activated slag/fly ash blend (SSASF) cements were exposed for 15 days to CO_2-laden H_2SO_4 (pH 1.1) at 90 C. The magnitude of acid erosion was estimated from the weight loss and the volume expansion by in-situ growth of the crystalline bassanite, $CaSO_4.1/2 H_2O$ phase which is a corrosion product of the cement. Using sodium silicate with mol. ratio of 2.50, The activated SSASF cement was better at mitigating acid erosion when the mole ratio of the SiO_2/Na_2O in the sodium silicate was 2.50 rather than 3.22. More importantly, the incorporation of more fly ash into the blend cement lowered the extent of acid erosion. Consequently, the most effective SSASF formulation for withstanding the attack of hot acid was the combination of 2.50 SiO_2/Na_2O ratio sodium silicate activator and 50/50 slag/fly ash ratio cement. There was less than 7 % weight loss by acid erosion in this effective cement after it was autoclaved at temperatures up to 300 C.

The phase composition study revealed that the cementitious structure of 100°C-autoclaved cement was comprised of an unknown amorphous compound and semi-crystalline $CaO.SiO_2.nH_2O$ (CSH) phase. Raising the temperature to 200°C led to the incorporation of the analcime phase in the zeolitic mineral family in the cementitious structure, while CSH phase was transferred to the tobermorite $(5CaO.6SiO_2.xH_2O)$ phase. At 300°C, analcime became the major crystalline phase coexisting with Na-P type zeolites along with the tobermorite   xonotlite phase transition. Hence, the NaOH hydrolysate derived from the hydrolysis of sodium silicate not only acted as the alkali activator, but also played an essential role in assembling the zeolite structure. Assuming that the amorphous phase formed in the 100°C cement is associated with sodium aluminosilicate-related hydrates, we believe that the zeolite-related phases contributed significantly to enhancing the resistance to hot acid of SSASF cement autoclaved at temperatures up to 300°C[10, 11]. In contrast, the efficacy of calcium silicate-related hydrates in mitigating the acid erosion of cement seems to be little, if any.

Furthermore our unpublished, most resent study to synthesize and characterize the polymeric zeolite called "geopolymer" revealed that the geopolymer cement prepared through the three step-reaction shown in Figure 2 has better performance in resisting acid erosion than that of the analcim and Na-P type zeolite.

Step 1. Dissolution of Al_2O_3-/SiO_2-enriched reactants in alkali medium

$$-[NaO.SiO_2-]_n- + nH_2O \longrightarrow xNa^+ + yH_3SiO_4^- + zOH^-$$
sodium silicate

$$3Al_2O_3.2SiO_2 + Al_2O_3.2SiO_2 + zOH^- \longrightarrow xAl(OH)_4^- + yH_3SiO_4^-$$
mullite in fly ash   metakaolin

Step 2. Reaction between these dissociated ionic species

$$Na^+ + H_3SiO_4^- + 2Al(OH)_4^- \longrightarrow \begin{array}{c} OH \ \ OH \ \ OH \\ | \ \ \ | \ \ \ | \\ Na\text{-}O\text{-}Al\text{-}O\text{-}Si\text{-}O\text{-}Al\text{-}OH + 3H_2O \\ | \ \ \ | \\ OH \ \ O^- \end{array}$$
Pre-geopolymer

Step 3. Self-polycondensation between pre-polymers

$$\begin{array}{c} OH \ \ OH \ \ OH \\ | \ \ \ | \ \ \ | \\ Na\text{-}O\text{-}Al\text{-}O\text{-}Si\text{-}O\text{-}Al\text{-}OH \\ | \ \ \ | \\ OH \ \ O^- \end{array} + \begin{array}{c} OH \ \ OH \ \ OH \\ | \ \ \ | \ \ \ | \\ Na\text{-}O\text{-}Al\text{-}O\text{-}Si\text{-}O\text{-}Al\text{-}OH \\ | \ \ \ | \\ OH \ \ O^- \end{array}$$

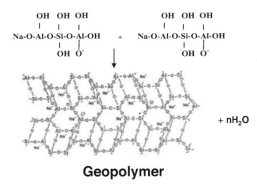

$+ nH_2O$

Geopolymer

Figure 2. Synthesis route of geopolymer cement

MATERIALS R&D IN ENHANCED GEOTHERMAL SYSTEMS (EGS)

Conventional hydrothermal technology requires natural hydrothermal reservoirs at locations of ~ 10 Km from the surface. The major advantage of new generation geothermal technology (Enhanced Geothermal Systems, EGS) is that hot, dry non-porous rock reservoirs can be located at depths of up to ~ 5 Km. To create the hydrothermal resources, the hot dry rock zone must be stimulated by pumping highly pressurized cold water through the injection well. Such a hydraulic stimulation leads to the generation and propagation of multiple fractures in the rock structure. The resulting network structure of fractures is used as hydrothermal resources. Production wells are installed in the fractured zone and will take up the steam generated after pressurized water from injection well passes through the hot fracture region.

Additionally, supercritical carbon dioxide ($scCO_2$) is being considered as a working fluid for an enhanced geothermal system. The supercritical CO_2 possesses improved heat transfer, and larger compressibility and expansibility, compared with that of water. The $scCO_2$ may also react with rocks in the reservoir, effectively sequestering the CO_2. However, the carbonation of the rock and clay minerals in the well may affect the structural integrity of the well.

Based upon the concept described, our current work involves two material-related R&D programs. One is to develop the geopolymer-based cementitious materials for use as temporary sealing materials and the other is related to the study of carbonation mechanisms of reservoir rock and clay by $scCO_2$.

Temporary geopolymer sealing materials

One pivotal factor governing cost reduction for geothermal drilling operations is lost circulation of drilling mud because of the formation of natural and low-pressure generated fractures in underground foundation structures during drilling. To deal with this problem, lost circulation zones must be sealed or plugged by appropriate materials. When the convention lost circulation materials are inadequate, Portland cement-based cementitious materials are commonly used. However, although the completely cured cementitious sealing materials are reliable for stabilizing the lost circulation zones, they must be modified with set-retarding reagents to avoid curing during down-hole pumping in high-temperature geothermal wells. Correspondingly, once the colloidal cementitious sealers are treated with the set retarder, it is very difficult to control the setting in the well. In addition to the material costs due to lost circulation, drilling rig and crew cost for unwanted waiting time for setting of sealers emplaced in fractures is enormously expensive. There is also the uncertainty of how much sealer needs pumping. In addition, the dynamic wellbore fluids and water frequently act to retard the curing of cements and cause uncured cements to wash away.

Thus, the sealing materials required to penetrate and pervade any size fracture and seal adequately at temperatures up to 250°C without disintegrating or washing out during the drilling. They must also decompose and be eliminated by the injection of pressurized water to ensure that the sealed fractures readily reopen after drilling operations.

At present, our focus is on developing a geopolymer that meets the following EGS sealing material criteria: 1) a proper Na_2O/SiO_2 mole ratio in sodium silicate reactant to provide a precise setting time for temperatures between 50° to 250°C; 2) self-degradable properties so that the sealed fracture zones are readily reopened by injection water; and 3) swelling and expansion properties to improve the adherence of the sealer to interior surfaces of fractures.

Carbonation mechanisms of reservoir rock by supercritical carbon dioxide

The structural integrity of the rock formations subjected to $scCO_2$ is being studied by: 1) chemical analyses of bore core samples taken from reservoir rock, 2) phase identification of carbonated minerals in aqueous and non-aqueous environments using National Synchrotron Light Source (NSLS) at BNL, and 3) changes in mechanical strength of carbonated minerals. The data will be used to model CO_2 and reservoir rock interactions, which could provide information on whether it is feasible to use $scCO_2$ as the working fluid for EGS systems.

ADVANCED COATINGS

Inexpensive carbon steel components coated with anticorrosion and antifouling materials can be used to reduce capital investment and the cost of operating and maintaining geothermal power plants. These components would replace very expensive stainless steel, titanium alloy, and Inconel components that commonly are employed to reduce corrosion in plants with brine temperatures up to 250 °C. The coatings play a pivotal role in extending the lifetime of carbon steel components such as heat exchangers, wellheads, and condensers and can lower the cost of the electricity generated by these plants.

Heat exchangers (HXs)

The economic utilization of binary working fluid geothermal energy conversion cycles would dramatically increase the size of the exploitable hydrothermal resource. The components of the HX, such as tubes, shell, and sheet are a major contributer to the cost of a binary plant. The stainless steel and titanium alloy HX tubes presently used in such binary-cycle plants offer great protection against corrosion caused by hot brine (Figure 3). These expensive, high-grade metal alloy tubes form corrosion-preventing passive oxide layers on their surface. The oxide layers cause the tubes' surfaces to become more susceptible to the reactions with silicate and silica. These surfaces eventually develop strongly bonded scales. The accumulation of multiple layers of scales impairs the HX's function and efficacy. Reusing the tubes is very costly because a time consuming cleaning operation is required. Thus, if inexpensive carbon-steel tubes could be coated with a thermally conductive material that resists corrosion, oxidation, and fouling, there would be a marked reduction in both the capital investment and the maintenance costs of the heat exchanger which can contain an average of 800 tubes.

Figure 3. Typical titanium alloy-based heat exchangers

Based upon the data we obtained from 1998 through 2006, the criteria for the liner systems developed in this program were: 1) continuous operation at hydrothermal temperature of 200°C; 2) thermal conductivity for the HX > 1.0 kcal/hr.m°C; 3) ionic impedance of lining film > 1 x 10^8 ohm-cm^2after 15-day-exposure to 200°C brine; 4) oxidation rate (O/C atomic ratio) of liners' surfaces < 0.05 after 15-day-exposure to 200°C brine; 5) abrasive wear rate of the liners' surfaces by SiO_2 grit (particle size of 15 m, under 150 m/s velocity and 0.6 MPa pressure) < 0.1 m/min after 15-day-exposure to 200°C brine; 6) bond strength of liner to tube >5.0 MPa; 7) tensile strength of lining film before exposure > 60.0 MPa; 8) low surface energy of liner/contact angle > 80° of water droplet on liner surfaces; and, 9) self-repairing properties. To meet these material criteria, BNL developed smart, high-performance polyphenylenesulfide (PPS)-based composite coating systems consisting of PPS as the matrix, polytetrafluoroethylene (PTFE) as the antioxidant[12,13], micro-scale carbon fiber as the thermal conductor and reinforcement[14], dicalcium aluminate powder as the self-repairing filler[15], nanoscale boehmite crystal as the wear resistant filler[16], and crystalline zinc phosphate as the primer[17]. Importantly, the zinc phosphate primer improved the adherence of carbon steel to the coating and in mitigated cathodic corrosion of the underlying steel[18].

Wellhead components

The wellhead components encounter very harsh environments with a flow velocity of ~ 3 m/sec of brine at 250°C. If their components were made of inexpensive carbon steel instead of titanium alloy-based metals, coatings would be needed to protect the component's surface against corrosion.

Since the melting temperature of PPS is around 250°C, it was assumed that this material can withstand the brine temperatures up to 200°C. Increasing the melting temperature of PPS is an important step towards developing a material to protect carbon steel components at 250°C against corrosion and scale deposition. To increase the melting point of PPS, polymer/clay nanocomposite technology was adapted by using montomorillonite (MMT) clay as the alternative nanoscale filler. The MMT nanofiller conferred three advanced properties on the semi-crystalline PPS. First, it raised its melting point by nearly 40 C to ~ 290 C (Figure 4). Second, it increased its crystallization energy, implying excellent adherence of the nanofillers' surfaces to PPS through a strong interfacial bond. Third, it abated the degree of hydrothermal oxidation due to sulfide sulfite linkage transformations. When advanced PPS nanocomposite was used as a corrosion-preventing coating for carbon steel in a simulated geothermal environment at 300 C, a coating of ~ 150 m thickness adequately protected the steel against hot brine corrosion[19].

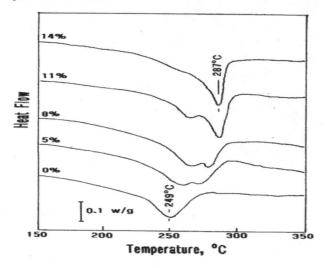

Figure 4. Different scanning calorimeter (DSC) curves for MMT nanofiller-filled PPS coatings

Air-cooling aluminum-finned condensers

The demand for electricity from geothermal binary plants increases during the summer. However, plant efficiency is decreased 30% compared to the winter because of the air-cooled condensers. A simple method to deal with this problem is to spray inexpensive, relatively clean, cool geothermal brine over the surface of the aluminum-finned steel tubing condenser. Although this method is very attractive, spraying brine on the condenser's components, aluminum fins and carbon steel tubes could cause corrosion and deposits geothermal brine-induced mineral scales. To deal with this problem, anti-corrosion and anti-fouling coatings are needed. The aluminum fins circling the surface of steel tube are of many different shapes including zigzag- or wave-shaped. Furthermore, the distance between the fins is less than 3 mm, while the height of the fin attached to the steel tube is ~ 15 mm.

Based upon the unique features of condensers and information obtained from field and in-house tests over the past three years, the following design criteria for the coatings were developed for this program: 1) low surface tension of < 60 dynes/cm, allowing it to easily permeate and wick between the fins; 2) excellent wetting behavior on the surfaces of both aluminum and steel/contact angle of < 70° of precursor solution droplet on both metal surfaces; 3) hydrophobic property of coating's surface to confer water-repellent and -shedding /low surface energy of < 30 mJ/m; 4) thin coating film of < 100 m to retain its thermal conductivity and to reduce material cost; 5) ASTM salt-spray resistance of > 1000 hours; 6) good adherence of coating to aluminum and steel (required strength of interfacial bond was unknown) 7) durability of coating to withstand 10,000 brine wet/dry cycles; 8) chemical inertness of coating surfaces to brine minerals to prevent scale deposition; and 9) low conductivity of corrosive ions/ionic impedance of $> 1 \times 10^5$ ohm-cm^2.

In an effort to design coatings to meet these material criteria, BNL succeeded in developing a new self-assembly, nanosynthesis technology that assembled a nanocomposite structure consisting of nanoscale rare-earth metal oxides as the corrosion inhibitors and water-based organometallic polymer (OMP) as the hydrophobic matrix. The environmentally benign cerium (Ce) oxide was chosen from the metal oxides and employed in this nanocomposite system. Using this synthesis technology involving three spontaneous reactions, condensation, amidation, and acetoxylation, between the Ce acetate dopant and aminopropylsilane triol (APST) as the film-forming precursor aqueous solution. The synthetic OMP material was composed of Ce oxide as the nanoscale filler and poly-acetamide-acetoxyl methyl-propylsiloxane (PAAMPS) polymer in a family of OMP (Figure 5). This nanocomposite coating extended the useful lifetime of steel exposed in a salt-fog chamber at 35 C from ~ 10 hours to ~ 768 hours. Furthermore, this coating system was far better at protecting an aluminum substrate than steel. The salt-spray resistance of the film-covered aluminum panels was extended to more than 1440 hours compared with ~ 40 hours for bare aluminum[20,21].

Under an extremely harsh environment of a 24,500 brine wet/dry cycles field test conduced by the National Renewable Energy Laboratory (NREL) at the Mammoth power plant, the Ce oxide/PAAMPS nanocomposite coatings displayed very promising results for protecting the aluminum fin and carbon steel of air-cooled condensers against corrosion and in minimizing the deposition of scales. However, two critical issues for further improvement of coating performance were addressed. One was the poor coverage of the coating over the sharp edges of ring-shaped fins. The other issue was to develop more effective nanoscale rare earth metal oxides for inhibiting the cathodic corrosion reaction at metal surfaces underneath the coating, instead of using Ce oxide.

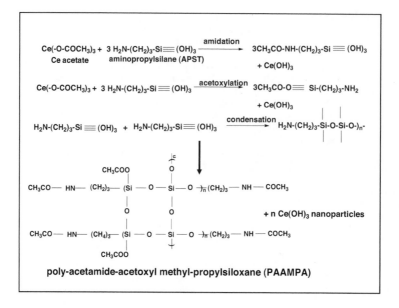

Figure 5 Synthesis of self-assembling PAAMPS nanocomposite coating containing $Ce(OH)_3$ nanoscale corrosion inhibiter

CONCLUSIONS

1. The three crystalline hydrothermal reaction products, hydroxyapatite $[Ca_5(PO_4)_3(OH)]$, boehmite (-AlOOH), and hydrogarnet $(3CaO.Al_2O_3.6H_2O)$ phases, were responsible for strengthening and densifying the CaP cements, as well as conferring on them resistance to CO_2. The mechanism underlying CO_2-resistance was the replacement of the OH groups within the hydroxyapatite phase by CO_3^{2-}, thereby leading to the formation of CO_3- intercalated and – sequestered hydroxyapatite; $2Ca_5(PO_4)_3(OH) + CO_3^{2-}$ $2Ca_5(PO_4)_3.(CO_3)$, while the boehmite phase displayed chemical inertness to CO_2. Although the hydrogarnet phase showed some sensitivity to carbonation, the hydroxyapatite $2Ca_5(PO_4)_3.CO_3$ phase transition occurring without any destruction of their structures together with the CO_2-inert boehmite phase were the major reasons why the CaP cements had an excellent resistance to CO_2. Also, the CaP cement possessed the following four important properties: 1) The maintenance of pumpability in the field; 2) low density slurry; 3) toughness; and 4) excellent bond durability to the casing pipe's surface.

2. Sodium silicate-activated slag/Class F fly ash (SSASF) blend cements were prepared by varying two parameters: the SiO_2/Na_2O mol. ratio in the sodium silicate activator and the slag/fly ash weight ratio. The cements were then autoclaved at 100 , 200 , and 300 C. The

usefulness of the autoclaved SSASF cements as acid-resistant geothermal well cements was evaluated by exposing them for 15 days to 90 C CO_2 -laden H_2SO_4 (pH 1.1). Among the combinations of the two parameters, the most effective one in inhibiting acid erosion consisted of 50/50 slag/fly ash ratio cement and 2.5/1.0 SiO_2/Na_2O mol. ratio activator. The weight loss by acid erosion of this cement was less than 7 %. The major contributor to such minimum acid erosion was the zeolite phase formed by interactions between the mullite in fly ash and the Na ion liberated from the activator. Further characteristics of this cement included a long setting time for its slurry (~ 1770 min at room temperature), a water permeability ranging from 2.4 x 10^{-4} to 1.9 x 10^{-5} Darcy, and a compressive strength from 4.0 to 53.4 MPa after autoclaving at 100 -300 C for 24 hours.

3. Enhanced Geothermal Systems propose using supercritical carbon dioxide as a working fluid because it possesses higher heat transfer and larger compressibility and expansibility, compared with that of water. Two material-related R&D programs are currently underway. One, is to develop geopolymer cement as a temporary sealing material in drilling operations. The other determines and models carbonation mechanisms of reservoir rock and clay minerals by supercritical carbon dioxide.

4. The internal surfaces of carbon steel heat exchanger (HX) tubes in a geothermal binary-cycle (Mammoth) power plant operating at a brine temperature of 160 C were lined with two thermally conductive high-temperature performance material systems to protect against corrosion and fouling by calcium silicate hydrate and silica scaling. The lined tubes (20-ft.-long, ~ 1.0 in. outside diameter) were tested for two years at the Mammoth power plant site to ensure that these lining systems performed adequately in service. One of these material systems was comprised of three different lining layers, zinc phosphate (ZnPh) as the primer, micro-scale carbon fiber thermal conductor-filled polyphenylenesulfide (PPS) as the intermediate layer, and polytetrafluoroethylene (PTFE)-blended PPS as the top surface layer. The other consisted of the ZnPh primer and nanoscale boehmite-filled PPS layer. Uncoated AISA AL-6XN stainless steel tubes also were used as the reference steel. The unlined stainless steel HX tube is well protected against corrosion by passive Cr, Fe, and Mn oxide layers on the outermost surface. However, calcium silicate hydrate and silica scales deposited on these oxide layers and strongly adhered to the tubes. This strong bond is reflected in the requirement for high-pressure hydroblasting at 55.1 MPa to scour the tube's surfaces. But, even after hydroblasting, ~ 2.0 m thick scales still remain on the oxide layer.

By contrast, the surface of a PPS top layer modified with PTFE as an anti-oxidant additive significantly retarded the hydrothermal oxidation of the liner. Such an anti-oxidant surface not only minimized the rate of the scale deposition, but also made it inert to reactions with the scales. Thus, all the scales deposited on the liner's surfaces were easily removed by hydroblasting with only ~ 18.0 MPa pressure. In addition, the PPS withstood a 160 C brine temperature and displayed great resistance to the permeation of brine through the liner, providing outstanding performance in protecting the tubes against corrosion.

The boehmite-filled PPS liner's surfaces without the PTFE suffered some degree of oxidation, so that the remnant of a few silica scales remained on the hydroblasted liner's surfaces. However, there was no internal delamination of the PPS liner from the ZnPh primer, nor were any blisters generated in the critical interfacial boundary region between the PPS and the ZnPh primer. Furthermore, brine-related elements were not detected in a ~ 5 m thick, superficial layer, strongly demonstrating that although the boehmite filler was incorporated, this liner adequately prevented corrosion of the tubes.

5. Nanoscale montomorillonite (MMT) clay fillers became dispersed in a polyphenylenesulfied (PPS) matrix through the processes of octadecylamine (ODA) intercalation   molten PPS co-intercalation   exfoliation. Cooling this molten exfoliated material led to the formation of a PPS/MMT nanocomposite. The MMT nanofiller conferred three advanced properties on the semi-crystalline PPS. First, it raised its melting point by nearly 40 C to 290 C. Second, it increased its crystallization energy, implying excellent adherence of the nanofillers' surfaces to PPS in through a strong interfacial bond. Third, it abated the hydrothermal oxidation due to sulfide   sulfite linkage transformations. When this advanced PPS nanocomposite was used as a corrosion-preventing coating for carbon steel in a simulated geothermal environment at 300 C, a coating of ~ 150   m thickness adequately protected the steel against hot brine corrosion. In contrast, an MMT-free PPS coating of similar thickness was not as effective in mitigating corrosion as the nanocompsite. In fact, the uptake of corrosive ionic electrolyte by the unmodified coating increased with exposure time.

6. Self-assembly nanocomposite synthesis technology was used to make $Ce(OH)_3$- dispersed poly-acetamide-acetoxyl methyl-propylsiloxane (PAAMPA) organometallic polymer. Three spontaneous reactions were involved: condensation, amidation, and acetoxylation, between the Ce acetate and aminopropylsilane triol (APST) at 150°C. An increase in temperature to 200°C led to the *in-situ* phase transformation of $Ce(OH)_3$ into Ce_2O_3 in the PAAMPA matrix. A further increase to 250 C caused oxidative degradation of the PAAMPA and extensive cracking in the composite. We assessed the potential of $Ce(OH)_3$/ and Ce_2O_3/ PAAMPA composite materials as corrosion-preventing coatings for carbon steel and aluminum. The Ce_2O_3 composite coating displayed better performance in protecting both metals against NaCl-caused corrosion than did the $Ce(OH)_3$ composite. Using the coating formed at 200°C, we demonstrated that the following four factors played an essential role in mitigating the corrosion of the metals: First, the coating's surface was minimally susceptibility to moisture; Second, the coating layer density was increased; Third, the cathodic oxygen reduction reaction was reduced due to the Ce_2O_3 which formed a passive film over the metal's surface; and Fourth, the coating had good adherence to metals. The last two factors contributed to minimizing the cathodic delamination of the coating from the metal's surface. We also noted that the affinity of the composite with the surface of aluminum was much stronger than with steel. Correspondingly, the rate of corrosion of aluminum was reduced by as much as two orders of magnitude for a nanoscale thick coating. In contrast, its ability to reduce the corrosion rate of steel was lower than one order of magnitude.

REFERENCES
[1] T.Sugama, U.S. Patent (Number: 5,246,496) entitled "Phosphate-bonded Calcium Aluminate Cements," (1995).
[2] T. Sugama and N.R. Carciello, Strength Development in Phosphate-bonded Calcium Aluminate Cements, *J. Am. Ceram. Soc.*, **74**, 1023-30 (1991).
[3] T. Sugama and J.M. Hill, Calcium Phosphate Cements Prepared by Acid-Base Reactions, *J. Am. Ceram. Soc.*, **75**, 2076-87 (1992).

[4] T. Sugama and L. Brothers, Calcium Aluminate Cements in Calcium Aluminate Blend Phosphate Cement Systems: Their Role in Inhibiting Carbonation and Acid Corrosion at Low Hydrothermal Temperature, *J. Mater. Sci.,* **37**, 3163-73 (2002).

[5] M. Nagai, T. Saeki, and T. Nishino, Carbon Dioxide Sensor Mechanism of Porous Hydroxyapatite Ceramics, *J. Am. Ceram. Soc.,***73**, 1456- (1990).

[6] T. Sugma and N.R. Carciello, Carbonation of calcium phosphate cement after long-term exposure to Na_2CO_3-laden water at 250°C, *Cem. Concr. Res.,* **23**, 1409-17 (1993).

[7] T. Sugama, Citric Acid as a Set Retarder for Calcium Aluminate Phosphate Cements, *J. Adv. in Cem. Res.,* **18**, 47-57 (2006).

[8] T. Sugama, L. Weber, and L. Brothers, Ceramic Fiber-reinforced Calcium Aluminate/Fly Ash/Polyphosphate Cements as a Hydrothermal Temperature of 280°C, *J. Adv. in Cem. Res.,* **14**, 25-34 (2002).

[9] T. Sugama, L. Brothers, and T. Van de Putte, Air-foamed Calcium Aluminate Phosphate Cement for Geothermal Wells, *Cem. Concr. Comp,* **27**, 758-768 (2005).

[10] T. Sugama and L. Brothers L, Sodium-silicate-activated Slag for Acid-resistant Geothermal Well Cements, *J. Adv. in Cem. Res.,* **16**, 77-87 (2004).

[11] T. Sugama, L. Brothers, and T. Van de Putte, Acid-resistant Cements for Geothermal Wells: Sodium Silicate Activated Slag/Fly Ash Blends, *J. Adv. in Cem. Res.,* **17**, 65-75 (2005).

[12] T. Sugama, D. Elling, and K. Gawlik, Poly(phenylenesulfide)-based Coatings for Carbon Steel Heat Exchanger Tubes in Geothermal Environments, *J. Mater. Sci.,* **37**, 4871-4800 (2002).

[13] T. Sugama and K. Gawlik, Anti-silica Fouling Coatings in Geothermal Environments, *Mater. Lett,* **57**, 666-673 (2002).

[14] T. Sugama and K. Gawlik, Milled Carbon Microfiber-reinforced Poly(phenylenesulfide) Coatings for Abating Corrosion of Carbon Stee, *Polymers & Polym. Comp.,* **11**, 161-170 (2003).

[15] T. Sugama and K. Gawlik, Self-repairing Poly(phenylenesulfide) Coatings in Hydrothermal Environments at 200°C, *Mater. Lett.,* **57**, 4282-4290 (2003).

[16] T. Sugama and K. Gawlik, Nanoscale Boehmite Filler for Corrosion- and Wear-resistant Polyphenylenesulfide Coatings, *Polymers & Polym. Comp.,* **12**, 153-167 (2004).

[17] T. Sugama and N. R. Carciello, Corrosion Protection of Steel and Bond Durability at Polyphenylene Sulfide-to-Anhydrous Zinc Phosphate Interfaces, *J. Appl. Polym. Sci.,* **45**, 1291-1301 (1992).

[18] T. Sugama and N.R. Carciello, Interfaces of Polyphenyletheretherketone (PEEK) and Polyphenylenesulfide (PPS) Coated Zinc Phosphated Steels after Heating-cooling Cycles in a Wet, Harsh Environment, *J. Coatings Tech.,* **66**, 43-54 (1994).

[19] T. Sugama, Polyphenylenesulfied/Montomorillonite Clay Nanocomposite Coatings: Their Efficacy in Protecting Steel Against Corrosion, *Mater. Lett.,* **34**, 332-339 (2006).

[20] T. Sugama, Cerium Acetate-modified Aminopropylsilane Triol: A Precursor of Corrosion-preventing Coating for Aluminum-finned Condensers, *J. Coat. Tech.,* **2**, 649-659 (2005).

[21] T. Sugama, U.S. Patent (Number: 7,507,480) entitled "Corrosion-Resistant Metal Surfaces," (1998).

NOVEL HIGH-TEMPERATURE MATERIALS ENABLING OPERATION OF EQUIPMENT IN ENHANCED GEOTHERMAL SYSTEMS

Matthew W. Hooker, Craig S. Hazelton, Kimiko S. Kano, Larry G. Adams, and Michael L. Tupper
Composite Technology Development, Inc.
Lafayette, CO 80026

Steven Breit
Wood Group ESP
Oklahoma City, OK 73135

ABSTRACT
Geothermal energy is a key renewable source of energy, which unlike wind and solar power, is not affected by changing weather and is always available to meet power demands. For geothermal energy to be widely utilized, deeper wells to reach hot, dry rock, located beneath the Earth's surface are required. Water will be introduced to create a geothermal reservoir, or an Enhanced Geothermal System (EGS). EGS reservoirs are at depths of up to 10 km, and the temperatures at these depths are limiting the use of available downhole equipment. Downhole submersible pumps are a key component for large-scale power generation from geothermal resources. Both Hydrothermal and Enhanced Geothermal Systems require a robust serviceable pump capable of bringing heat to the surface. The critical attributes of such a pump are temperature tolerance and low service needs. Development of geothermal will depend on the flow of high temperature resources. At that point, reliability and robustness will be key operating features. Existing pumps are limited to 190°C, preventing high value hydrothermal resources in the 190 to 220°C range from being developed. Novel high-temperature composite materials are being developed and demonstrated for use in new ESPs that can meet the challenging demands of EGS wells and can accelerate the exploitation of this large renewable source of energy.

INTRODUCTION

Geothermal energy is a key renewable energy resource that is currently being advanced in several nations. Unlike wind and solar power, geothermal energy is not affected by changing weather and is therefore always available to meet base-load power demands. Presently, commercial geothermal reservoirs are naturally formed by the combination of molten magma far beneath the Earth's surface and rainwater that seeps through the soil and is converted to steam when it reaches to the magma. Power plants that utilize these naturally-occurring geothermal energy resources are often referred to as hydrothermal systems.

Geothermal power in the United States, generated using hydrothermal systems, is currently produced from relatively shallow wells located primarily in California and Nevada. In these locations, geothermal energy is produced under nearly ideal circumstances, which include porous rock and an ample supply of sub-surface water. While these resources are currently being utilized, they are not very prevalent in the United States. However, there are significant domestic thermal resources, commonly referred to as "hot, dry rock", located at depths of 3 to 10 km below the Earth's surface. For geothermal energy to be more widely utilized, and to tap into the large potential offered by generating power from the heat of the earth, deeper wells will be necessary to reach these resources. In addition, water will be injected into these deep wells to create geothermal reservoirs. When heated by the hot rock, the resulting steam is then used to generate power in a manner similar to that of conventional hydrothermal processes. Geothermal reservoirs produced using this approach are commonly referred to as Enhanced Geothermal Systems (EGS).[1,2] An illustration of an EGS power plant is shown in Figure 1.[3]

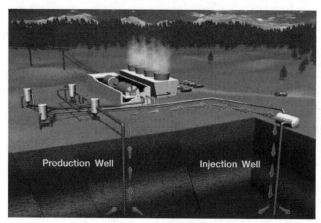

Figure 1. Illustration of an Enhanced Geothermal System.[3]

Because EGS reservoirs are typically located at depths of 3 to 10 km, electric submersible pumps (ESP's) will be needed to transport the heated fluids to the surface. While artificial lift is regularly used in the Oil and Gas industry, the application of this technology in geothermal energy is challenging due to the increased temperatures of the wells. In fact, the need for advanced pumping technologies was recently identified in the U.S. Department of Energy's Geothermal Technologies Program Multi-Year Research, Development, and Demonstration Plan.[2]

A key need in the adaptation of ESP's for use in high-temperature wells is high-temperature electrical insulations for use in the motor windings, as well as the cables used to deliver power to downhole equipment. As shown in Table I, more than half of all failures in existing ESP systems occur in either the motor (32%) or cable (21%), with the other failures occurring in various mechanical components.[4]

Table I. Component Failures in ESP Systems.

ESP System Component (Primary Failed Item)	Percentage of Total Failures (%)
Assembly (non-specific)	1
Cable	21
Sensor	1
Gas Handler	1
Motor	32
Pump	30
Intake	4
Seal/Protector	10
Other	1

Because these devices are operated at high voltages (typically 3 to 5 kV) and elevated temperatures, the performance of downhole ESP's is often de-rated to prolong their use. In most electrical systems, the insulation is often the limiting factor in the performance of the device. An electrical insulation commonly used in current ESP systems is Poly Ether Ether Ketone, or PEEK. PEEK is a thermoplastic polymer and is applied to solid wires using a co-extrusion process. While this material can be readily applied to wires, its resistivity decreases at temperatures in excess of 170°C which can lead to premature failures in high-temperature applications such as those anticipated in Enhanced Geothermal Systems.

To address the need for reliable, high-temperature motors and cables, CTD is currently developing composite insulations comprised of fiberglass reinforcements and inorganic polymers. These insulations can be applied to wires using scalable manufacturing methods, and are based on a technology that has already been demonstrated in the production high-field magnets[5,6] and high-temperature heaters for the in-situ production of oil shale.[7]

For EGS applications, the electrical insulation must withstand temperatures on the order of 200 to 250°C in the near term, with longer-term goals of 300°C operation. For motors, the insulated wire is tightly wound within the stator. This will result in relatively high strains at the turns, but the windings will be in a fixed position during its use. Alternatively, cable insulations must withstand relatively low strains during the manufacture and deployment, but may be subject to these strains at elevated temperatures as the cables deliver power to the system. In each instance, the insulation must provide stable, high-voltage performance at elevated temperatures. This report describes recent results on the production and testing of high-temperature wire insulations for use in both of these applications.

EXPERIMENTAL PROCEDURE

Candidate insulation materials were tested in both their bulk forms, as well as on wires and integrated into sub-scale motors and cables. In each instance the candidate insulation is a glass-fiber reinforced composite with an inorganic polymer matrix. Flat-plate composite laminates, as well as composite-insulated wires, were produced using S2-glass fabric and CTD-1200-series resins. Once produced, the results of the testing are related to PEEK to provide a relative basis for the performance of these new materials.

Fabrication and Testing of Insulation Materials

Because these materials will be subjected to bending strains during manufacture and deployment, mechanical testing was performed to assess their stress-strain behaviors at both 25 and 250°C. For this testing, composite laminates were fabricated using a wet lay up process and cured at 150°C under 3.5 MPa of pressure. Next, the specimens were machined so that the fiber reinforcement has a ±45 orientation within the test article. This orientation approximates the fiber angles that will be used to produce the insulated wires, and is therefore more representative of the wire insulation used in motors and cables. The stress-strain characteristics of the composites were then characterized using a four-point bending test.[8]

The insulated wires were produced by applying an S2-glass braid onto copper wires, and then applying one of the candidate resins to the glass braid. Once produced, the dielectric breakdown strength and insulation resistance of each insulation material was tested at 25 and 250°C. Dielectric breakdown testing is a destructive procedure in which an increasing DC voltage is applied across the thickness of the insulator until failure occurs. Next, the Dielectric Breakdown Strength (in kV/mm) and Electrical Strength Constant (in kV/mm$^{1/2}$) were calculated using the measured breakdown voltage and insulation thickness.[9] Electrical resistivity (in Ω–cm) was calculated by measuring the leakage current as a function of applied voltage across the thickness of the specimens. In this measurement, the voltage was increased from 0 to 6 kV and the current was measured at each 1 kV increment.

Resistivity was then calculated using the measured current-voltage relationship and the specimen geometry.

Fabrication of Prototype Devices
In addition to testing the performance of the insulation in a bulk form, these materials were also applied to long lengths of copper wire so that sub-scale assemblies could be produced. This work included the fabrication of a 100-meter, three-phase cable as well as small scale motor windings (or statorettes). Illustrations of each component are shown in Figure 2. The cable design utilized 4-AWG wires, whereas the statorettes use 8 to 12 AWG wire.

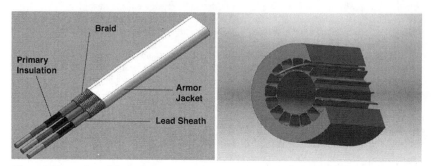

Figure 2. Illustration of three-phase motor lead extension (left) and statorette for testing insulated motor wires (right).

As seen above, the motor lead extension is comprised of three wires with a primary composite insulation. A lead sheath is then applied over the insulation, followed by a braid fabric and an armor jacket. In this work, the composite insulation was applied by wrapping pre-impregnated glass tapes along the length of the wires. The insulation was applied using a continuous process in which the inorganic polymer was also cured and spooled. The application of the tape, as well as the completed spool of insulated wire, is shown in Figure 3.

Figure 3. Application of prepreg tape insulation to copper wire (left) and spool of insulated wire (right).

To produce wire for the statorettes, S2-glass was braided onto 12-AWG copper wires and the resin was applied using a continuous, reel-to-reel type process. The details of the insulation application process are given elsewhere.[10] The statorettes were then wound with the candidate insulations to produce an assembly that approximates that of an ESP motor. In this work, the statorette diameter is the same as that of the motor. This approach allows for the production of test pieces that can be readily fabricated and tested, while also simulating the configuration of the wires within a motor.

RESULTS AND DISCUSSION
Insulation Materials and Performance

Mechanical and electrical testing of the various composite insulations showed that these materials exhibit several useful characteristics for use in downhole geothermal equipment. Initially, the stress-strain behaviors of candidate insulation materials were tested at 25 and 250°C. As seen in Figure 4, the composite insulations exhibit lower induced stresses than PEEK when mechanically loaded in a four-point bending configuration at 25°C. This finding is especially important for cable applications because these wires are not typically exposed to high degrees of strain, but the insulation must remain robust during the various spooling and deployment operations to which the cables are subjected. As expected, the stresses were lower at 250°C for all of the materials tested (Figure 5). Most notably, PEEK went from the highest stress to the lowest stress when the temperature was increased. This is due to the thermoplastic transition this material exhibits, and at 250°C the material readily flows under modest forces. The stress-strain behavior of the CTD-1200-series insulations at 250°C is also noteworthy because the stresses generated at a given strain are essentially the same as those induced at room temperature. Thus, these stress-strain characteristics of these materials will be consistent during both cable manufacture and deployment into geothermal wells.

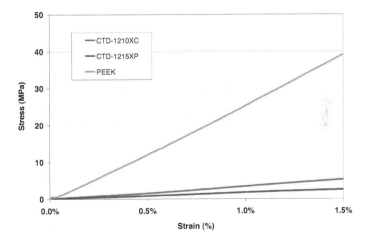

Figure 4. Stress-Strain behavior of candidate insulations at 25°C.

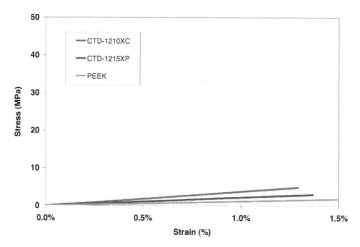

Figure 5. Stress-Strain behavior of candidate insulations at 250°C.

In addition to their mechanical performance, the dielectric properties of the composite insulations also compare favorably to PEEK at 250°C. As shown in Table II, all of the composite materials possess higher dielectric breakdown strengths and electrical resistivities at elevated temperatures than PEEK. This improved performance is critical for future geothermal applications because the operating temperatures of the downhole equipment will continue to increase, and as the temperature increases the equipment must otherwise be de-rated to compensate for the limits of the electrical insulation used to construct the downhole tools.

Table II. Dielectric Properties of Candidate Insulations on Copper Wire at 250°C.

	PEEK	CTD-1202	CTD-1210XC	CTD-1215XPC
Breakdown Voltage (kV)	22.0	22.2	37.7	35.1
Insulation Thickness (mm)	0.84	0.56	0.45	0.44
Dielectric Strength (kV/mm)	26.3	40.7	61.3	79.1
Electrical Strength Constant (kV/mm$^{1/2}$)	24.1	30.0	41.2	52.7
Resistivity @ 5 kV (GΩ-cm)	16.1	286	234	467

As seen in Table II, CTD-1210XC and CTD-1215XPC possess dielectric breakdown strengths that are two to three times that of PEEK, while also providing resistivities that are more than an order of magnitude higher at 250°C. These improvements in electrical performance are important for both motors and cables, and will be important in the design of next-generation downhole equipment.

Fabrication of Prototype Cables and Motor Windings

In addition to establishing the baseline properties of the inorganic insulation systems, both prototype cables and motor windings were fabricated to demonstrate the feasibility of using these new materials in future energy systems. As seen in Figure 6 (left), a three-phase motor lead extension was produced using a tape-wrapped insulation. The photo below shows the primary composite insulation layer, the lead sheath for environmental protection, the braided sleeve over the lead layer, and the armor that protects the cable. As a part of an ongoing cable design optimization, braided fiber reinforcements are also being considered for use in these cables. An example of a fiberglass braid on a 4-AWG (5.2-mm diameter) wire can be seen at the right in Figure 6. As previously discussed, the wires used in motor lead extensions are somewhat larger than those used in motor windings and the relative sizes of the various wires for cable and motor applications can also be seen in this photo.

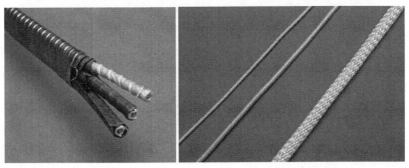

Figure 6. Three-phase cable with tape-wrapped composite insulation (left) and braided insulations to be considered in future work with both motors and cables (right).

Sub-scale motor windings are also being produced so that the performance of the insulation can be assessed in a configuration that approximates the design of an ESP motor, and after being subjected to an assembly process that is similar to what the wire will undergo during motor fabrication. As seen in Figure 7 (left), the statorettes have the same cross-section as a full-scale motor, but the length of the unit is shorter to allow for the lab-scale testing of the windings. Figure 7 (left) shows a statorette wound with PEEK-insulated wires, but all insulation materials are tested using a similar configuration. After winding, the electrical properties of the statorettes are measured under a variety of operational conditions. This includes room-temperature testing, as well as testing at elevated temperatures and after thermal conditioning. High-temperature testing and thermal conditioning are performed by placing the statorettes in a test chamber with controlled temperature and pressure. Figure 7 (right) shows the introduction of a statorette with various insulation materials unto the test chamber.

As seen in Figure 8, preliminary test data for the composite insulation materials compares favorably to that of PEEK. In this example, the leakage current of the various insulations was measured at 250°C at 3 kV after conditioning at 300°C for 168 hours. This testing showed that the leakage current of the PEEK is approximately 20 μA at 250°C, whereas the CTD-1200-series composite insulations exhibited leakage current values on the order of 2 to 3 µA. This finding is important because the higher leakage currents indicate a lower resistivity within the insulation, and thus reduces the volumetric efficiency of the motors. Additional testing is currently ongoing to further characterize the long-term performance of the composite insulations for use in future geothermal systems.

Figure 7.   Example of statorette wound with PEEK-insulated wire (left) and statorette being placed inside a test vessel (right).

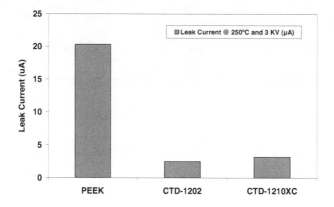

Figure 8. Leakage current values measured at 250°C and 3 kV for candidate wires after 168-hours of thermal conditioning at 300°C.

FUTURE WORK
    The results presented herein appear to indicate that fiber-reinforced inorganic insulation materials developed by CTD hold great promise for improving the performance of ESP motors for use in geothermal and other applications. CTD will work with the ESP industry to qualify these materials for use in ESP motors and to introduce them into the market. CTD also believes that the high-

temperature and high-voltage capabilities of these insulation materials can provide value for other motor and electrical machine applications. It seems to follow that operating a motor at higher temperatures and higher voltages offers the potential to reduce the size and weight of the motor, improve the volumetric performance and efficiency, and reduce the cost of these motors due to lower quantities of materials required to build smaller motors and reduced manufacturing time, if the number of turns required in the stator coils can be reduced.

CONCLUSIONS

In this work, inorganic composite insulations were fabricated and tested relative to the known requirements for use in ESP equipment. Use of these new materials in both motor windings and motor lead extensions were considered. In each instance, the insulations were found to provide improved performance to thermoplastic insulations at temperatures up to 250°C, and prototype components have been produced to demonstrate their applicability for use in these systems. These findings are important for future EGS applications because of the high temperatures present in these wells, and the need to ensure the reliability of the downhole equipment used in the generation of power from these resources.

ACKNOWLEDGEMENTS

The work described herein was funded, in part, by the U.S. Department of Energy through grant numbers DE-FG02-08ER85138 and DE-FG36-08GO18183. The authors gratefully acknowledge this support.

REFERENCES
[1] The Future of Geothermal Energy: Impact of Enhanced Geothermal Systems (EGS) on the United States in the 21st Century, prepared by the Massachusetts Institute of Technology under DOE Contract DE-AC07-05ID14517, 2006.
[2] Geothermal Technologies Program – Multi-Year Research, Development, and Demonstration Plan, United States Department of Energy, 2008.
[3] Introduction to Geothermal Energy, Geothermal Education Office, 2000.
[4] N. Griffiths and S. Breit, "The World's First Wireline Retrievable Electric Submersible Pumping System," presented at the European Artificial Lift Forum, Aberdeen, Scotland, February 28, 2008.
[5] J.A. Rice, P.E. Fabian, and C.S. Hazelton, "Wrappable Ceramic Insulation for Superconducting Magnets," *Advances in Cryogenic Engineering,* Vol. 46, pp. 267-273 (2000).
[6] J.A. Rice, C.S. Hazelton, and P.E. Fabian, "Ceramic Electrical Insulation for Electrical Coils, Transformers, and Magnets," U.S. Patent 6,407,339, June 18, 2002.
[7] M.W. Hooker, M.W. Stewart, C.S. Hazelton, and M. Tupper, "High-Performance Downhole Heater Cables," presented at the Rocky Mountain Section of the American Association of Petroleum Geologists," June 2006.
[8] ASTM Test Standard D6272–02, "Standard Test Method for Flexural Properties of Unreinforced and Reinforced Plastics and Electrical Insulating Materials by Four-Point Bending," ASTM International, West Conshohocken, PA, 2002.
[9] ASTM Standard Test Method D3755, "Standard Test Method for Dielectric Breakdown Voltage and Dielectric Strength of Solid Electrical Insulating Materials Under Direct-Voltage Stress," ASTM International, West Conshohocken, PA, 2004.
[10] M.W. Hooker, C.S. Hazelton, K.S. Kano, M.L. Tupper, and S. Breit, "High-Temperature Electrical Insulations for EGS Downhole Equipment," Proceedings of the 35th Workshop on Geothermal Reservoir Engineering, Stanford University, Stanford, CA, February 1-3, 2010.

Author Index